Functional Properties of Food Source Probiotics

Functional Properties of Food Source Probiotics

Guest Editor

Jiajia Song

Basel • Beijing • Wuhan • Barcelona • Belgrade • Novi Sad • Cluj • Manchester

Guest Editor
Jiajia Song
College of Food Science
Southwest University
Chongqing
China

Editorial Office
MDPI AG
Grosspeteranlage 5
4052 Basel, Switzerland

This is a reprint of the Special Issue, published open access by the journal *Foods* (ISSN 2304-8158), freely accessible at: https://www.mdpi.com/journal/foods/special_issues/CB2TO57JWN.

For citation purposes, cite each article independently as indicated on the article page online and as indicated below:

Lastname, A.A.; Lastname, B.B. Article Title. *Journal Name* **Year**, *Volume Number*, Page Range.

ISBN 978-3-7258-6952-7 (Hbk)
ISBN 978-3-7258-6953-4 (PDF)
https://doi.org/10.3390/books978-3-7258-6953-4

Contents

About the Editor

Jiajia Song

Jiajia Song is an Associate Professor at the College of Food Science, Southwest University. He received his PhD from China Agricultural University and completed postdoctoral research at the University of Alabama at Birmingham. His research focuses on the bioactivity, delivery, and application of functional factors in dairy systems, including probiotics, proteins, and peptides. He has served as Principal Investigator for multiple national, provincial, and municipal research projects, as well as industry funded technology development programs. He has published more than 100 peer-reviewed articles, with over 2000 citations. His honors include the First Prize of the Chongqing Science and Technology Progress Award as a main contributor and recognition as Chongqing D Category Talent. He currently serves as an Associate Editor for *Frontiers in Immunology* and *Frontiers in Nutrition*, Academic Editor for *International Journal of Food Science*, and as an Editorial Board Member for *Journal of Future Foods*, *Foods*, and *Scientific Reports*, as well as a Young Editorial Board Member for *eFood* and *Food Science of Animal Products*.

Preface

The rapid development of microbiome research has highlighted the crucial role of food source probiotics in human health, nutrition, and food innovation. Increasing evidence indicates that probiotics, prebiotics, synbiotics, and postbiotics derived from fermented foods and natural food matrices can modulate gut microbiota composition and exert beneficial effects on metabolic health, immune regulation, and disease prevention. These advances have drawn growing interest from researchers in food science, nutrition, and related disciplines.

This Reprint was conceived to provide a focused overview of recent scientific progress on the functional properties of food source probiotics and their practical applications. The collected articles address a wide range of topics, including the biological mechanisms underlying probiotic and postbiotic activity, the development of functional fermented foods, and the evaluation of probiotic safety for food use. Together, they reflect current research trends and methodological approaches in this rapidly evolving field.

The aim of this Reprint is to serve as a useful reference for researchers, food technologists, and students who are interested in functional foods and microbiome-based health strategies. By bringing together experimental studies and reviews from diverse perspectives, this Reprint sought to promote interdisciplinary exchange and inspire further research into the effective and responsible use of food source probiotics.

Jiajia Song
Guest Editor

foods

Editorial

Functional Properties of Probiotics in Food Sources

Jiajia Song

College of Food Science, Southwest University, Chongqing 400715, China; jiajias@swu.edu.cn

1. Introduction

Probiotics are live microorganisms that, when consumed in sufficient quantities, offer numerous health benefits to the host. Dairy products, fermented foods, fruits, and vegetables are prime sources of these beneficial microbes. Research has shown that probiotics from food sources exhibit various bioactivities, such as improving gut health, mitigating food protein allergies, and enhancing exercise performance [1–3]. They also play a role in preventing disorders beyond the intestines, including metabolic disorders [4], hyperuricemia [5], polycystic ovarian syndrome [6], and neurodegenerative diseases [7]. In addition, prebiotics are bioactive compounds that are either added to foods or naturally present in them, serving as substrates for specific gut microorganisms and providing health benefits [8]. Synbiotics, which combine probiotic bacteria with prebiotic ingredients, exhibit even stronger bioactivities, such as reducing lipid accumulation and alleviating autism spectrum disorder-like behaviors, compared to probiotics alone [9,10]. Postbiotics are defined as preparations of non-viable microorganisms and/or their cellular components that provide health benefits to the host, with obesity, immune system function, and microbiota being key targets [11,12]. Probiotics are also widely used in fermented foods. For instance, *Lactobacillus casei* Zhang has been found to enhance the properties of stirred yogurt by increasing gel strength and exopolysaccharide content and improving various technological aspects [13]. Similarly, *Lactobacillus paraplantarum* LS-5 has been reported to boost the antioxidant activity of fermented sauerkraut [14].

This Special Issue, titled "Functional Properties of Food Source Probiotics", in *Foods* includes eleven articles in total: ten research articles and one review article. The content is primarily focused on the following three areas: the biological functions of probiotics, postbiotics, and prebiotics; the characteristics and flavors of fermented foods, such as fruit wines, fruit juices, and yogurts; and the safety evaluation of probiotic strains.

2. An Overview of Published Articles

Tan et al. isolated *Lactiplantibacillus plantarum* HFY11 (LP-HFY11) from naturally fermented yak yogurt and investigated its effects on oxazolone-induced colitis in mice [Contribution 1]. Their study shows that LP-HFY11 significantly alleviates colitis symptoms and modulates inflammation- and immunity-related factors in the colon and serum. Additionally, LP-HFY11 was found to influence gut microbiota composition. These findings suggest that LP-HFY11 has potential as a probiotic and may be effective in treating oxazolone-induced colitis. In a separate study, Lee et al. demonstrated for the first time that *Lactiplantibacillus brevis* GKEX, sourced from douchi, in combination with resistance exercise training, can enhance muscle mass, thickness, strength, and power, while reducing body fat percentage over a six-week period in a double-blind randomized trial [Contribution 2]. Niu et al. explored the effects of heat-killed *Bifidobacterium longum* BBMN68 in pasteurized yogurt on alleviating mugwort pollen-induced allergic inflammation. Their findings indicate that the pasteurized yogurt containing heat-killed BBMN68 reduces allergic airway inflammation by maintaining the Th1/Th2 immune balance and altering gut microbiota structure and function [Contribution 3]. Similarly, Shiu et al. examined the effects of various inactivated forms of *Lactobacillus gasseri* HM1, with or without recombinant lactoferrin (LF), on metabolic disorders in obese mice [Contribution 4]. Their results suggest that different deactivation methods influence the efficacy of probiotics in improving metabolic indices. Specifically, sonication-killed HM1 (SK-HM1) is most effective in reducing inflammation in perirenal and epididymal fat tissues, while SK-HM1/LF shows significant improvement in glucose tolerance, highlighting the potential benefits of recombinant LF for glucose metabolism. Additionally, the exopolysaccharide from *Lactiplantibacillus plantarum* HMX2, isolated from Chinese northeast sauerkraut, was supplemented in the diet of juvenile turbot. This addition enhances growth performance, boosts antioxidant, digestive, and immune-related enzyme activities, improves survival rates, and regulates the intestinal microbiota, indicating its positive impact on juvenile turbot cultivation [Contribution 5]. Deng et al. reviewed the mechanisms by which probiotics and their components and metabolites mitigate colon cancer, providing a comprehensive overview of their effects on the disease [Contribution 6]. Furthermore, water-insoluble polysaccharide from Poria cocos, a type of prebiotic, was found to restore intestinal barrier function, increase short-chain fatty acids, reduce inflammatory cytokines, and modify gut microbiota structure, effectively alleviating antibiotic-associated diarrhea in mice [Contribution 7].

In fruit wine production, *Saccharomyces cerevisiae* is commonly used, but the application of non-Saccharomyces strains can introduce new and desirable flavors. Compared to mulberry wine fermented solely with *S. cerevisiae*, co-fermentation with *P. kluyveri* significantly reduces the levels of 2-methylpropyl acetate, allyl isothiocyanate, ethyl crotonate, isobutyl propanoate, butyl 2-methylbutanoate, alcohols, aldehydes, ketones, and organic acids, suggesting that *P. kluyveri* can create a novel flavor profile for mulberry wine [Contribution 8]. Balmori et al. investigated five pomelo cultivars for probiotic beverage production using *Lacticaseibacillus paracasei* and identified Tubtim Siam as the most suitable cultivar due to its high *L. paracasei* survival, antioxidant properties, bile acid binding, lipase activity inhibition, and cholesterol micellization disruption [Contribution 9]. He et al. studied the effects of konjac glucomannan (KGM) on the growth of *Lactiplantibacillus plantarum* SHY130 and evaluated the synbiotic combination of *L. plantarum* SHY130 and KGM on the physicochemical, antioxidant, and sensory properties of stirred yogurt. Their research offers new insights into developing synbiotic yogurt products [Contribution 10].

To be considered a probiotic, a bacterial strain must have a well-defined safety profile. Ma et al. assessed the safety of *Akkermansia muciniphila* PROBIO, demonstrating that it lacks mutagenic or clastogenic properties, does not spread antibiotic resistance genes, shows no toxicological effects in acute and sub-chronic toxicity studies, lacks potential virulence factors or pathogenic properties, and its moxifloxacin resistance does not transfer. These findings confirm the safety of *A. muciniphila* PROBIO for food applications [Contribution 11].

3. Conclusions

The articles in this Special Issue reveal the effects and mechanisms of probiotics, postbiotics, and prebiotics in improving constipation, diarrhea, colon cancer, athletic performance, allergic reactions, metabolic syndrome, and animal growth, as well as their safety. Additionally, some strains were explored for their applications in enhancing the quality and flavor of fermented foods. The findings in this Special Issue emphasize the significance of probiotics in food sources and further encourage research development in this area. In the future, research on the action mechanisms of probiotics should focus on identifying key molecular targets and fundamental substances, while the evaluation of effects in human trials should also be considered.

Conflicts of Interest: The author declare no conflicts of interest.

List of Contributions:

1. Tan, F.; Zhou, X.; Ren, L.; Kong, C.-S. Effect of *Lactiplantibacillus plantatum* HFY11 on colitis in mice. *Foods* **2024**, *13*, 1496.

2. Lee, M.-C.; Hsu, Y.-J.; Ho, C.-S.; Tsai, Y.-S.; Chen, C.-C.; Huang, C.-C. Supplementation with *Lactiplantibacillus brevis* GKEX combined with resistance exercise training improves muscle mass, strength performance, and body fat condition in healthy humans. *Foods* **2024**, *13*, 1030.
3. Niu, X.; Yin, X.; Wu, X.; Zhang, Q.; Jiang, Y.; He, J.; Zhao, Y.; Zhang, C.; Ren, Y.; Lai, M. Heat-killed *Bifidobacterium longum* BBMN68 in pasteurized yogurt alleviates mugwort pollen-induced allergic airway responses through gut microbiota modulation in a Murine model. *Foods* **2023**, *12*, 2049.
4. Shiu, W.-C.; Liu, Z.-S.; Chen, B.-Y.; Ku, Y.-W.; Chen, P.-W. Evaluation of a standard dietary regimen combined with heat-inactivated *Lactobacillus gasseri* HM1, lactoferrin-producing HM1, and their sonication-inactivated variants in the management of metabolic disorders in an obesity mouse model. *Foods* **2024**, *13*, 1079.
5. Hu, G.; Wang, Y.; Xue, R.; Liu, T.; Zhou, Z.; Yang, Z. Effects of the exopolysaccharide from *Lactiplantibacillus plantarum* HMX2 on the growth performance, immune response, and intestinal microbiota of juvenile turbot, *Scophthalmus maximus*. *Foods* **2023**, *12*, 2051.
6. Deng, X.; Yang, J.; Zhang, Y.; Chen, X.; Wang, C.; Suo, H.; Song, J. An update on the pivotal roles of probiotics, their components, and metabolites in preventing colon cancer. *Foods* **2023**, *12*, 3706.
7. Lai, Y.; Deng, H.; Fang, Q.; Ma, L.; Lei, H.; Guo, X.; Chen, Y.; Song, C. Water-insoluble polysaccharide extracted from *Poria cocos* alleviates antibiotic-associated diarrhea based on regulating the gut microbiota in mice. *Foods* **2023**, *12*, 3080.
8. Ding, B.; Zhao, S.; Zhang, W.; Lin, Y.; Xiong, L. The effect of co-culture with different *Pichia kluyveri* and *Saccharomyces cerevisiae* on volatile compound and characteristic fingerprints of mulberry wine. *Foods* **2024**, *13*, 422.
9. Balmori, V.; Marnpae, M.; Chusak, C.; Kamonsuwan, K.; Katelakha, K.; Charoensiddhi, S.; Adisakwattana, S. Enhancing phytochemical compounds, functional properties, and volatile flavor profiles of pomelo (*Citrus grandis* (L.) Osbeck) juices from different cultivars through fermentation with *Lacticaseibacillus paracasei*. *Foods* **2023**, *12*, 4278.
10. He, P.; Wang, Y.; Yang, J.; Suo, H.; Song, J. The effect of *Lactiplantibacillus plantarum* SHY130 and konjac glucomannan on the physicochemical, antioxidant, and sensory properties of stirred yogurt. *Foods* **2024**, *13*, 2324.
11. Ma, X.; Tian, M.; Yu, X.; Liu, M.; Li, B.; Ren, D.; Wang, W. Characterization and preliminary safety evaluation of *Akkermansia muciniphila* PROBIO. *Foods* **2024**, *13*, 442.

References

1. Gyawali, I.; Zhou, G.; Xu, G.; Li, G.; Wang, Y.; Zeng, Y.; Li, J.; Zhou, J.; Zhu, C.; Shu, G. Supplementation of microencapsulated probiotics modulates gut health and intestinal microbiota. *Food Sci. Nutr.* **2023**, *11*, 4547–4561. [CrossRef] [PubMed]
2. Tian, X.; Liang, X.; He, H.; Cui, Q.; Liu, Q.; Fan, R.; Liu, T.; Yi, H.; Gong, P.; Wang, Q. Probiotics alleviate food protein allergy in mice by activating TLR4 signaling pathway. *Mol. Nutr. Food Res.* **2023**, *67*, 2200579. [CrossRef] [PubMed]
3. Chen, Z.-X.; Yang, J.; Song, J.-J.; Suo, H.-Y. Research progress on the mechanism of exercise fatigue and the role of probiotics in counteracting it. *Food Health* **2024**, *6*, 14. [CrossRef]
4. Wang, Y.; Ai, Z.; Xing, X.; Fan, Y.; Zhang, Y.; Nan, B.; Li, X.; Wang, Y.; Liu, J. The ameliorative effect of probiotics on diet-induced lipid metabolism disorders: A review. *Crit. Rev. Food Sci. Nutr.* **2024**, *64*, 3556–3572. [CrossRef] [PubMed]
5. Sun, L.; Ni, C.; Zhao, J.; Wang, G.; Chen, W. Probiotics, bioactive compounds and dietary patterns for the effective management of hyperuricemia: A review. *Crit. Rev. Food Sci. Nutr.* **2024**, *64*, 2016–2031. [CrossRef] [PubMed]
6. Calcaterra, V.; Rossi, V.; Massini, G.; Casini, F.; Zuccotti, G.; Fabiano, V. Probiotics and polycystic ovary syndrome: A perspective for management in adolescents with obesity. *Nutrients* **2023**, *15*, 3144. [CrossRef] [PubMed]
7. Ojha, S.; Patil, N.; Jain, M.; Kole, C.; Kaushik, P. Probiotics for neurodegenerative diseases: A systemic review. *Microorganisms* **2023**, *11*, 1083. [CrossRef] [PubMed]
8. Ferreira, V.C.; Barroso, T.L.C.T.; Castro, L.E.N.; Da Rosa, R.G.; de Siqueira Oliveira, L. An overview of prebiotics and their applications in the food industry. *Eur. Food Res. Technol.* **2023**, *249*, 2957–2976. [CrossRef]
9. Tang, T.; Song, J.; Li, J.; Wang, H.; Zhang, Y.; Suo, H. A synbiotic consisting of *Lactobacillus plantarum* S58 and hull-less barley β-glucan ameliorates lipid accumulation in mice fed with a high-fat diet by activating AMPK signaling and modulating the gut microbiota. *Carbohydr. Polym.* **2020**, *243*, 116398. [CrossRef] [PubMed]
10. Wang, C.; Chen, W.; Jiang, Y.; Xiao, X.; Zou, Q.; Liang, J.; Zhao, Y.; Wang, Q.; Yuan, T.; Guo, R. A synbiotic formulation of *Lactobacillus reuteri* and inulin alleviates ASD-like behaviors in a mouse model: The mediating role of the gut–brain axis. *Food Funct.* **2024**, *15*, 387–400. [CrossRef] [PubMed]

11. Pimentel, T.C.; Cruz, A.G.; Pereira, E.; da Costa, W.K.A.; da Silva Rocha, R.; de Souza Pedrosa, G.T.; dos Santos Rocha, C.; Alves, J.M.; Alvarenga, V.O.; Sant'Ana, A.S. Postbiotics: An overview of concepts, inactivation technologies, health effects, and driver trends. *Trends Food Sci. Technol.* **2023**, *138*, 199–214. [CrossRef]
12. Salminen, S.; Collado, M.C.; Endo, A.; Hill, C.; Lebeer, S.; Quigley, E.M.; Sanders, M.E.; Shamir, R.; Swann, J.R.; Szajewska, H. The International Scientific Association of Probiotics and Prebiotics (ISAPP) consensus statement on the definition and scope of postbiotics. *Nat. Rev. Gastroenterol. Hepatol.* **2021**, *18*, 649–667. [CrossRef] [PubMed]
13. Bai, M.; Huang, T.; Guo, S.; Wang, Y.; Wang, J.; Kwok, L.-Y.; Dan, T.; Zhang, H.; Bilige, M. Probiotic *Lactobacillus casei* Zhang improved the properties of stirred yogurt. *Food Biosci.* **2020**, *37*, 100718. [CrossRef]
14. ZhoZhou, H.; Wang, S.; Liu, W.; Chang, L.; Zhu, X.; Mu, G.; Qian, F. Probiotic properties of *Lactobacillus paraplantarum* LS-5 and its effect on antioxidant activity of fermented sauerkraut. *Food Biosci.* **2023**, *52*, 102489.

foods

MDPI

Article

Heat-Killed *Bifidobacterium longum* BBMN68 in Pasteurized Yogurt Alleviates Mugwort Pollen-Induced Allergic Airway Responses through Gut Microbiota Modulation in a Murine Model

Xiaokang Niu [1,†], Xindi Yin [1,†], Xiuying Wu [2], Qi Zhang [1], Yunyun Jiang [2], Jingjing He [1], Yuyang Zhao [3], Chao Zhang [1], Yimei Ren [3], Mengxuan Lai [2], Yue Sang [4] and Ran Wang [1,*]

1 Key Laboratory of Functional Dairy, Department of Nutrition and Health, China Agricultural University, Beijing 100190, China
2 Inner Mongolia Mengniu Dairy (Group) Co., Ltd., Hohhot 011500, China
3 College of Food Science & Nutritional Engineering, China Agricultural University, Beijing 100083, China
4 Hebei Engineering Research Center of Animal Product, Sanhe 065200, China
* Correspondence: wangran@cau.edu.cn
† These authors contributed equally to this work.

Abstract: Many probiotic bacteria have been proven to prevent allergic airway responses through immunomodulation. This study was conducted to evaluate the effects of heat-killed *Bifidobacterium longum* BBMN68 (BBMN68) in pasteurized yogurt on the alleviation of mugwort pollen (MP)-induced allergic inflammation. BALB/c mice aged 5–6 weeks were randomly assigned and fed pasteurized yogurt containing heat-killed BBMN68 for 27 days, followed by allergic sensitization and challenge with MP extract. The allergic mice that received pasteurized yogurt containing heat-killed BBMN68 had improved immune status, including a lower serum IgE level, decreased serum interleukin (IL)-4, IL-5, and IL-13 concentrations, and alleviated airway inflammation manifested by increased macrophage and decreased eosinophil and neutrophil counts in BALF, as well as airway remodeling and suppressed peribronchial cellular infiltration. Moreover, oral administration of pasteurized yogurt containing heat-killed BBMN68 significantly modulated gut microbiota composition by influencing the proportion of beneficial genera associated with inflammation and immunity, such as *Lactobacillus*, *Candidatus_Saccharimonas*, *Odoribacter*, and *Parabacteroides*, which also negatively correlated with serum IgE and Th2 cytokine levels. These results demonstrated that pasteurized yogurt containing heat-killed BBMN68 had mitigative effects on allergic airway inflammation, likely through maintaining the systemic Th1/Th2 immune balance by altering the structure and function of the gut microbiota.

Keywords: airway allergic inflammation; *Bifidobacterium longum* BBMN68; heat-killed bacteria; gut microbiota; gut-lung axis; pollen allergy; postbiotics

1. Introduction

Allergic airway disease is characterized by airway structure and function deterioration induced by T helper 2 (Th2) responses derived from adaptive immune normality due to harmless environmental stimuli such as pollen [1]. Mugwort pollen (MP) is one of the most common aeroallergens [2]. The dominant allergenic protein Art v 1 of mugwort pollen triggers pro-inflammatory Th2-associated cytokines that result in Th1/Th2 imbalance and the following allergic symptoms such as IgE secretion, mucus hypersecretion, eosinophilic inflammation, and airway hyperresponsiveness (AHR), leading to life quality impairment and extra economic costs [3]. The development of allergic airway disease is usually accompanied by gut microbiome shifts [4]. Therefore, preventive and treatment approaches through modulation of gut microbiota have attracted increasing attention.

Probiotics are living microorganisms that benefit the hosts gut microbiome [5]. They are able to enhance the gut microbiome's resilience as well as local and systemic immune function [1,6–8]. *Lactobacillus* and *Bifidobacterium* are the most well-studied probiotics with strain-specific properties of immunomodulation and immunostimulation. Research on supplementation of live *Lactobacillus rhamnosus* GG (LGG) has verified its potential as a promising strategy for allergy prevention by decreasing serum IgE levels, lung eosinophil counts, lung Th2 cytokines, AHR, lung matrix metalloproteinase 9 expression, inflammatory cell infiltration, and the concentration of exhaled nitric oxide [9–11]. In recent years, postbiotics, defined as "preparation of inanimate microorganisms and/or their components that confers a health benefit on the host", have been employed to mimic the effects of probiotics and overcome the challenges in their application, such as the instability and safety issues of live cells [12]. Both live and heat-killed LGG have shown anti-inflammatory effects in different models [13,14]. Administration of heat-killed *Clostridium butyricum* CGMCC0313-1 through aerosol inhalation was shown to adjust the Th1/Th2 balance and suppress the NF-κB/NLRP3 inflammatory pathway in the lungs of allergic mice [15]. Heat-killed *Lactobacillus plantarum* and *Lactobacillus curvatus* inhibited AHR by reducing cytokine levels and inducing regulatory factor Foxp3 expression [16]. The immunological effects of heat-killed probiotic bacteria suggest their important roles in relieving pollen allergies.

Bifidobacterium longum BBMN68 (BBMN68) is a probiotic strain isolated from healthy Chinese centenarians [17]. It has previously been shown to boost innate and acquired immune functions, including enhancement of macrophages and natural killer cell activities, induction of CD11c+ CD103+ dendritic cells (DC) and semi-mature DCs, and proliferation of splenic lymphocytes [18–20]. Moreover, the immunomodulatory benefits of BBMN68 are primarily derived from the cell wall fraction [19]. Therefore, it is proposed that heat-killed BBMN68 has potential for pollen allergy alleviation by modulating immune function. In this study, we investigated the effects of oral administration of heat-killed BBMN68 in pasteurized yogurt on the mitigation of allergic symptoms through interventions on the immune system and gut microbiota in MP-induced mice.

2. Materials and Methods

2.1. Preparation of the Pasteurized Yogurt Containing Heat-Killed BBMN68

B. longum BBMN68 (CGMCC 2265) was obtained from the China General Microbiological Culture Collection (CGMCC, Beijing, China). As a comparison, *Lactobacillus rhamnosus* GG (LGG, ATCC 53103) was purchased from the American Tissue Culture Collection (ATCC, Virginia, MA, USA). Both strains were cultured in MRS (De Man, Rogosa, and Sharpe) broth (Land Bridge Technology Co., Ltd., Beijing, China) at 37 °C for 24 h. Bacterial cells were centrifuged and washed before heat-inactivation in a water bath at 85 °C for 30 min.

All test yogurts were fermented with *Lactobacillus bulgaricus* and *Streptococcus thermophilus* based on a common procedure for conventional production and then pasteurized at 75 °C for 1 min. The heat-killed BBMN68 or LGG were additionally supplemented into the pasteurized yogurt at a concentration of 10^8 cells/mL. It contains 78 kcal, 5.3 g of carbohydrate, 3.5 g of fat, and 3.0 g of protein per 100 g of yogurt.

2.2. Animals and Experimental Design

A total of 40 male BALB/c mice (Vital River Laboratories, Beijing, China) at an age of 5–6 weeks were randomly assigned to five groups, with 8 mice in each group. Mice were housed in a specific-pathogen-free (SPF) facility with free access to standard pelleted laboratory chow and water under standard conditions (12 h light/dark cycle, 20 ± 2 °C, 45–50% humidity). The animal management and experimental procedures were approved by the Animal Care and Use Ethics Committee (PONY-2021-FL-72).

The procedures for establishing the model of mugwort pollen (MP)-induced allergic sensitization in mice are shown in Figure 1, as previously described with slight modifications [2]. Briefly, mice received four intraperitoneal injections of 0.1 mL MP

(Aileji Biotechnology Co., Ltd., Hangzhou, China) at 2 mg/mL in the presence of 5% aluminum hydroxide (Thermo Fisher Scientific, Waltham, MA, USA) on days 0, 6, 12, and 18 for sensitization. Then mice were challenged intranasally with 40 µL MP at 8 mg/mL on days 23, 24, 25, 26, and 27. Mice in the control group received an equal amount of saline for sensitization and provocation and had daily administration of 0.2 mL of phosphate buffered saline (PBS, Solarbio, Beijing, China) during the whole experimental period (control group). Mice in the other four MP-induced allergic groups received 0.2 mL of PBS (MP group), pasteurized plain yogurts (Y + MP group), pasteurized yogurt containing heat-killed BBMN68 (BBMN68 + MP group), or pasteurized yogurt containing heat-killed LGG (LGG + MP group) intragastrically.

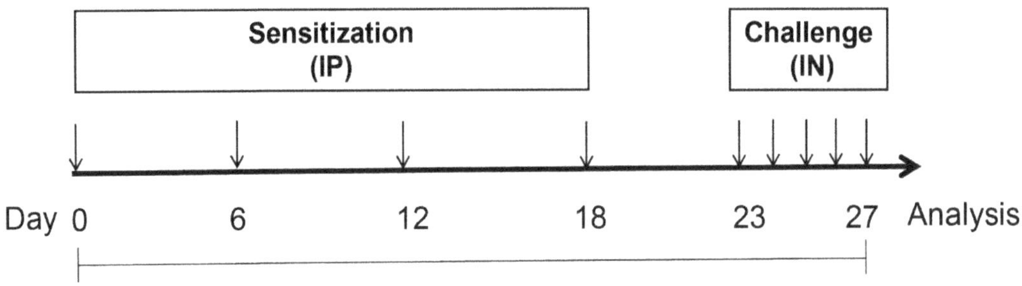

Figure 1. Procedures for establishing the model of mugwort pollen (MP)-induced allergic sensitization in mice. Mice received intraperitoneal (IP) injections of 0.2 mg MP for sensitization. The challenge was performed intranasally (IN) with 0.32 mg MP. Mice in the control group received saline for sensitization and challenge.

On day 28, mice were fasted for 12 h and anesthetized through an intraperitoneal injection of 10% chloral hydrate. Bronchoalveolar lavage fluid (BALF) was acquired by lavaging the lungs three times with 0.8 mL of sterile PBS. Blood and lung tissue samples were collected. Fecal pellets were freshly collected and stored on ice, then immediately transported to the laboratory for 16S rRNA gene sequencing.

2.3. Determination of Serum Immunoglobulins and Cytokines

Total serum IgE, IgG, Th2-associated cytokines IL-4, IL-5, IL-13, Th1-associated cytokine IFN-γ, and Th17-associated cytokine IL-17 were determined by enzyme-linked immunosorbent assay (ELISA) using commercial kits (Neobioscience Technology, Shenzhen, China; Multisciences Biotech, Hangzhou, China). All experimental procedures were conducted following the manufacturer's instructions.

2.4. The Differential Cell Counts in BALF

After centrifugation, the precipitated BALF was resuspended with PBS. The cell counts of macrophages, eosinophils, neutrophils, and lymphocytes in BALF were observed on a glass slide stained with hematoxylin and Congo Red (Solarbio, Beijing, China) using a Leica DM 2500 microscope (Leica, Wetzlar, Germany). The percentage of differential cells was calculated for 200 cells.

2.5. Lung Histology and Hematoxylin-Eosin Staining

The left lung was fixed with 4% paraformaldehyde for 24 h and embedded in paraffin. The tissue sections were cut into 5 µm thick using a microtome and stained with hematoxylin and eosin (HE). The inflammatory cell infiltration was assessed at 20× magnification under a light microscope (Leica, Wetzlar, Germany). Then the semi-quantitative analysis of inflammation severity was scored based on the area of infiltration sites in the lung tissue sections: score 0 (no inflammation); score 1 ($\leq$1/3 tissue involvement); score 2 (1/2 tissue involvement); and score 3 ($\geq$2/3 tissue involvement) [21].

2.6. Fecal Microbiota Analysis

The genomic DNA of fecal samples was extracted using the MOBIO Power Fecal DNA Isolation Kit (Qiagen, Stockach, Germany) according to the instructions. The amplification of the V3–V4 region of the 16S rRNA gene was performed with the PCR bar-coded primer pair 338F (5′-ACTCCTACGGGAGGCAGCAG-3′) and 806R (5′-GGACTACHVGGGTWTCTAAT-3′). PCR products were purified with the AxyPrep DNA Gel Extraction Kit (Axygen Biosciences, Union City, CA, USA) after agarose gels electrophoresis. The purified amplicons were subsequently pooled in equimolar amounts and paired-end sequenced on an Illumina MiSeq PE300 platform (Illumina, San Diego, CA, USA).

The sequencing raw data were demultiplexed, quality-filtered by fastp (version 0.20.0), and merged by FLASH (version 1.2.7). Operational taxonomic units (OTUs) with 97% similarity were clustered using UPARSE (version 7.1), and taxonomies were analyzed using the RDP Classifier (version 2.2). The rarefaction analysis, including the Shannon, Ace, and Chao indices, was performed to demonstrate the α-diversity. Principal coordinate analysis (PCoA) was performed with the Quantitative Insights into Microbial Ecology (QIIME) pipeline (version 1.9.0). The correlation between gut microbiota and allergic sensitization parameters in serum was revealed using Pearson correlation. A differential abundance of taxa was identified using the linear discriminant analysis effect size (LEfSe).

2.7. Statistical Analysis

Statistical analysis was performed using a one-way ANOVA followed by Tukey's multiple comparison test to compare groups on the SPSS software (version 25.0, IBM Corp., Armonk, NY, USA). Data were plotted with the GraphPad Prism 8.0.1 software (GraphPad Software Inc., Boston, MA, USA) and expressed as mean ± standard error of the mean (SEM). A p-value < 0.05 was considered significant.

3. Results

3.1. Protective Effects of Heat-Killed BBMN68 in Pasteurized Yogurt on Allergic Airway Responses

The protective effects of pasteurized yogurt containing heat-killed bacteria were tested using a MP-induced allergic sensitization mouse model, as depicted in Figure 1. After sensitization and challenge, the serum antibody IgE and IgG levels were measured (Figure 2). Compared with the control group, the total IgE and IgG in the serum of MP-exposed mice were increased ($p < 0.05$), indicating that MP induced the shift of the systemic immune balance toward Th2 immune responses. The allergic mice that received pasteurized plain and BBMN68 yogurt had lower ($p < 0.05$) serum IgE levels compared to the MP group, whereas the effect was not observed in the LGG + MP group.

The Th2-associated cytokines IL-4, IL-5, IL-13, Th1-associated cytokine IFN-γ, and Th17-associated cytokine IL-17 in serum were also determined (Figure 3A–E). The IL-4, IL-5, and IL-13 levels in the serum of MP-exposed mice were significantly higher than those in the control group ($p < 0.05$). In comparison with the MP group treated with no yogurt, oral administration of heat-killed BBMN68 or LGG pasteurized yogurt reduced ($p < 0.05$) serum IL-4 and IL-5 levels. In addition, as shown in Figure 3C, pasteurized yogurt with and without BBMN68 or LGG all played a significant role in decreasing ($p < 0.05$) the IL-13 level in allergic mice. No significant difference was identified in serum IFN-γ and IL-17 levels.

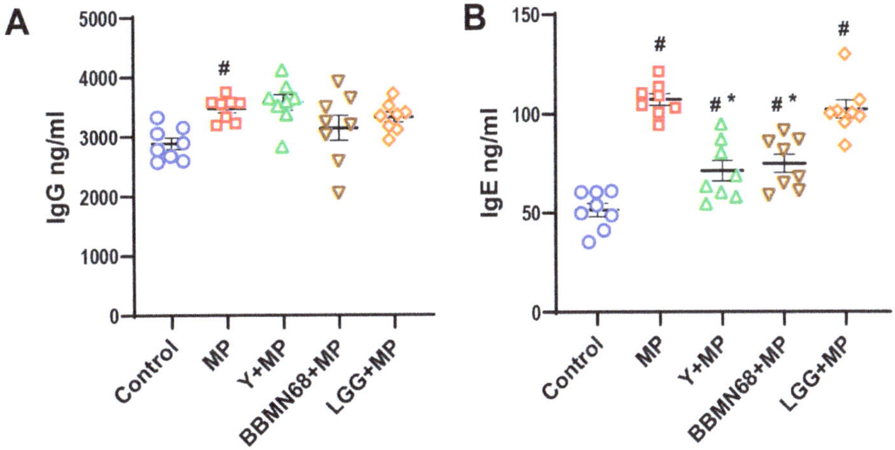

Figure 2. Serum IgG (**A**) and IgE (**B**) antibody levels in mice treated with pasteurized plain and heat-killed BBMN68 or LGG yogurt exposed to mugwort pollen (MP). Data are expressed as mean ± SEM, different color graphics in each group represent duplicates. #: $p < 0.05$ compared to the control group; *: $p < 0.05$ compared to the MP group; $n = 8$.

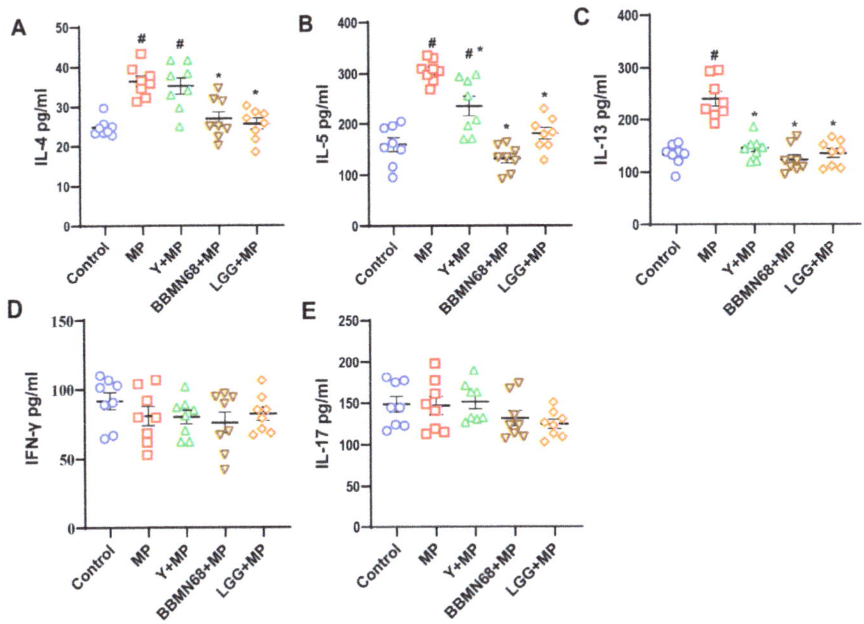

Figure 3. Serum cytokines of mice treated with pasteurized plain and heat-killed BBMN68 or LGG yogurt exposed to mugwort pollen (MP). (**A**) IL-4, (**B**) IL-5, (**C**) IL-13, (**D**) IFN-γ, (**E**) IL-17. Data are expressed as mean ± SEM, different color graphics in each group represent duplicates. #: $p < 0.05$ compared to the control group; *: $p < 0.05$ compared to the MP group; $n = 8$.

The effects of heat-killed BBMN68 yogurt on allergic airway inflammation were evaluated by BALF differential cell counts and lung histology (Figure 4A–E). The MP-induced allergic mice had lower ($p < 0.05$) macrophage and higher ($p < 0.05$) eosinophil and neutrophil counts than mice in the control group. The significantly increased ($p < 0.05$)

macrophage counts were observed in allergic mice that received pasteurized plain and BBMN68 yogurt compared to mice in the MP group; among them, the macrophage count was highest in the BBMN68 + MP group. Both neutrophil and eosinophil counts decreased ($p < 0.05$) in allergic mice that received pasteurized plain, BBMN68, or LGG yogurt. Moreover, the HE staining of left lung tissue sections indicated different pasteurized yogurt treatments in allergic mice promoted airway remodeling and suppressed peribronchial cellular infiltration, manifested by a lower inflammation score ($p < 0.05$), especially in the mice gavaged with heat-killed BBMN68 yogurt (Figure 4E).

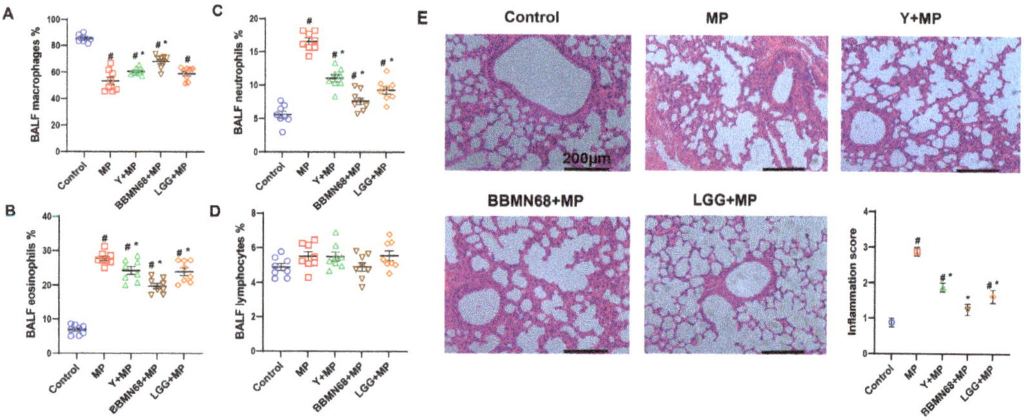

Figure 4. The BALF differential cell counts (%) of mice treated with pasteurized plain and heat-killed BBMN68 or LGG yogurt exposed to mugwort pollen (MP). (**A**) macrophages, (**B**) eosinophils, (**C**) neutrophils, (**D**) lymphocytes per 200 cells; (**E**) representative images of left lung stained with H and E, scale bars are 200μm; semi-quantitative scoring of inflammation in lung tissues. Data are expressed as mean ± SEM, different color graphics in each group represent duplicates. #: $p < 0.05$ compared to the control group; *: $p < 0.05$ compared to the MP group; $n = 8$.

3.2. Intervention of Heat-Killed BBMN68 in Pasteurized Yogurt on Gut Microbiota Composition Systemically Correlates with Allergic Airway Response

Herein, we analyzed the effects of different heat-killed probiotic bacteria in pasteurized yogurt on gut microbiota composition in MP-sensitized mice (Figure 5). The diversity of gut microbiota was decreased ($p < 0.05$) in allergic mice of the MP group, as indicated by the Ace and Chao index, in contrast to control mice (Figure 5A). Even though the diversity was also lower in allergic mice receiving pasteurized plain and heat-killed BBMN68 pasteurized yogurt, the proportions of beneficial *Spirillum*, *Mycobacterium*, *Lactobacillus*, and *Prevotella* were higher than those in the MP group, as shown in Figure 5C. The PCoA plot based on the unweighted-unifrac distances of the OTUs was presented (Figure 5B). The MP group and the control group were separated, while the BBMN68 + MP group was distinctly separated from the MP group. In Figure 5C, the bacterial community structure showed that Bacteroidetes and Firmicutes were the major bacterial phyla, and Bacteroidetes were more dominant in the MP and all pasteurized yogurt treatment groups than the control group. At the genus level, the proportion of harmful bacteria, including *Clostridium* and *Cholephila* increased while beneficial *Spirillum*, *Mycobacterium*, and *Lactobacillus* dropped in the gut microbiota of mice in the MP group in contrast to the control group. The MP induced a gut environment favoring Th1/Th2 imbalance and inflammation by altering microbiota structure in mice; supplementation of pasteurized yogurt with and without heat-killed probiotics obviously mitigated this deleterious effect via increasing health-promoting bacteria proportion.

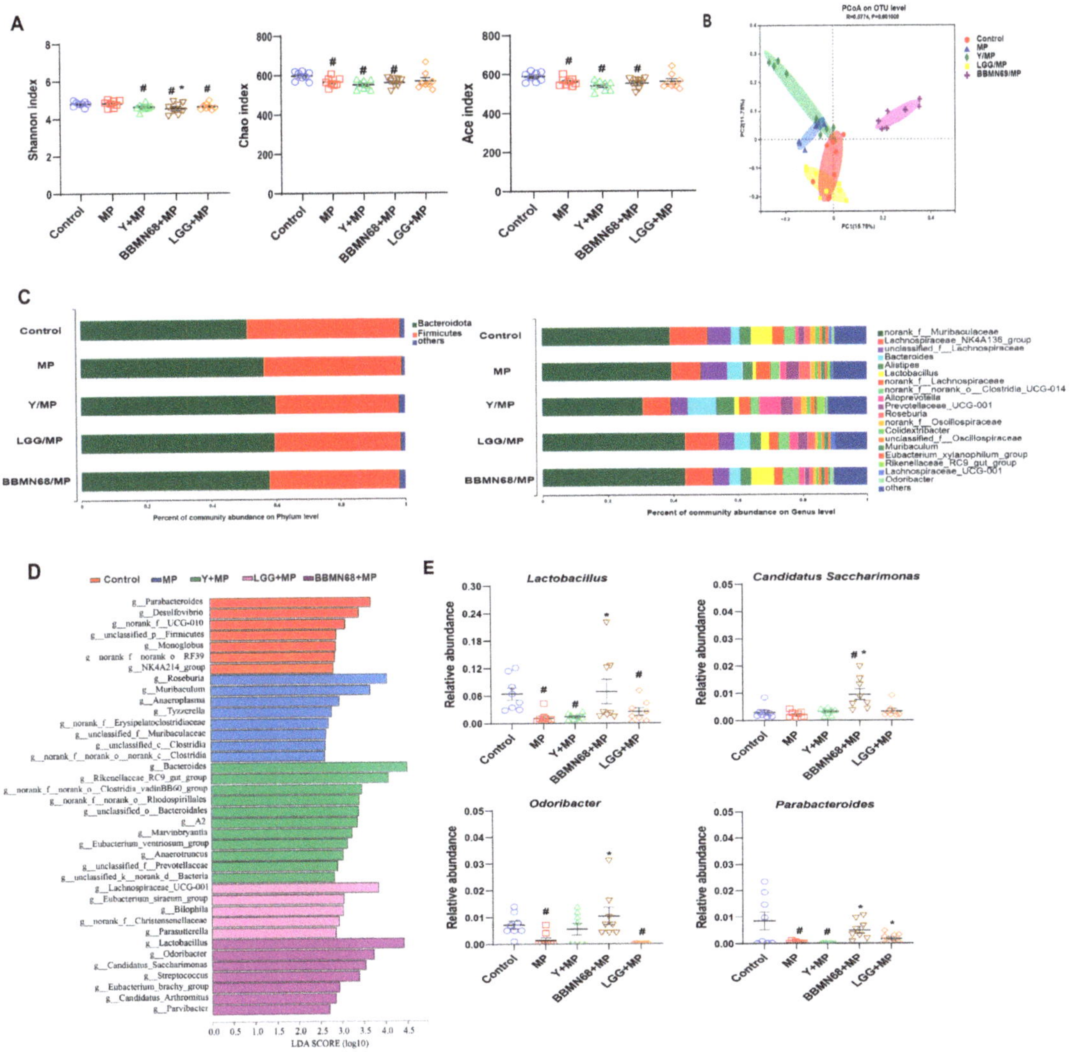

Figure 5. Effects of oral administration of pasteurized plain and heat-killed BBMN68 or LGG fermented yogurt on the fecal microbiota of allergic mice exposed to mugwort pollen (MP). (**A**) Alpha-diversity was calculated with the Shannon index, Chao index, and Ace index. (**B**) PCoA plot based on the unweighted-unifrac distances of the OTUs. (**C**) Percentage of community abundance on phylum and genus levels. (**D**) The differentially abundant taxa identified by LEfSe with LDA values (LDA score > 2.5). (**E**) Comparison of relative abundances of significantly different representative beneficial microbial taxa at genus level. Data are expressed as mean ± SEM, different color graphics in each group represent duplicates. #: $p < 0.05$ compared to the control group; *: $p < 0.05$ compared to the MP group; $n = 6–8$.

An LEfSE analysis was performed to compare the relative contributions of different taxa among different groups (Figure 5D). A total of 38 significant different taxa at the genus level among groups were determined. Compared with the control group, increased pro-inflammatory bacteria (*Tyzzerella*, *Clostridia*, and *Anaeroplasma*) were found in the intestinal tract of allergic mice in the MP group. The anti-inflammatory bacteria (e.g., *gRhodospirillales* and *Prevotellaceae*) were enriched in the Y + MP group. Beneficial bacteria *Lachnospiraceae* and *Eubacterium_siraeum*_group were detected in the LGG + MP group. Meanwhile, higher abundances of anti-inflammatory beneficial bacteria *Lactobacillus*, *Odoribacter*, *Eubacterium_brachy*_group, and *Parvibacter* were detected in the BBMN68 + MP group. A comparison of representative microbial taxa from these discriminative genera is shown in Figure 5E. Allergic mice receiving pasteurized yogurt containing heat-killed BBMN68 had significantly higher abundances ($p < 0.05$) of beneficial *Lactobacillus*, *CandidatusSaccharimonas*, *Odoribacter*, and *Parabacteroides* than those in the MP group.

The Pearson correlation between gut microbial composition and serum IgE and Th2 cytokines was analyzed (Figure 6). Significant negative correlations were detected between allergic responses and genus involved with immune function, such as *Lactobacillus* and serum IgE, IL-4, IL-5, and IL-13; *Candidatus_Saccharimonas* and serum IL-5, IL-13; *Parabacteroides* and IgE, IL-4, IL-5; *Enterorhabdus* and IgE, IL-4; *Odoribacter* and IgE; as well as *norank_f__Muribaculaceae* and IL-4. In particular, *Lactobacillus*, *Candidatus_Saccharimonas*, *Odoribacter*, and *Parabacteroides* were enriched in the gut of allergic mice after heat-killed BBMN68 pasteurized yogurt intake (Figure 5E).

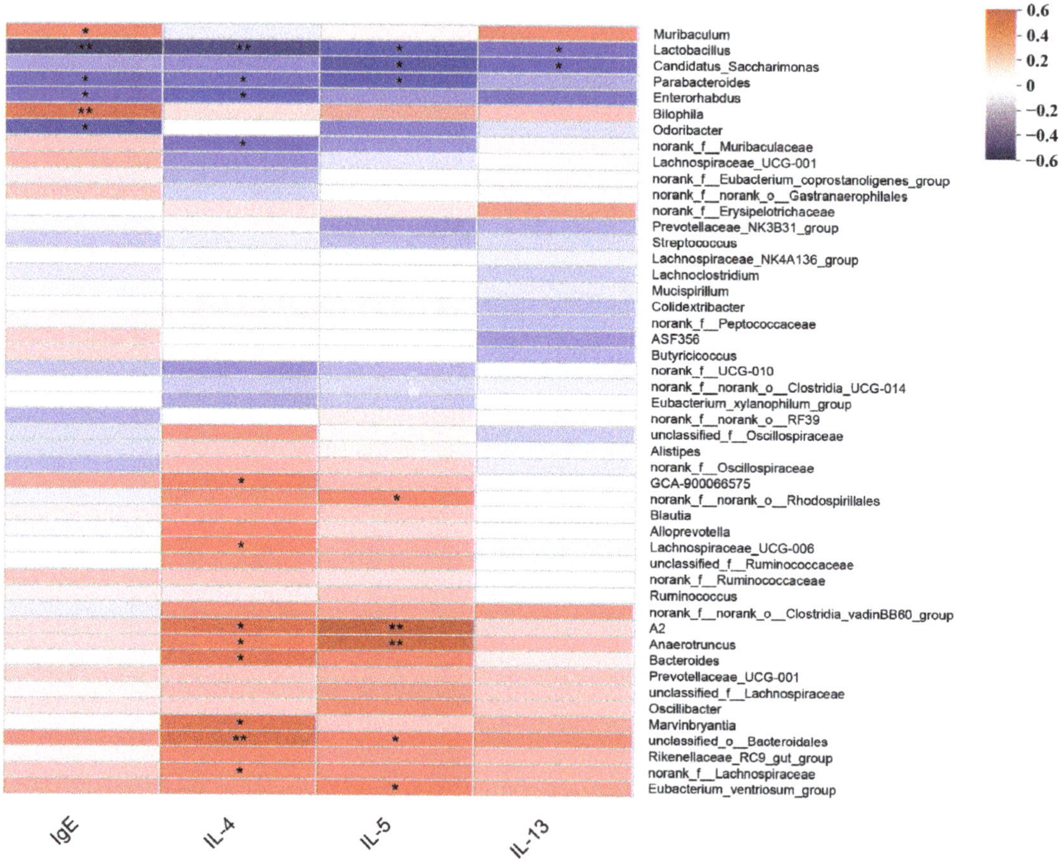

Figure 6. Pearson correlations between the relative abundance of gut microbiota and serum IgE, IL-4, IL-5, and IL-13 levels of allergic mice treated with pasteurized plain and heat-killed BBMN68 or LGG yogurt. * $p < 0.05$, ** $p < 0.01$.

4. Discussion

The rising prevalence of allergic airway disease has become a threat to public health that affects millions of people and causes high economic costs in many countries. Exposure to airborne allergen sources is inevitable and brings challenges to life quality improvement. In this study, we found the protective effects of heat-killed *B. longum* BBMN68 in pasteurized yogurt in a MP-induced allergic sensitization mouse model. The results demonstrated that oral supplementation of heat-killed BBMN68 decreased serum IgE levels, adjusted Th1/Th2 immune balance, alleviated airway inflammation, promoted airway remodeling, and reshaped the gut microbiota composition without undesirable side effects.

As a typical source of good bacteria and nutrients, yogurt possesses multiple health-promoting effects, including immunologic and metabolic benefits [22–24]. It has been shown that fresh and pasteurized yogurt had no significant differences in immunological parameters [25]. Pasteurized yogurt has many advantages; for example, it can be temporarily stored at room temperature, which greatly expands its potential use. Moreover, yogurts supplemented with additional probiotics and nutrients such as vitamin D display combined effects to enhance positive outcomes via modulating gut microbiota [26]. In this study, oral supplementation of pasteurized yogurt with heat-killed BBMN68 exhibited improved protection against MP-induced allergies compared with pasteurized plain yogurt, which indicated that yogurt can be a nutritious delivery medium for postbiotics.

Heat-killed probiotics, also known as postbiotics, have been used as a novel strategy for the prevention and treatment of many immune and gastrointestinal disorders, such as allergic airway disease, food allergies, and necrotizing enterocolitis. Oral administration of heat-inactivated *Lactobacillus plantarum* K37 showed anti-allergic effects by modulating airway hyperresponsiveness in ovalbumin-sensitized BALB/c mice [27]. Heat-killed *E. faecalis* YM-73 and *L. salivarius* AP-32 switched the immune response from a Th2 toward a Th1 response via increasing Th1-associated cytokines and reducing Th2-associated cytokines in Caco-2 cells [28]. Our results showed that heat-killed LGG significantly downregulated the Th2 cytokines but not the total IgE level, which is consistent with a previous study [11]. Numerous studies suggest that the major health-promoting roles of beneficial microbes, for example LGG, are linked with their non-viable byproducts, including cell-wall composition (mainly peptidoglycan and lipoteichoic acids), exopolysaccharides, surface-layer proteins, cell-free supernatants, bacteriocins, and metabolites such as short-chain fatty acids (SCFAs). These by products have the potential to facilitate the growth of beneficial commensal bacteria, hinder the growth and adherence of pathogenic bacteria, and induce anti-inflammatory and/or antioxidant activity [10,29–33]. Even though the functional constituents and their biological activities of *B. longum* BBMN68 as a postbiotic have yet to be fully elucidated, a previous study indicated that its cell wall fraction is the major effective component compared to the cell-free extract in a model of LPS-stimulated macrophage-like cells [19]. In this study, the protective effects of heat-killed LGG were less pronounced than those of heat-killed BBMN68, which may be due to their strain-specific immunomodulation and conclusion that heat-killed BBMN68 has a better capacity for reducing sensitization severity. The temperature and duration of heat treatment have a critical impact on the immune-related properties of probiotics cell surfaces. Li et al. found that heat treatment of LGG at 80 °C for 20 min decreased LPS-induced proinflammatory mediators and increased anti-inflammatory mediators in rats [14]. Another study on *Bifidobacterium longum* indicated that heat treatment of CECT-7347 at 121 °C for 20 min exhibited antioxidant, anti-inflammatory, and gut-barrier protection properties [34]. Scanning electron microscopy analysis of the LGG cell wall after spray-drying revealed that the pili were removed after heating, which resulted in reduced adherence capacity [35]. Further research is needed to fully understand the impact of heat treatment on the immunomodulatory properties of LGG and BBMN68 and determine the optimal conditions for preserving probiotic properties.

Eosinophilic airway inflammation is predominant in both allergic and non-allergic sensitization [36]. The oral administration of heat-killed BBMN68 in pasteurized yogurt significantly reduced BALF eosinophils, largely due to the decreased Th2-associated cytokines IL-4, IL-5, and IL-13 levels in serum, and therefore may modulate the contraction and length adaptation of airway smooth muscle cells, contributing to the improvement of diminished airflow obstruction. Meanwhile, Th1 and Th17 related immune responses were not observed in this research, which implicates that BBMN68 maintains systemic immune balance primarily through downregulating the Th2 response. The suppression of IL-4 and IL-13 also accounts for lower IgE production [1]. The serum IgG production was not altered with postbiotic treatments; similar results have been reported on probiotics using different allergic models [4,11]. It was speculated that successful treatment of allergic disease does not necessarily depend on decreased allergen-specific IgG1 or IgE antibody levels [4,37]. In addition, systemic total IgG is not representative of IgG subclass profiles; the MP allergenic protein Art v 1-specific IgG should be determined in future studies. In recent years, the role of eosinophils in the pathogenesis of severe allergic inflammation has been recognized, which makes anti-eosinophil therapy a promising target [38]. Our result could lead to the development of new therapies using postbiotics to target eosinophilic airway inflammation. In addition, the correlations between immune reactions, inflammation, and specific strains of gut bacteria were explored. The links were found between specific genus and serum cytokine levels, which provide more evidence for effective anti-IL-5 or anti-IL-4 approaches by precisely targeting gut microbiota for allergic airway disease prevention and treatment.

The phenomenon in this study also supported the idea of a "diet-gut-lung axis". Emerging evidence has revealed the communication between allergic inflammation and microbiota in the gut and lung after probiotic and prebiotic administration via oral or intranasal routes, thus leading to a reduction in severity and susceptibility in animals and humans [11,39,40]. The microbiota, both in the gut and lung, plays important roles in the development and regulation of allergic inflammation by influencing the maturation and function of the systematic immune system [41,42]. The constituents and metabolites derived from the gut microbiota can be the key signaling molecules of the immune system. Among them, SCFAs (e.g., acetate, butyrate, and propionate) are recognized as the most important mediators [43]. They can bind to G-protein-coupled receptors (GPCR) such as 41 and 43 located on the surface of intestinal epithelial cells and therefore activate regulatory T cells (Treg) and inhibit NF-κB activity, leading to anti-inflammatory effects and gut homeostasis maintenance [44]. For example, butyrate is reported to induce IL-10 and IL-18 production through GPR109A to inhibit colonic inflammation and inflammation-associated cancers [45]. Moreover, the SCFAs produced in the gut can be transported to the blood stream and influence lung and bone marrow homeostasis. SCFAs in the lung inhibit the activity of histone deacetylase and promote the mTOR-S6K pathway to regulate T cell differentiation and cytokine production to combat allergic airway diseases [46]. SCFAs in the bone marrow, a major site of innate and adaptive immune cell development, can promote the differentiation of hematopoietic stem cells and the production of monocyte and DC progenitors (MDPs) and common DC precursors (CDPs), which can migrate to the lung to mitigate Th2 cell-mediated inflammation [47,48]. Herein, a negative correlation was found between the genus involved with SCFA-producing bacteria *Lactobacillus*, *Parabacteroides*, and *Odoribacter* and serum antibodies and Th2 cytokines, indicating that heat-killed BBMN68 may influence the systemic immune system by favoring SCFA-producing bacteria growth. Previous studies have shown that dysbiosis of the gut microbiome stimulates local and systematic inflammatory pathways and leads to airway hyperresponsiveness [42]. In this study, even though lower bacterial community diversity was observed, the yogurt containing heat-killed BBMN68 modified the gut microbiota composition by promoting the abundance of beneficial bacteria such as *Spirillum*, *Mycobacterium*, *Lactobacillus*, and *Prevotella*. Overall, the improved inflammatory conditions in the lung may result from the gut homeostasis maintained by inactivated *B. longum* BBMN68 intake. The influence of heat-killed BBMN68 on the lung microbiome should be assessed in future studies.

This research provides evidence that pasteurized yogurt containing BBMN68 exhibits immunomodulatory activity in a mouse model and highlights its potential application in the protection of pollen-induced allergic airway inflammation. However, the applicability of such findings to humans is uncertain and requires further research. More clinically relevant parameters of allergic inflammation-associated diseases typically observed in humans should be determined to improve the translational value of data from animal models. Based on that, clinical trials of different human life stages, especially early life, deserve further exploration to determine the effects of pasteurized yogurt containing BBMN68 on allergic airway diseases.

In conclusion, this study testified that treatment of allergic sensitization by oral administration of yogurt containing heat-killed BBMN68 is valid in an MP-induced allergic mouse model and that heat-killed BBMN68 can be used as a postbiotic. The optimal dose and most effective mode of administration via nasal or oral routes still require further investigation. More importantly, future studies on the functional constituents of BBMN68 and the mechanisms of their biological activities are essential for its potential application.

Author Contributions: The authors' responsibilities were as follows: R.W., X.W. and Y.J.: conceptualization; X.N. and X.Y.: data curation, formal analysis; R.W. and J.H.: funding acquisition, supervision; X.N., X.Y., Q.Z., Y.Z., C.Z. and Y.R.: methodology; X.N. and X.Y.: visualization; X.N. and X.Y.: original draft writing; X.N., X.Y., J.H., M.L., Y.S. and R.W.: review and editing. All authors have read and agreed to the published version of the manuscript.

Funding: This study is financially supported by the National Key R&D Program of China (No. 2021YFD1600204).

Institutional Review Board Statement: The animal study protocol was approved by the Ethics Committee of Pony Testing International Group Co., Ltd. (Beijing, China) (Approval Number: PONY-2021-FL-72).

Data Availability Statement: All related data and methods are presented in this paper. Additional inquiries should be addressed to the corresponding author.

Acknowledgments: The graphical abstract was created with BioRender.com.

Conflicts of Interest: The authors declare no conflict of interest related to this work. The participation of the author from the Inner Mongolia Mengniu Dairy (Group) Co. did not affect the authenticity and objectivity of the experiment.

References

1. Finkelman, F.D.; Hogan, S.P.; Hershey, G.K.K.; Rothenberg, M.E.; Wills-Karp, M. Importance of cytokines in murine allergic airway disease and human asthma. *J. Immunol.* **2010**, *184*, 1663–1674. [CrossRef] [PubMed]
2. Ma, W.J.; Bao, M.J.; Zhu, J.P.; Yao, H.Y.; Xie, Y.C.; Guan, Y.; Li, F.F.; Dong, X.W.; Zheng, Y.M.; Xie, Q.M. Oral administration of allergen extracts from mugwort pollen desensitizes specific allergen-induced allergy in mice. *Vaccine* **2012**, *30*, 1437–1444. [CrossRef] [PubMed]
3. Meadows, A.; Kaambwa, B.; Novielli, N.; Huissoon, A.; Fry-Smith, A.; Meads, C.; Barton, P.; Dretzke, J. A systematic review and economic evaluation of subcutaneous and sublingual allergen immunotherapy in adults and children with seasonal allergic rhinitis. *Health Technol. Assess* **2013**, *17*, 1–322. [CrossRef]
4. Spacova, I.; van Beeck, W.; Seys, S.; Devos, F.; Vanoirbeek, J.; Vanderleyden, J.; Ceuppens, J.; Petrova, M.; Lebeer, S. Lactobacillus rhamnosus probiotic prevents airway function deterioration and promotes gut microbiome resilience in a murine asthma model. *Gut Microbes* **2020**, *11*, 1729–1744. [CrossRef]
5. Gibson, G.R.; Hutkins, R.; Sanders, M.E.; Prescott, S.L.; Reimer, R.A.; Salminen, S.J.; Scott, K.; Stanton, C.; Swanson, K.S.; Cani, P.D.; et al. The International Scientifc Association for Probiotics and Prebiotics (ISAPP) consensus statement on the definition and scope of prebiotics. *Nat. Rev. Gastroenterol. Hepatol.* **2017**, *14*, 491–502. [CrossRef]
6. Kalliomäki, M.; Antoine, J.M.; Herz, U.; Rijkers, G.T.; Wells, J.M.; Mercenier, A. Guidance for substantiating the evidence for benefcial efects of probiotics: Prevention and management of allergic diseases by probiotics. *J. Nutr.* **2010**, *140*, S713–S721. [CrossRef]
7. Mennini, M.; Dahdah, L.; Artesani, M.C.; Fiocchi, A.; Martelli, A. Probiotics in asthma and allergy prevention. *Front. Pediatr.* **2017**, *5*, 165. [CrossRef]
8. West, C.E.; Dzidic, M.; Prescott, S.L.; Jenmalm, M.C. Bugging allergy; role of pre-, pro- and synbiotics in allergy prevention. *Allergol. Int.* **2017**, *66*, 529–538. [CrossRef]
9. Kalliomäki, M.; Salminen, S.; Poussa, T.; Arvilommi, H.; Isolauri, E. Probiotics and prevention of atopic disease: 4-year follow-up of a randomised placebo-controlled trial. *Lancet* **2003**, *361*, 1869–1871. [CrossRef]

10. Wu, Z.; Pan, D.; Guo, Y.; Sun, Y.; Zeng, X. Peptidoglycan diversity and anti-inflammatory capacity in Lactobacillus strains. *Carbohydr. Polym.* **2015**, *128*, 130–137. [CrossRef]
11. Spacova, I.; Petrova, M.I.; Fremau, A.; Pollaris, L.; Vanoirbeek, J.; Ceuppens, J.L.; Seys, S.; Lebeer, S. Intranasal administration of probiotic Lactobacillus rhamnosus GG prevents birch pollen-induced allergic asthma in a murine model. *Allergy* **2019**, *74*, 100–110. [CrossRef] [PubMed]
12. Salminen, S.; Collado, M.C.; Endo, A.; Hill, C.; Lebeer, S.; Quigley, E.M.M.; Sanders, M.E.; Shamir, R.; Swann, J.R.; Szajewska, H.; et al. The International Scientific Association of Probiotics and Prebiotics (ISAPP) consensus statement on the definition and scope of postbiotics. *Nat. Rev. Gastroenterol. Hepatol.* **2021**, *18*, 649–667. [CrossRef] [PubMed]
13. Zhang, L.; Li, N.; Caicedo, R.; Neu, J. Alive and Dead Lactobacillus rhamnosus GG Decrease Tumor Necrosis Factor-α–Induced Interleukin-8 Production in Caco-2 Cells. *J. Nutr.* **2005**, *135*, 1752–1756. [CrossRef] [PubMed]
14. Li, N.; Russell, W.M.; Douglas-Escobar, M.; Hauser, N.; Lopez, M.; Neu, J. Live and heat-killed Lactobacillus rhamnosus GG: Effects on proinflammatory and anti-inflammatory cytokines/chemokines in gastrostomy-fed infant rats. *Pediatr. Res.* **2009**, *66*, 203–207. [CrossRef]
15. Li, L.; Sun, Q.; Xiao, H.; Zhang, Q.; Xu, S.; Lai, L.; Li, Z.; Li, C. Aerosol inhalation of heat-killed clostridium butyricum CGMCC0313-1 alleviates allergic airway inflammation in mice. *J. Immunol. Res.* **2022**, *2022*, 8447603. [CrossRef]
16. Hong, H.J.; Kim, E.; Cho, D.; Kim, T.S. Differential suppression of heat-killed lactobacilli isolated from kimchi, a Korean traditional food, on airway hyper-responsiveness in mice. *J. Clin. Immunol.* **2010**, *30*, 449–458. [CrossRef]
17. Hao, Y.; Huang, D.; Guo, H.; Xiao, M.; An, H.; Zhao, L.; Zuo, F.; Zhang, B.; Hu, S.; Song, S.; et al. Complete genome sequence of Bifidobacterium longum subsp. longum BBMN68, a new strain from a healthy chinese centenarian. *J. Bacteriol.* **2011**, *193*, 787–788. [CrossRef]
18. Yang, H.Y.; Liu, S.L.; Ibrahim, S.A.; Zhao, L.; Jiang, J.L.; Sun, W.F.; Ren, F.Z. Oral administration of live Bifidobacterium substrains isolated from healthy centenarians enhanced immune function in BALB/c mice. *Nutr. Res.* **2009**, *29*, 281–289. [CrossRef]
19. Zhu, J.; Zhao, L.; Guo, H.; Jiang, L.; Ren, F. Immunomodulatory effects of novel bifidobacterium and lactobacillus strains on murine macrophage cells. *Afr. J. Microbiol. Res.* **2011**, *5*, 8–15.
20. Yang, J.; Zhang, H.; Jiang, L.; Guo, H.; Luo, X.; Ren, F. Bifidobacterium longum BBMN 68-specific modulated dendritic cells alleviate allergic responses to bovine β-lactoglobulin in mice. *J. Appl. Microbiol.* **2015**, *119*, 1127–1137. [CrossRef]
21. Zhang, C.; Li, W.; Li, X.; Wan, D.; Mack, S.; Zhang, J.; Wagner, K.; Wang, C.; Tan, B.; Chen, J.; et al. Novel aerosol treatment of airway hyper-reactivity and inflammation in a murine model of asthma with a soluble epoxide hydrolase inhibitor. *PLoS ONE* **2022**, *17*, e0266608. [CrossRef] [PubMed]
22. Kok, C.R.; Hutkins, R. Yogurt and other fermented foods as sources of health-promoting bacteria. *Nutr. Rev.* **2018**, *76*, 4–15. [CrossRef] [PubMed]
23. Daniel, N.; Nachbar, R.T.; Tran, T.T.T.; Ouellette, A.; Varin, T.V.; Cotillard, A.; Quinquis, L.; Gagné, A.; St-Pierre, P.; Trottier, J.; et al. Gut microbiota and fermentation-derived branched chain hydroxy acids mediate health benefits of yogurt consumption in obese mice. *Nat. Commun.* **2022**, *13*, 1343. [CrossRef]
24. Meydani, S.N.; Ha, W.K. Immunologic effects of yogurt. *Am. J. Clin. Nutr.* **2000**, *71*, 861–872. [CrossRef]
25. Ballesta, S.; Velasco, C.; Borobio, M.V.; Arguelles, F.; Perea, E.J. Fresh versus pasteurized yogurt: Comparative study of the effects on microbiological and immunological parameters, and gastrointestinal comfort. *Enferm. Infecc. Microbiol. Clin.* **2008**, *26*, 552–557. [CrossRef]
26. Gasparri, C.; Perna, S.; Spadaccini, D.; Alalwan, T.; Girometta, C.; Infantino, V.; Rondanelli, M. Is vitamin D-fortified yogurt a value-added strategy for improving human health? A systematic review and meta-analysis of randomized trials. *J. Dairy Sci.* **2019**, *102*, 8587–8603. [CrossRef]
27. Liu, Y.W.; Liao, T.W.; Chen, Y.H.; Chiang, Y.C.; Tsai, Y.C. Oral administration of heat-inactivated Lactobacillus plantarum K37 modulated airway hyperresponsiveness in ovalbumin-sensitized BALB/c mice. *PLoS ONE* **2014**, *9*, e100105. [CrossRef]
28. Ou, C.C.; Lin, S.L.; Tsai, J.J.; Lin, M.Y. Heat-killed lactic acid bacteria enhance immunomodulatory potential by skewing the immune response toward Th1 polarization. *J. Food Sci.* **2011**, *76*, M260–M267. [CrossRef]
29. Johnson-Henry, K.C.; Hagen, K.E.; Gordonpour, M.; Tompkins, T.A.; Sherman, P.M. Surface-layer protein extracts from Lactobacillus helveticus inhibit enterohaemorrhagic Escherichia coli O157:H7 adhesion to epithelial cells. *Cell Microbiol.* **2007**, *9*, 356–367. [CrossRef]
30. Yan, F.; Cao, H.; Cover, T.L.; Whitehead, R.; Washington, M.K.; Polk, D.B. Soluble proteins produced by probiotic bacteria regulate intestinal epithelial cell survival and growth. *Gastroenterology* **2007**, *132*, 562–575. [CrossRef]
31. Wu, M.H.; Pan, T.M.; Wu, Y.J.; Chang, S.J.; Chang, M.S.; Hu, C.Y. Exopolysaccharide activities from probiotic bifidobacterium: Immunomodulatory effects (on J774A.1 macrophages) and antimicrobial properties. *Int. J. Food Microbiol.* **2010**, *144*, 104–110. [CrossRef] [PubMed]
32. Bermudez-Brito, M.; Plaza-Díaz, J.; Muñoz-Quezada, S.; Gómez-Llorente, C.; Gil, A. Probiotic mechanisms of action. *Ann. Nutr. Metab.* **2012**, *61*, 160–174. [CrossRef] [PubMed]
33. Liu, Z.; Zhang, Z.; Qiu, L.; Zhang, F.; Xu, X.; Wei, H.; Tao, X. Characterization and bioactivities of the exopolysaccharide from a probiotic strain of Lactobacillus plantarum WLPL04. *J. Dairy Sci.* **2017**, *100*, 6895–6905. [CrossRef] [PubMed]

34. Martorell, P.; Alvarez, B.; Llopis, S.; Navarro, V.; Ortiz, P.; Gonzalez, N.; Balaguer, F.; Rojas, A.; Chenoll, E.; Ramón, D.; et al. Heat-treated Bifidobacterium longum CECT-7347: A whole-cell postbiotic with antioxidant, anti-Inflammatory, and gut-barrier protection properties. *Antioxidants* **2021**, *10*, 536. [CrossRef] [PubMed]
35. Kiekens, S.; Vandenheuvel, D.; Broeckx, G.; Claes, I.; Allonsius, C.; De Boeck, I.; Thys, S.; Timmermans, J.P.; Kiekens, F.; Lebeer, S. Impact of spray-drying on the pili of Lactobacillus rhamnosus GG. *Microb. Biotechnol.* **2019**, *12*, 849–855. [CrossRef]
36. George, L.; Brightling, C.E. Eosinophilic airway inflammation: Role in asthma and chronic obstructive pulmonary disease. *Ther. Adv. Chronic Dis.* **2016**, *7*, 34–51. [CrossRef]
37. Pellaton, C.; Nutten, S.; Thierry, A.C.; Boudousquié, C.; Barbier, N.; Blanchard, C.; Corthésy, B.; Mercenier, A.; Spertini, F. Intragastric and intranasal administration of Lactobacillus paracasei NCC2461 modulates allergic airway inflammation in mice. *Int. J. Inflam.* **2012**, *2012*, 686739.
38. Vatrella, A.; Maglio, A.; Pelaia, C.; Ciampo, L.; Pelaia, G.; Vitale, C. Eosinophilic inflammation: An appealing target for pharmacologic treatments in severe asthma. *Biomedicines* **2022**, *10*, 2181. [CrossRef]
39. McLoughlin, R.; Berthon, B.S.; Rogers, G.B.; Baines, K.J.; Leong, L.E.X.; Gibson, P.G.; Williams, E.J.; Wood, L.G. Soluble fibre supplementation with and without a probiotic in adults with asthma: A 7-day randomised, double blind, three way cross-over trial. *eBioMedicine* **2019**, *46*, 473–485. [CrossRef]
40. Zhang, J.; Ma, J.; Li, Q.; Su, H.; Sun, X. Exploration of the effect of mixed probiotics on microbiota of allergic asthma mice. *Cell Immunol.* **2021**, *367*, 104399. [CrossRef]
41. Barcik, W.; Boutin, R.C.T.; Sokolowska, M.; Finlay, B.B. The role of lung and gut microbiota in the pathology of asthma. *Immunity* **2020**, *52*, 241–255. [CrossRef] [PubMed]
42. Hufnagl, K.; Pali-Schöll, I.; Roth-Walter, F.; Jensen-Jarolim, E. Dysbiosis of the gut and lung microbiome has a role in asthma. *Semin. Immunopathol.* **2020**, *42*, 75–93. [CrossRef] [PubMed]
43. Dang, A.T.; Marsland, B.J. Microbes, metabolites, and the gut–lung axis. *Mucosal. Immunol.* **2019**, *12*, 843–850. [CrossRef] [PubMed]
44. Le Poul, E.; Loison, C.; Struyf, S.; Springael, J.Y.; Lannoy, V.; Decobecq, M.E.; Brezillon, S.; Dupriez, V.; Vassart, G.; van Damme, J.; et al. Functional characterization of human receptors for short chain fatty acids and their role in polymorphonuclear cell activation. *J. Biol. Chem.* **2003**, *278*, 25481–25489. [CrossRef]
45. Singh, N.; Gurav, A.; Sivaprakasam, S.; Brady, E.; Padia, R.; Shi, H.; Ganapathy, V. Activation of Gpr109a, receptor for niacin and the commensal metabolite butyrate, suppresses colonic inflammation and carcinogenesis. *Immunity* **2014**, *40*, 128–139. [CrossRef]
46. Park, J.; Kim, M.; Kang, S.G.; Jannasch, A.H.; Cooper, B.; Patterson, J.; Kim, C.H. Short-chain fatty acids induce both effector and regulatory T cells by suppression of histone deacetylases and regulation of the mTOR–S6K pathway. *Mucosal Immunol.* **2015**, *8*, 80–93. [CrossRef]
47. Desch, A.N.; Henson, P.M.; Jakubzick, C.V. Pulmonary dendritic cell development and antigen acquisition. *Immunol. Res.* **2013**, *55*, 178–186. [CrossRef]
48. Kopf, M.; Schneider, C.; Nobs, S.P. The development and function of lung-resident macrophages and dendritic cells. *Nat. Immunol.* **2015**, *16*, 36–44. [CrossRef]

foods

MDPI

Article

Effects of the Exopolysaccharide from *Lactiplantibacillus plantarum* HMX2 on the Growth Performance, Immune Response, and Intestinal Microbiota of Juvenile Turbot, *Scophthalmus maximus*

Gege Hu, Yihui Wang, Rui Xue, Tongji Liu, Zengjia Zhou and Zhennai Yang *

Key Laboratory of Geriatric Nutrition and Health of Ministry of Education, Beijing Advanced Innovation Center for Food Nutrition and Human Health, Beijing Engineering and Technology Research Center of Food Additives, Beijing Technology and Business University, No. 11 Fu-Cheng Road, Hai-Dian District, Beijing 100048, China; hugege0819@163.com (G.H.); wangyihui202103@163.com (Y.W.); 2130021028@st.btbu.edu.cn (R.X.); 2130021006@st.btbu.edu.cn (T.L.); m15255825668@163.com (Z.Z.)

* Correspondence: yangzhennai@th.btbu.edu.cn; Tel.: +86-10-68984870

Abstract: In this study, the exopolysaccharide (EPS) from *Lactiplantibacillus plantarum* (HMX2) was isolated from Chinese Northeast Sauerkraut. Its effects on juvenile turbot were investigated by adding different concentrations of HMX2-EPS (C: 0 mg/kg, H1: 100 mg/kg, H2: 500 mg/kg) to the feed. Compared with the control group, HMX2-EPS significantly improved the growth performance of juvenile turbot. The activities of antioxidant enzymes, digestive enzymes, and immune-related enzymes were significantly increased. HMX2-EPS could also increase the secretion of inflammatory factors and enhance the immune response of turbot by regulating the IFN signal transduction pathway and exhibit stronger survival rates after the *A. hydrophila* challenge. Moreover, HMX2-EPS could improve the diversity of intestinal microbiota in juvenile fish, increase the abundance of potential probiotics, and reduce the abundance of pathogenic bacteria. The function of gut microbes in metabolism and the immune system could also be improved. All results showed better effects with high concentrations of HMX2-EPS. These results indicated that HMX2-EPS supplementation in the diet could promote growth, improve antioxidant activity, digestive capacity, and immunity capacity, and actively regulate the intestinal microbiota of juvenile turbot. In conclusion, this study might provide basic technical and scientific support for the application of *L. plantarum* in aquatic feed.

Keywords: *Lactiplantibacillus plantarum*; exopolysaccharide (EPS); immunity; intestinal health; *Scophthalmus maximus*

1. Introduction

Probiotics are widely used in aquaculture because of their diverse beneficial properties, as they have the capability to improve growth performance and antioxidant levels and enhance innate immune response [1]. Probiotics can effectively reduce the number of antibiotics in aquaculture, thereby reducing the production of drug-resistant bacteria, and preventing side effects such as drug residues [2]. Therefore, the use of probiotics on aquatic animals has become a very effective measure to replace antibiotics [3]. Recently, more and more researchers have paid attention to the secondary metabolites of probiotics, such as exopolysaccharide (EPS). EPSs have unique physicochemical properties and biological activities, so they are applied in food, medicine, and other industrial products, which has stimulated interest in exploring and discovering new EPSs [4].

EPS is a high molecular weight long-chain polymer that is mainly produced by the metabolism of microorganisms, including cyanobacteria, fungi, and bacteria [5]. Lactic acid bacteria (LAB) have attracted unique research interest among EPS-producing bacteria due to their beneficial properties of recognized safety [6]. As an important secondary metabolite of LAB, EPS has many biological functions, making it one of the most potential substitutes for immunomodulators. The modulatory effects of LAB-derived EPS on host immunity, antioxidant activity, and intestinal microbiota have been extensively studied. EPS from *Lactobacillus casei* WXD030 could enhance RAW264.7 cell proliferation and phagocytic activity, as well as induce NO, TNF-α, IL-1β, and IL-6 production to play a role in regulating immune responses [7]. EPS produced by *Lactobacillus plantarum* RS20D could stimulate macrophages to release NO and up-regulate gene expression of pro-inflammatory cytokines at the mRNA level [8]. Liu et al. found that EPS from *Lactobacillus* Y42 exerted antioxidant function by up-regulating the expression and activities of antioxidant enzyme lines such as catalase (CAT), superoxide dismutase (SOD), and glutathione peroxidase (GSH-PX) in HT-29 cells [9]. EPS produced by *Lactobacillus plantarum* YW11 effectively recovered the intestinal microbial diversity in mice and increased the abundance of Roseburia, Ruminococcus, and Blautia with an increased content of butyric acid [10]. Many in vitro research results showed that EPS could effectively inhibit the growth of harmful bacteria such as *Enterobacter sakazakii*, *Escherichia coli*, *Listeria monocytogenes*, *Staphylococcus aureus*, *Candida albicans*, and *Salmonella typhimurium* [11]. It has been reported that some EPSs of LAB can be effective stimulants for improving the growth performance and immunity of fish and have a promising application in aquaculture due to their biological activities, such as antibacterial, antioxidant, anticancer, and modulating immunity. Feng et al. found that EPS of *Lactococcus lactis* Z-2 isolated from healthy carp could enhance its immune function, antioxidant capacity, and disease resistance [12]. However, studies on the growth performance, immune response, disease resistance, and intestinal microbiota effects of EPS from food source probiotics on aquatic animals have been reported relatively little.

In this study, *Scophthalmus maximus* is an economically important species of flounder known as turbot. Its meat is delicious and rich in nutritional value, and it is widely cultivated in Europe and China [13]. At present, there is little research that has reported on food-derived probiotics exopolysaccharides as feed additives for aquatic animals. We isolated a strain of *Lactiplantibacillus plantarum* HMX2 that was isolated from Chinese Northeast Sauerkraut. EPS was extracted to investigate the effects on growth performance, antioxidant activities, digestive enzyme activities, immune capacity, and intestinal microbiota of juvenile turbot. The microbiome technology was used to reveal the regulatory mechanism of HMX2-EPS on the composition and diversity of core taxa in turbot to clarify the prebiotic effect of the EPS and provide the basic data for the application of *L. plantarum* in turbot farming. This study could further reveal the role of LAB in aquatic feed, which had positive significance for the comprehensive development and utilization of LAB.

2. Materials and Methods

2.1. Preparation of EPS Samples

Lactiplantibacillus plantarum HMX2 was provided by the Dairy Laboratory at Beijing Technology and Business University of China. HMX2 was cultured in an SDM medium adjusted to pH 6.6 with 1 M acetic acid [14]. The EPS produced by the strain was then isolated and prepared using the method previously reported [15]. A total of 80% (wt/vol) trichloroacetic acid (TCA) (Shanghai Aladdin Biochemical Technology Co., Ltd., Shanghai, China) was added to the fermentation broth to achieve the 4% (wt/vol) trichloroacetic acid concentration. After stirring for 2 h at room temperature, followed by centrifugation at 4 °C for 30 min, 2 volumes of ethanol (Shanghai Aladdin Biochemical Technology Co., Ltd., Shanghai, China) were added to the supernatant and then kept overnight at 4 °C. The pellet of crude HMX2-EPS was collected after centrifugation at 12,000× *g* for 30 min at 4 °C. The precipitate was dialyzed and lyophilized with a lyophilizer (Christ, Hagen, Germany). Then, the protein was removed by the Sevage method, and the pigment was removed by activated carbon [16] to obtain the HMX2-EPS.

2.2. Experimental Design and Daily Management

The healthy juvenile *S. maximus* used in the experiment were purchased from a mariculture company in Yantai City, China. The experimental fish were temporarily maintained in aerated artificial seawater for 7 days, siphoning 30% of the water every day. During the temporary cultivation, the water temperature was controlled at (17 ± 1) °C, and the basic feed was fed after changing the water every day (Yantai Penganyuan Marine Food Co., Ltd., Yantai, China). According to our experimental arrangement, 300 juvenile turbot were randomly divided into 3 groups with 100 fish in each group and repeated four times, which were then divided into a control group and two experimental groups. The control group (C) was fed with basic feed, while the experimental group was fed with basic feed supplemented with 100 mg/kg (H1) and 500 mg/kg (H2) HMX2-EPS, respectively, and each group had three replicates. The culture experiment was conducted for 6 weeks, during which the water was changed once a day, with 1/3 to 1/2 water each time. The residual bait and feces were removed, and the health status of the fish was recorded. After the aquaculture experiment, the experimental fish were fasted for 24 h. All fish in the tank were fished out, and the quantity, length, and weight were recorded, respectively.

2.3. Sample Collection

Ten fish were randomly collected, and the intestines were quickly dissected and separated with 100 mg/L MS 222 (Beijing Green Hengxing Biotechnology Co., Ltd., Beijing, China). The intestines were washed with sterilized 0.85% normal saline and placed in RNA-free centrifuge tubes and frozen storage tubes and placed in liquid nitrogen. The intestines were then placed in an RNA-free centrifuge tube and a cryopreservation tube and placed in liquid nitrogen. The samples were stored in the refrigerator at −80 °C for the subsequent experiments.

2.4. Growth Performance Measurement

Equations below

$$\text{Weight gain rate (WGR, \%)} = (Wi - W_0)/W_0 \times 100\%;$$

$$\text{Specific growth rate (SGR, \%/d)} = (\text{Ln } Wi - \text{Ln } W_0)/t \times 100\%;$$

$$\text{Feed conversion ratio (FCR)} = Wf/(Wi - W_0);$$

$$\text{Survival rate (SR, \%)} = Ni/N_0 \times 100\%.$$

where W_0 is the initial body mass of the fish (g), Wi is the final body mass (g), Wf is the feed intake (g), N_0 is the initial number (tail), Ni is the final number (tail), and t is the number of days for the breeding experiment (d).

2.5. Enzymatic Assay

The digestive enzymes, antioxidant enzymes, and immune enzymes were assayed with corresponding commercial detection kits (C016-1-1, A054-1-1, A007-1-1, A001-3-2, A001-3-1, A060-2-1, Nanjing Jiancheng Bioengineering Institute, Nanjing, China). These digestive enzymes included α-amylase (AMS) and lipase (LPS), the antioxidant enzymes included catalase (CAT), superoxide dismutase (SOD), and the immune enzymes included acid phosphatase (ACP), and alkaline phosphatase (AKP). A bicinchoninic acid (BCA) protein assay kit (Beyotime Biotechnology, Haimen, China) was used to determine the protein concentration.

2.6. qPCR Analysis of Immune-Related Genes and Antioxidant Genes

The total RNA was extracted from intestinal cells by the TRIzol method, and RNA was examined by 0.1% agarose gel electrophoresis and a Nanodrop 2000 spectrophotometer (Thermo Fisher, Waltham, MA, USA). RNA was used as a template for reverse transcription to synthesize cDNA. Based on an ABI 7500 gene quantitative real-time detection system

(Thermo Fisher, USA), relative expression of the immune genes interleukin-1beta (IL-1β), interleukin-10 (IL-10), Myeloid differentiation primary response gene 88 (MyD88), type I interferon (IFN1), toll-like receptor 2 (TLR-2), superoxide dismutase (SOD), catalase (CAT), and glutathione peroxidase 1 (GPX1) were determined by SYBR green I method. Specific primers for genes are demonstrated in Table 1. The reaction system for real-time qPCR was 20 μL, including 0.4 μL upstream and downstream primers (10 μmol/L), 6 μL cDNA template, 10 μL SYBR Mix I, and 3.2 μL diethylpyrocarbonate (DEPC) water. The program was 50 °C for 20 s, 95 °C for 7 min, followed by 40 cycles of 95 °C for 10 s and 60 °C for 30 s. The relative mRNA expression levels were detected by the $2^{-\triangle\triangle CT}$ method.

Table 1. Primers of internal reference and target genes.

Gene	Primer Name	Sequence (5′->3′)	Genebank ID
β-actin	β-actin F β-actin R	GCTGTCTTCCCTTCTATCGTCG TCCATGTCATCCCAGTTGGTC	XM_035614477
IL-1β	IL-1β F IL-1β R	ATGGAGTGCAACATGAGCGA GAGCAGGTTTTCGTCCCTGA	XM_035640817
IL-10	IL-10 F IL-10 R	AGCTCAAGTCCGATGTCAGC TCAAGAGCTGGGTGATGCAC	XM_035632547
IFN-1	IFN-1 F IFN-1 R	CCGACGGGCATTATGGGTAG CGAGTATCCCTTCGTCCCAC	XM_035615472
MyD88	MyD88 F MyD88 R	CCCAATGGTAGCCCTGAGAT CATCTCGGTCGAACACACAC	XM_035618686
TLR-2	TLR-2 F TLR-2 R	CTTCGAGCCAGGTAAACCC AGAAAGACCAGGATCAGCACG	XM_035606298
SOD	SOD F SOD R	TATCAAGAGGCGCTGGCAAA GCTTCCATTAGCTCCCCCTG	XM_035617519
CAT	CAT F CAT R	CCAGAAACCCAGCCTCACTT GAAGGCACGGACCTGTGTAA	XM_035643471
GPX1	GPX1 F GPX1 R	TTCTGCCAAGGGACTCGTTG TCAAAGCCATTCCCTGGACG	XM_035631441

2.7. Infection Experiment of Aeromonas Hydrophila

After the sample collection, the remaining 20 fish in each group were used for the challenge experiment. Each fish was intraperitoneally injected with A. hydrophila suspension with a concentration of (3.3×10^7) cfu/mL 100 μL in the experimental group, while the control group was injected intraperitoneally with 0.65% normal saline 100 μL. The death of fish in each group was observed and recorded, and the survival rate (SR) was calculated according to the following formula.

$$\text{Survival rate (\%)} = \text{Final survival mantissa/initial mantissa} \times 100$$

2.8. Intestinal Microbial Composition

The total DNA of microbes in the intestine was extracted directly with the E.Z.N.A. Stool DNA Kit (OMEGA, Stamford, CT, USA). Amplification of the hypervariable region (V4) of the bacterial 16S DNA gene was performed for sequencing analysis and species identification. The PCR products obtained by amplification were detected by 1% agarose gel electrophoresis, cut and recovered, and sequenced and analyzed on Novogene's Illumina MiSeq platform (Tianjin, China) after accurate quantification. The method of sequencing and analysis referred to Yang et al. [17].

2.9. Statistical Analysis

All the data were presented as mean ± standard deviation (mean ± S.D.). Statistical analysis of experimental data was performed using SPSS 26.0 software and one-way ANOVA followed by LSD and Duncan's test for multiple comparisons of significant differences between groups, with $p < 0.05$ considered the significant level.

3. Results

3.1. Growth Performance of Juvenile Turbot

The growth performance of juvenile turbot is shown in Table 2. After the culture experiment, there was no significant difference in survival rates of turbot among all groups ($p > 0.05$). The FBL, FBW, SGR, and FCR of the H1 and H2 groups were significantly higher than those of the C group ($p < 0.05$), and there was no significant difference between the two EPS groups. The WGR of the H2 group was the highest, which was significantly higher than those of the C and H1 groups ($p < 0.05$). Therefore, the addition of HMX2-EPS had a significant effect on the growth of juvenile turbot, while the addition concentration had little effect on the growth of turbot.

Table 2. Effects of HMX2-EPS supplementation on the growth performance of juvenile turbot. IBL, initial body length; IBW, initial body weight; FBL, final body length; FBW, final body weight; WGR, weight gain rate; SGR, specific growth rate; FCR, feed conversion ratio; SR, survival rate. C, 0 mg/kg; H1, 100 mg/kg; H2, 500 mg/kg. Different superscripts in each line indicate significant differences ($p < 0.05$).

Items	Treatments		
	C	H1	H2
IBL (cm)	5.94 ± 0.07	5.85 ± 0.12	5.93 ± 0.14
IBW (g)	6.11 ± 0.60	6.15 ± 0.45	6.16 ± 0.31
FBL (cm)	7.54 ± 0.21 ^{b}	8.36 ± 0.11 ^{a}	8.44 ± 0.07 ^{a}
FBW (g)	21.07 ± 0.34 ^{b}	23.29 ± 0.59 ^{a}	24.15 ± 0.38 ^{a}
WGR (%)	244.84 ± 3.72 ^{c}	278.70 ± 8.76 ^{b}	292.05 ± 6.68 ^{a}
SGR (%/d)	2.95 ± 0.1 ^{b}	3.17 ± 0.09 ^{a}	3.25 ± 0.11 ^{a}
FCR	1.04 ± 0.08 ^{b}	1.17 ± 0.04 ^{a}	1.14 ± 0.05 ^{a}
SR (%)	98.33 ± 2.36	100.00	98.33 ± 2.36

3.2. Digestive and Immune-Related Enzyme Activities in Juvenile Turbot

As shown in Table 3, the LPS activity in the H2 group was the highest and significantly higher than that in the H1 and C groups ($p < 0.05$). The AMS activity in the H1 and H2 groups was significantly higher than that of the C group ($p < 0.05$). In addition, the AKP and ACP activities of the groups H1 and H2 were significantly higher than that of the C group ($p < 0.05$), and there was no significant difference between the two groups ($p > 0.05$). The activity of CAT in the H1 and H2 groups was significantly higher than that in group C ($p < 0.05$), while SOD activity in the H2 group was the highest, and there was a significant difference between H1 and H2 groups, which was significantly higher than that in group C ($p < 0.05$). The results showed that the addition of HMX2-EPS to the diet could improve the digestive enzyme activity and the non-specific immune enzyme activity in the intestine of turbot.

Table 3. Effects of HMX2-EPS supplementation on the digestive enzyme activities and immune-related enzyme activities of juvenile turbot. C, 0 mg/kg; H1, 100 mg/kg; H2, 500 mg/kg. Different superscripts in each line indicate significant differences ($p < 0.05$).

Items	Treatments		
	C	H1	H2
LPS (U/gprot)	23.68 ± 2.11 ^{b}	25.51 ± 1.23 ^{b}	35.13 ± 4.76 ^{a}
AMS (U/gprot)	0.42 ± 0.04 ^{b}	0.59 ± 0.06 ^{a}	0.64 ± 0.02 ^{a}
AKP (U/100mL)	79.23 ± 5.92 ^{b}	104.76 ± 10.64 ^{a}	98.23 ± 3.87 ^{a}
ACP (U/100mL)	90.58 ± 3.44 ^{b}	111.29 ± 13.13 ^{a}	108.32 ± 9.22 ^{a}
SOD (U/mL)	405.01 ± 18.98 ^{c}	533.12 ± 13.13 ^{b}	655.94 ± 24.22 ^{a}
CAT (U/mL)	13.39 ± 5.44 ^{b}	53.56 ± 9.33 ^{a}	48.38 ± 8.14 ^{a}

3.3. Expression of Immune-Related Genes and Antioxidant Genes

As revealed in Figure 1, the addition of HMX2-EPS had no significant effect on the expression of TLR-2 and MyD88 ($p > 0.05$). However, the gene expression of 1L-1β and 1L-10 in H1and H2 groups was significantly higher than that in the C group ($p < 0.05$), and the expression of IFN1 in the H2 group was the highest and significantly higher than that in the H1 and C groups ($p < 0.05$). Compared with the control group, the gene expression of CAT, SOD, and GPX1 was upregulated in the HMX2-EPS groups ($p < 0.05$). It showed that the addition of large doses of HMX2-EPS could enhance non-specific immunity and antioxidant capacity.

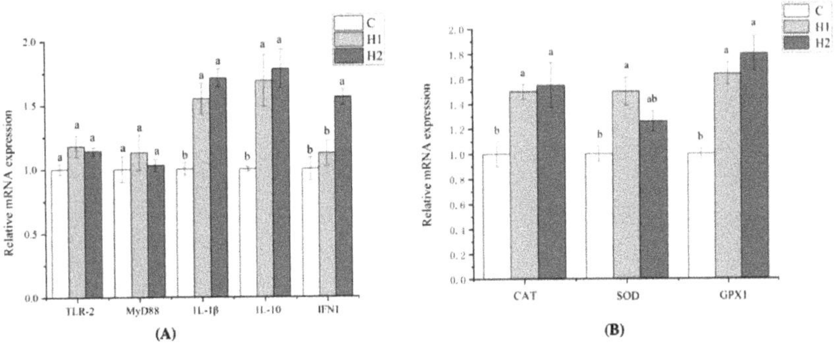

Figure 1. Effects of HMX2-EPS supplementation on the expression of immune-related genes in the intestines of juvenile turbot (**A**,**B**). C, 0 mg/kg; H1, 100 mg/kg; H2, 500 mg/kg. Different superscripts in each line indicate significant differences ($p < 0.05$).

3.4. Survival Rate after *A. hydrophila* Challenge

Different doses of HMX2-EPS could improve the survival rate of juvenile turbot after being poisoned by *A. hydrophila* (Figure 2). The highest survival rate was in the H2 group, which maintained 75% after 48 h, while the survival rate of juvenile turbot in the control group was only 20%.

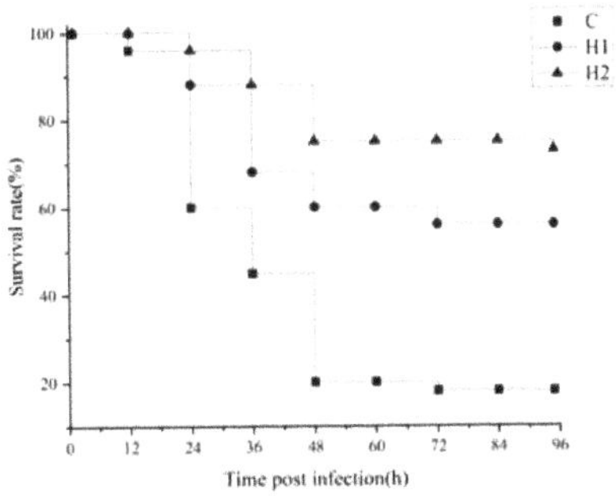

Figure 2. Effects of HMX2-EPS supplementation on the survival rate (%) of juvenile turbot challenged by *A. hydrophila*.

3.5. Intestinal Microbiota Regulation by HMX2-EPS

3.5.1. OUT Number and Alpha Diversity Analysis

A total of 60,352–68,279 effective sequences were obtained from the intestinal samples of juvenile turbot. Alpha diversity indices are presented in Table 4. Compared with the control group, the observed species, Shannon, Chao1, ACE, and PD whole tree indices, were significantly increased ($p < 0.05$) in H1 and H2 groups. However, there was no significant difference in the Simpson data among the three groups ($p > 0.05$). The results showed that HMX2 EPS supplementation could improve the abundance and evenness of microorganisms in the intestine.

Table 4. Effects of HMX2-EPS supplementation on the α diversity of the intestinal flora of juvenile turbot. Different superscripts in each line indicate significant differences ($p < 0.05$).

Items	Treatments		
	C	H1	H2
observed_species	990.91 ± 4.24 ^{b}	1297.26 ± 99.15 ^{a}	1406.81 ± 136.12 ^{a}
Shannon	6.88 ± 0.61 ^{b}	8.38 ± 0.12 ^{a}	9.21 ± 0.81 ^{a}
Simpson	0.91 ± 0.13	0.92 ± 0.67	0.95 ± 0.62
Chao1	1310.50 ± 14.33 ^{b}	1668.27 ± 16.14 ^{a}	1848.50 ± 26.17 ^{a}
ACE	810.18 ± 8.18 ^{b}	1863.23 ± 19.48 ^{a}	1987.27 ± 12.42 ^{a}
PD_whole_tree	92.51 ± 1.27 ^{b}	154.18 ± 1.36 ^{a}	174.79 ± 2.88 ^{a}

3.5.2. Beta Diversity Analysis

As shown in Figure 3A, the beta diversity analysis, according to the nonmetric multidimensional scaling (NMDS), showed that there was a difference between the C, H1, and H2 groups (stress = 0.019). The clustering results are presented at the phylum level (Figure 3B). H1 and H2 groups were clustered into a branch and then clustered with the C group into a branch. The results showed that the species composition structure of the H1 group and H2 group was similar, and the diversity of flora was higher than that of the C group.

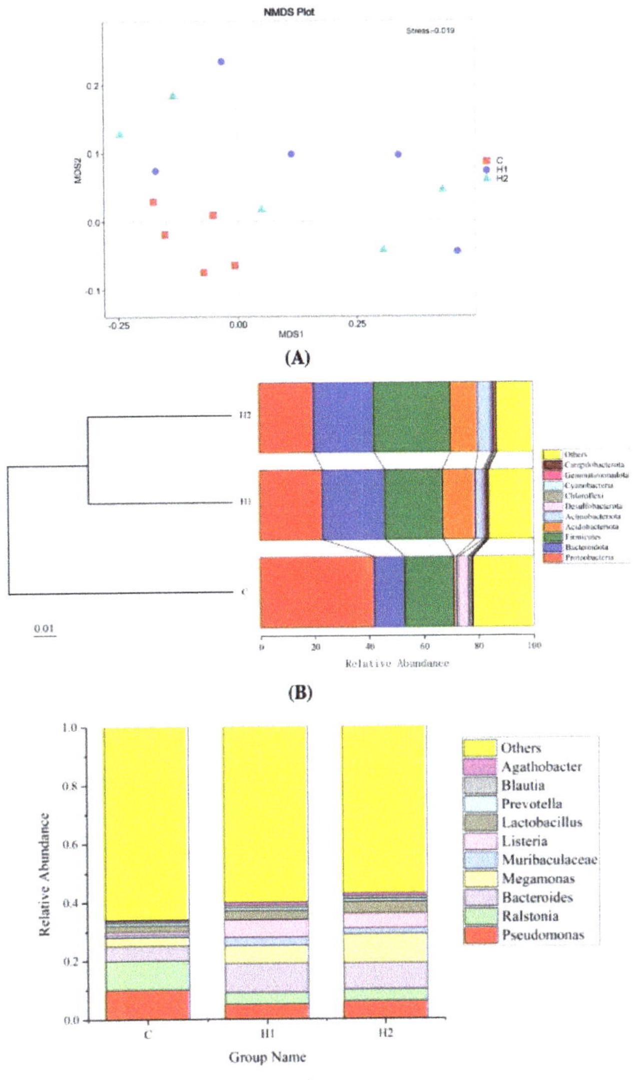

Figure 3. Nonmetric multidimensional scaling (NMDS) analysis of the intestinal bacterial community from juvenile turbot (**A**). Notes: NMDS can accurately reflect the degree of difference between samples when stress <0.2. Unweighted UniFrac distance matrix is used for UPGMA clustering analysis, and the clustering results are presented at the phylum level (**B**). Column charts of relative abundance of the top 10 gut microbiota at the genus level (**C**). C, 0 mg/kg; H1, 100 mg/kg; H2, 500 mg/kg.

3.5.3. Microbial Composition and LefSe Analysis

At the phylum level (Figure 3B), *Proteobacteria* (20.01–42.19%), *Firmicutes* (17.71–28.02%), and *Bacteroidota* (11.13–23.2%) formed the core microbiota. *Proteobacteria* was the dominant phylum in the control group, which was higher than that in the H1 and H2 groups. Compared with the control group, the abundance of *Bacteroidota*, *Firmicutes*, and *Acidobacteriota* significantly increased in the H1 and H2 groups. At the genus level

(Figure 3C), the abundance of potential probiotic-related genera, including *Bacteroides*, *Megamonas*, *Muribaculaceae*, and *Listeria* increased, and harmful bacteria such as *Ralstonia* and *Pseudomonas* decreased in the H1 and H2 groups.

To determine the difference in the abundance of species among populations, we performed the LEfSe analysis (Figure 4). We could see the analysis and the advantages of group C in phyla *Proteobacteria*, families *Psychromonadaceae*, *Pseudomonadaceae*, and *Methyloligellaceae*. The advantages of group H1 bacteria in phyla *Bacteroidota* and *Firmicutes*, families *Flavobacteriaceae*, *Streptococcaceae*, *Staphylococcaceae*, and *Enterococcaceae*. The advantages of group H2 bacteria in phyla *Firmicutes*, families *Bacillaceae* and *Lachnospiraceae*.

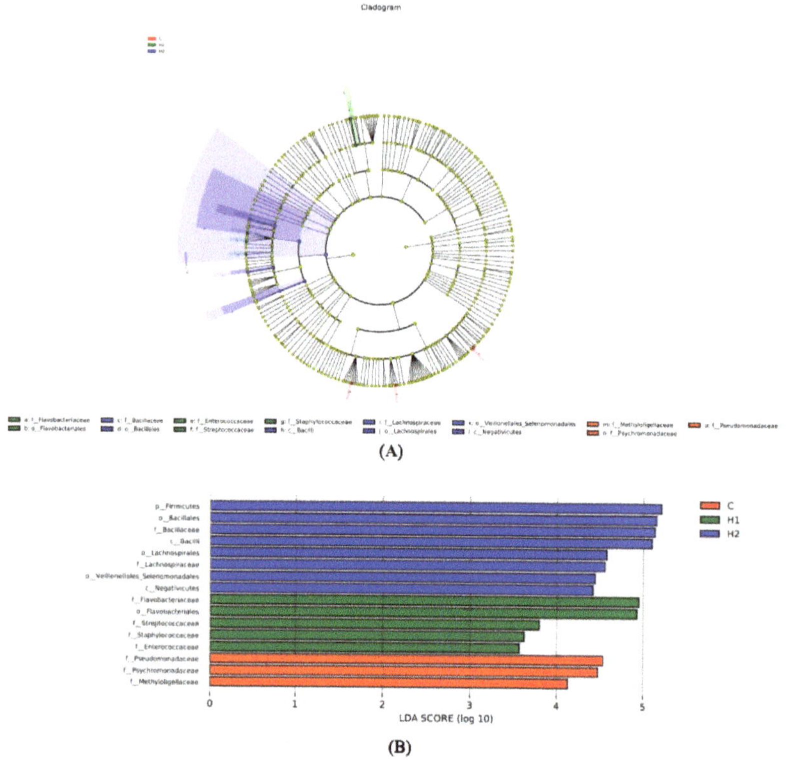

Figure 4. Linear discriminant analysis effect size (LEfSe) analysis of the gut microbiota fed with HMX2-EPS supplementation diets ($p \leq 0.05$ and linear discriminant analysis (LDA) cut-off > 4) (n = 6). (**A**): Cladogram showing differences in the abundance of taxa; (**B**): linear discriminant analysis scores of the abundance of taxa. C, 0 mg/kg; H1, 100 mg/kg; H2, 500 mg/kg.

3.5.4. Microbial Function Prediction

The abundance of functional categories by gut microbes at the second level (subfunction in each major type of metabolic pathway) is shown in Figure 5. Compared with the control group, the intestinal microorganisms in H1 and H2 groups were rich in enzyme families, immune systems, carbohydrate metabolism, and biosynthesis of other secondary metabolites. However, the abundance of some disease-related pathways was reduced, such as endocrine and metabolic diseases and infectious diseases.

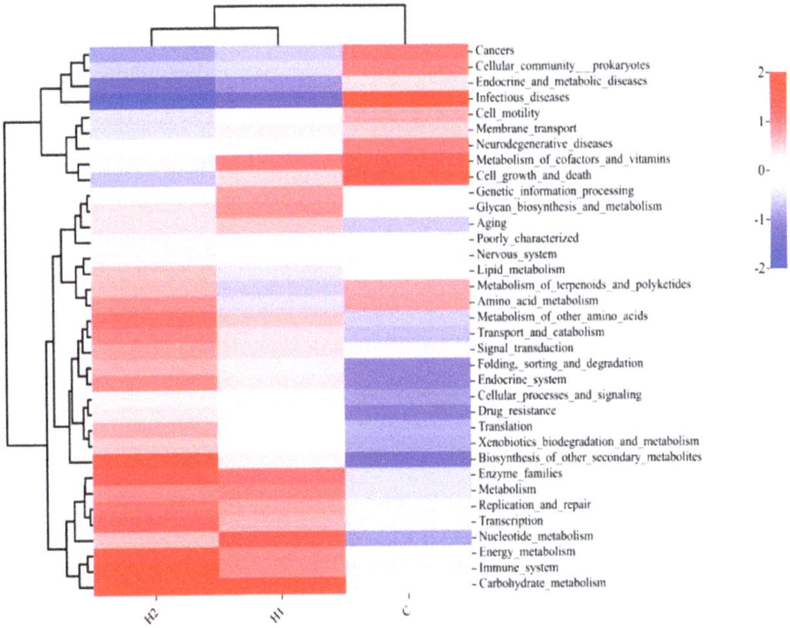

Figure 5. Heatmap of microbial functions at the second level (subfunction in each major type of metabolic pathway) of juvenile turbot. C, 0 mg/kg; H1, 100 mg/kg; H2, 500 mg/kg.

4. Discussion

EPS is the medium of interaction between probiotics and host organisms [7]. Although the effects of probiotics in improving the growth performance, immune response, and regulating intestinal flora of fish have been extensively studied in the past few decades [18], studies on the effects of EPS isolated from *Lactiplantibacillus plantarum* on juvenile turbot have not been reported. At present, we have isolated a potential probiotic called *L. plantarum* HMX2, which had high EPS production. Therefore, this experiment was conducted to investigate the effects of the EPS produced by *L. plantarum* HMX2 on the growth performance, antioxidation, digestion, immune ability, and intestinal flora of *S. maximus*.

Compared with EPS from animals and plants, EPS produced by bacteria had many advantages, such as strong operability, high reproductive ability, and advanced performance [11]. Li et al. [19] found that the appropriate EPS produced by Antarctic bacteria *Pseudoalteromonas* sp. 3-3-1-2 could effectively increase the body weight and body length of *Pseudopleuronectes yokohamae*. In this study, the addition of HMX2-EPS to the feed significantly improved the growth performance of juvenile turbot, although no significant difference was observed when different concentrations of EPS were added.

After ingestion of EPS, the first target immune reaction site was the intestinal mucosa. The intestine was considered the largest digestive, absorption, immune, and endocrine organ in animals [20]. Therefore, we used the intestine of the turbot juvenile to conduct the related experiments. Digestive enzymes were important indicators of digestive capacity and played an important role in the growth and development of fish. According to different digestive objects, it could be roughly divided into lipase, amylase, protease, and cellulase [21]. The addition of *Bifidobacterium longum* A17 EPS to the diet could improve the activities of digestive enzymes AMS, LPS, and protease in goldfish intestines under cadmium stress [22]. In this study, HMX2-EPS could improve the activities of digestive enzymes AMS and LPS in the intestine of juvenile turbot, and 500 mg/kg of HMX2-EPS showed the best effect. The results illustrated that EPS had a promoting effect on the digestive system, which ensured the intestinal absorption of nutrients and thus achieved the effect of improving the growth performance of fish.

The antioxidant capacity of the organism's defense system was closely related to health. Defense systems included both enzymatic and non-enzymatic systems, such as total antioxidant capacity (T-AOC), SOD, CAT, GPX, and glutathione peroxidase (GSH Px) belonged to the enzymatic system [23,24]. In this study, the addition of HMX2-EPS could improve the activity of SOD and CAT and promote the expression of antioxidant-related genes GPX1, CAT, and SOD in juvenile turbot. Chen et al. [25] found that the antioxidant capacity of *Cyprinus carpio L* was improved by the EPS from *Lactbacillu plantarum* HS-07 at different concentrations, and the activities of T-SOD, GSH Px, and CAT were significantly higher than those of the control group. Similar antioxidant capacity was demonstrated for EPS-9 from *Lactococcus lactis*, which significantly enhanced the activities of T-SOD, CAT, and GSH PX in the hepatopancreas and increased the concentrations of T-AOC and GSH in *Cyprinus carpio* [26]. The antioxidant activity of HMX2-EPS to improve the activity of antioxidant enzymes, as observed in this study, suggests the potential application of HMX2-EPS as an antioxidant agent to protect fish from body damage caused by oxidative stress.

ACP and AKP were not only important non-specific hydrolytic immune enzymes in the immune defense system but also important metabolic regulatory enzymes, which could be used as indicators to discriminate the body's immune competence in aquatic animals [27]. The activities of AKP, ACP, SOD, and CAT in red tilapia were significantly improved by *Shewanella frigiimarina* W32-2 EPS, and the addition of 3 mg/mL was optimal [28]. EPS 20 # 3 and 3-3-1-2 produced by polar bacteria *Pseudoalteromonas paragorgicola* could significantly improve the non-specific immunity of turbot with significantly increased activities of AKP, ACP, T-SOD, and CAT [29]. In this study, the addition of high concentration HMX2-EPS could significantly improve the activities of AKP and ACP in the intestine, indicating that the EPS could improve the activity of immune enzymes in fish.

Most LAB and their EPS could protect against pathogens by improving the phagocytic ability of monocytes, promoting the proliferation of lymphocytes, and inducing cytokines to regulate the immune system, thereby improving the body's immune defense ability [30]. Therefore, the detection of cytokines was often used to evaluate the immune response of the body. Zhang et al. [31] EPS from *Lactobacillus rhamnosus* LGG and *Lactobacillus casei* BL23 to the feed significantly increased the expression of pro-inflammatory factor (1L-6, TNF-a) and anti-inflammatory factor (1L-10) to improve the innate immunity of zebrafish. In this study, HMX2-EPS had no effect on the expression of TLR-2 and MyD88, indicating that HMX2-EPS did not activate the TLR-2/MyD88 pathway to induce inflammation. However, a high dose of HMX2-EPS significantly promoted the expression of 1L-1β, 1L-10, and IFN1. IFN1 is a crucial type I interferon for fish and provides the host with strong protection against pathogen infection. After pathogen infection was detected by the corresponding PRR, IFN was produced and secreted by multiple types of immune and non-immune cells, and the secreted IFN orchestrated innate and adaptive immunity through different mechanisms [32]. The results showed that HMX2-EPS could increase the secretion of inflammatory factors to trigger an inflammatory response, and it might enhance the immune response of turbot by regulating the IFN signal transduction pathway, thus promoting the immune response to invasive pathogens.

It was reported that the main pathogenic bacteria that could infect fish included *Vibrio anguillarum, Vibrio harveyi, Aeromonas hydrophila, Edwardia tarda,* and so on, which could cause symptoms such as body surface ulcers, ascites, and organ necrosis [33]. Wang [34] found that *Flavobacterium cloumnare* Pf1 EPS significantly improves the survival rate of grass carp infected by *F. cloumnare*. In this experiment, HMX2-EPS could improve the survival rate of juvenile turbot infected with *A. hydrophila*, perhaps because the high expression of IFN1 improves the disease resistance rate of turbot. The result showed that EPS could enhance innate immunity and had certain development value in aquatic production as an immune enhancer.

The intestine is a complex micro-ecosystem. The intestinal microbes of fish were closely related to the metabolism, nutrient absorption, growth and development, and immunity of the host, so maintaining the normal flora homeostasis in the intestine was essential for normal life activities [35]. The diversity of the gut microbiota correlated with the stability of the microbial community, and a decrease in diversity led to an increased risk of biological diseases [36]. In this study, the number of unique OUTs and alpha diversity indices in the EPS experimental groups was significantly increased compared with the control group, which indicated that the addition of HMX2-EPS in the diet significantly increased the abundance and evenness of intestinal microbiota of juvenile turbot. Beta diversity analysis found that EPS experimental group was significantly different from the control group, indicating that HMX2-EPS played a biological role and had an impact on the structure of intestinal flora. These results suggest that HMX2-EPS might be the key factor affecting intestinal flora.

Factors such as fish species, growth, habitat environment, feed species, and physiological characteristics of the gut all influence the amount and composition of the gut flora in fish, which could be both unified and interactive [35]. The gut microbiota is mainly composed of *Firmicutes, Bacteroidota, Proteobacteria,* and *Actinobacteriota*, which could effectively control the diversity and community structure of fish intestinal microorganisms [37]. There were two main ways for polysaccharides to regulate intestinal microorganisms. One was to regulate at the phylum level for *Firmicutes* and *Bacteroidota*. Another way was to promote the growth of beneficial bacteria and reduce the number of harmful bacteria, thus promoting intestinal health [38]. Both *Firmicutes* and *Bacteroidota* helped the host degrade many indigestible polysaccharides to provide energy and maintain the health of the animal intestines [38]. Generally, the imbalance of colonic flora and the increase of Proteobacteria abundance coexist. *Proteobacteria* could produce a large amount of endotoxin, which aggravates the structural destruction of intestinal epithelial tight junction and causes oxidative stress reaction and host metabolic disorder [39]. This study found that HMX2-EPS significantly increased the abundance of *Firmicutes* and *Bacteroidota* and decreased the abundance of *Proteobacteria. Lactobacillus rhamnosus* LGG-EPS could increase the abundance of *Firmicutes* and *Fusobacteria* and decrease the abundance of *Proteobacteria* and *Actinobacteria* in the zebrafish intestine to maintain the stability of the intestine flora structure [40], which was similar to our results. At the genus level, HMX2-EPS increased the reproduction of beneficial bacteria, such as *Bacteroides, Megamonas,* and *Muribaculaceae*, and decreased harmful bacteria, such as *Pseudomonas* and *Ralstonia*. Studies found that probiotics can colonize the intestine through the adhesion of EPS to intestinal epithelial cells. EPS could also promote probiotics' growth, improve their stress resistance and prebiotic potential, and form a protective layer to help probiotics evade immune surveillance and promote the colonization of probiotics in the intestine [41]. The polysaccharide utilization sites of *Bacteroides* provided a major proteinaceous mechanism for the initiation of the metabolism of many carbohydrates [42]. *Megamonas* could prevent and treat inflammation-related diseases, such as digestive inflammation-related diseases, rheumatoid arthritis, and cardiovascular diseases [43]. *Muribaculaceae* was related to the intestinal mucosal immune system and had the function of promoting intestinal metabolism [10,44]. *Colwellia* and *Ralstonia* produced corresponding toxins that reduced host resistance [45,46]. These results illustrated that HMX2-EPS could promote more kinds of probiotics to colonize and reduce the adhesive space of harmful intestinal bacteria, which enabled the healthy development of intestinal flora in juvenile turbot.

The Tax4Fun was performed to predict the functional profiles of gut microbiota. The intestinal microorganisms of juvenile turbot in HMX2-EPS groups were rich in immune system, enzyme families, carbohydrate metabolism, and biosynthesis of other secondary metabolites. The functions predicted corresponded to the composition and abundance of the above described intestinal flora. Ma et al. [47] found that metabolism and endocrine functions were more abundant in the *Lactobacillus* HNUB20-EPS group than in the control group. These results confirmed that EPS could serve as a carbon source for gut microbes, providing energy to microbes and participating in a variety of life activities [48,49]. EPS had better functional and spatial characteristics, which contributed to the rapid degradation, utilization, and maintenance of intestinal homeostasis of various nutrients. It might also be one of the reasons for the enhanced growth performance and immune function of juvenile turbot.

5. Conclusions

In the present study, supplementation with HMX2-EPS significantly improved the growth performance and activities of antioxidant enzymes, digestive enzymes, and immune-related enzymes in vivo of juvenile turbot. HMX2-EPS also significantly improved the survival rate of turbot by enhancing the immune response through regulating the IFN signal transduction pathway. Moreover, HMX2-EPS could improve the diversity of intestinal microbiota and improve the function of gut microbes in metabolism and the immune system. Therefore, there was a positive effect of HMX2-EPS supplementation in the feed on the cultivation of turbot juveniles.

Author Contributions: Conceptualization, G.H., Y.W., R.X., T.L., Z.Z. and Z.Y.; methodology, G.H. and Y.W.; software, R.X. and T.L.; validation, Z.Z.; formal analysis, G.H., Y.W., R.X., T.L., Z.Z. and Z.Y.; investigation, G.H. and Y.W.; resources, Z.Y.; data curation, T.L.; writing—original draft preparation, G.H. and R.X. writing—review and editing, Z.Y.; visualization, G.H., supervision, Z.Y.; project administration, Z.Y.; funding acquisition, Z.Y. All authors have read and agreed to the published version of the manuscript.

Funding: This study was supported by the National Natural Science Foundation of China (project no. 32272296).

Institutional Review Board Statement: This study was approved by the Scientific Research Ethics Committee of Beijing Technology and Business University (No. 3, 2023). All experiments were in line with the Management Rules of Laboratory Animals (Chinese Order No. 676 of the State Council, revised 1 March 2017).

Informed Consent Statement: Not applicable.

Data Availability Statement: The data presented in this study are available on request from the corresponding author.

Conflicts of Interest: The authors declare no conflict of interest.

References

1. Wang, A.; Ran, C.; Wang, Y.; Zhang, Z.; Ding, Q.; Yang, Y.; Olsen, R.E.; Ringø, E.; Bindelle, J.; Zhou, Z. Use of probiotics in aquaculture of China: A review of the past decade. *Fish Shellfish Immunol.* **2019**, *86*, 734–755. [CrossRef] [PubMed]
2. Vieco-Saiz, N.; Belguesmia, Y.; Raspoet, R.; Auclair, E.; Gancel, F.; Kempf, I.; Drider, D. Benefits and inputs from lactic acid bacteria and their bacteriocins as alternatives to antibiotic growth promoters during food-animal production. *Front. Microbiol.* **2019**, *10*, 57. [CrossRef] [PubMed]
3. Akhter, N.; Wu, B.; Memon, A.M.; Mohsin, M. Probiotics and prebiotics associated with aquaculture: A review. *Fish Shellfish Immunol.* **2015**, *45*, 733–741. [CrossRef] [PubMed]
4. Wang, Y.Q.; Wu, J.T.; Lv, M.; Shao, Z.; Hungwe, M.; Wang, J.; Bai, X.; Xie, J.; Wang, Y.; Geng, W. Metabolism characteristics of lactic acid bacteria and the expanding applications in food industry. *Front. Bioeng. Biotechnol.* **2021**, *9*, 612285. [CrossRef]
5. Malik, J.A. *Microalgae and Cyanobacteria: Potential Biofertilizers—Bioremediation and Phytoremediation Technologies in Sustainable Soil Management*; Apple Academic Press: Palm Bay, FL, USA, 2022; pp. 55–74.
6. Raman, J.; Kim, J.S.; Choi, K.R.; Eun, H.; Yang, D.; Ko, Y.-J.; Kim, S.-J. Application of lactic acid bacteria (LAB) in sustainable agriculture: Advantages and limitations. *Int. J. Mol. Sci.* **2022**, *23*, 7784. [CrossRef]
7. Xiu, L.; Zhang, H.; Hu, Z.; Liang, Y.; Guo, S.; Yang, M.; Du, R.; Wang, X. Immunostimulatory activity of exopolysaccharides from probiotic Lactobacillus casei WXD030 strain as a novel adjuvant in vitro and in vivo. *Food Agric. Immunol.* **2018**, *29*, 1086–1105. [CrossRef]
8. Zhu, Y.; Wang, X.; Pan, W.; Shen, X.; He, Y.; Yin, H.; Zhou, K.; Zou, L.; Chen, S.; Liu, S. Exopolysaccharides produced by yogurt-texture improving Lactobacillus plantarum RS20D and the immunoregulatory activity. *Int. J. Biol. Macromol.* **2019**, *121*, 342–349. [CrossRef]
9. Liu, Y.J.; Zhang, Y.Q.; Yuan, Y.; Li, H.Y.; Mu, G.Q.; Tuo, Y.F. Studies on the Antioxidative Activity of Exopolysaccharide from Lactobacillus Strains. *J. Chin. Inst. Food Sci. Technol.* **2019**, *19*, 21–35.

10. Zhang, M.; Hao, X.N.; Aziz, T.; Jian, Z.; Zhennai, Y. Exopolysaccharides from Lactobacillus plantarum YW11 improve immune response and ameliorate inflammatory bowel disease symptoms. *Acta Biochim. Pol.* **2020**, *67*, 485–493.
11. Li, J.Z.; Li, Q.K.; Shan, A.S. Biological Activities of Exopolysaccharides Produced by Lactic Acid Bacteria and Its Application Prospect in Livestock Production. *Chin. J. Anim. Nutr.* **2021**, *33*, 1901–1912.
12. Feng, J.C.; Cai, Z.L.; Chen, Y.Y.; Zhu, H.Y.; Chang, X.L.; Wang, X.F.; Liu, Z.; Zhang, J.X.; Nie, G.X. Effects of an exopolysaccharide from Lactococcus lactis Z-2 on innate immune response, antioxidant activity, and disease resistance against *Aeromonas hydrophila* in *Cyprinus carpio* L. *Fish Shellfish Immunol.* **2020**, *98*, 324–333. [CrossRef] [PubMed]
13. Xu, X.W.; Zheng, W.W.; Meng, Z.; Xu, W.; Liu, Y.; Chen, S. Identification of stress-related genes by co-expression network analysis based on the improved turbot genome. *Sci. Data* **2022**, *9*, 374. [CrossRef] [PubMed]
14. Wang, J.; Wu, T.; Fang, X.B.; Min, W.H.; Yang, Z.N. Characterization and immunomodulatory activity of an exopolysaccharide produced by *Lactobacillus plantarum* JLK0142 isolated from fermented dairy tofu. *Int. J. Biol. Macromol.* **2018**, *115*, 985–993. [CrossRef] [PubMed]
15. Wang, J.; Zhao, X.; Tian, Z.; Yang, Y.; Yang, Z. Characterization of an exopolysaccharide produced by *Lactobacillus plantarum* YW11 isolated from Tibet Kefir. *Carbohydr. Polym.* **2015**, *125*, 116–125. [CrossRef] [PubMed]
16. Yu, Y.; Chen, Z.; Wang, Y.F.; Zhang, M.Z.; Wang, Y.H.; Ai, N.; Shen, M.H.; Jiang, B. Preparation, structure and immuneomodulatory activity of exopolysaccharide from *Cordyceps militaris*. *Food Sci.* **2021**, *42*, 106–113.
17. Yang, G.; Tian, X.L.; Dong, S.L. *Bacillus cereus* and rhubarb regulate the intestinal microbiota of sea cucumber (*Apostichopus japonicus* Selenka): Species-species interaction, network, and stability. *Aquaculture* **2019**, *512*, 734284. [CrossRef]
18. Rohani, M.F.; Islam, S.M.M.; Hossain, M.K.; Ferdous, Z.; Siddik, M.A.B.; Nuruzzaman, M.; Pandeniya, U.; Brown, C.; Shahjhan. Probiotics, prebiotics and synbiotics improved the functionality of aquafeed: Upgrading growth, reproduction, immunity and disease resistance in fish. *Fish Shellfish Immunol.* **2022**, *120*, 569–589. [CrossRef]
19. Li, K.; Zheng, F.R.; Du, Z.J.; Wang, B.; Zhang, N.; Qi, L.; Lin, X.Z. Effects of exopolysaccharides produced by the Antarctic bacterium *Pseudoalteromonas* sp. 3-3-1-2 on the growth and immune activity of Pseudopleuronectes yokohamae. *Chin. Fish. Qual. Stand.* **2015**, *5*, 8–14.
20. Deng, Z.; Luo, X.M.; Liu, J.; Wang, H. Quorum sensing, biofilm, and intestinal mucosal barrier: Involvement the role of probiotic. *Front. Cell. Infect. Microbiol.* **2020**, *10*, 538077. [CrossRef]
21. Naeem, M.; Manzoor, S.; Abid, M.U.H.; Tareen, M.B.K.; Asad, M.; Mushtaq, S.; Ehsan, N.; Amna, D.; Xu, B.; Hazafa, A. Fungal proteases as emerging biocatalysts to meet the current challenges and recent developments in biomedical therapies: An updated review. *J. Fungi* **2022**, *8*, 109. [CrossRef]
22. Wang, Y. The Biosynthetic Pathway of Exopolysaccharides from *Bifidobacterium longum* and Its Protective Effect on Goldfish under Cadmium Stress. Master's Thesis, Inner Mongolia University, Hohhot, China, 2020.
23. Djordjevic, A.; Spasic, S.; Jovanovic-Galovic, A.; Djordjevic, R.; Grubor-Lajsic, G. Oxidative stress in diabetic pregnancy: SOD, CAT and GSH-Px activity and lipid peroxidation products. *J. Matern. Fetal Neonatal Med.* **2004**, *16*, 367–372. [CrossRef] [PubMed]
24. Arsova-Sarafinovska, Z.; Matevska, N.; Eken, A.; Petrovski, D.; Banev, S.; Dzikova, S.; Georgiev, V.; Sikole, A.; Erdem, O.; Sayal, A.; et al. Glutathione peroxidase 1 (GPX1) genetic polymorphism, erythrocyte GPX activity, and prostate cancer risk. *Int. Urol. Nephrol.* **2009**, *41*, 63–70. [CrossRef] [PubMed]
25. Chen, Y.Y.; Feng, J.G.; Li, M.; Liu, S.S.; Cai, Z.L.; Huang, M.Y.; Chang, X.L.; Zhang, J.X. Probiotic effect of fish-derived Lactobacillus plantarum HS-07 exopolysaccharide in common carp (*Cyprinus carpio* L.). *J. Fish. Sci. China* **2021**, *28*, 591–601.
26. Feng, J.; Cai, Z.; Chen, Y.; Zhu, H.; Chang, X.; Wang, X.; Liu, Z.; Zhang, J.; Nie, G. Effects of an exopolysaccharide from Lactococcus lactis Q-9 on innate immune response, antioxidant activity, and disease resistance against Aeromonas hydrophila in Cyprinus carpio. *J. Fish. China* **2020**, *44*, 1477–1487. [CrossRef] [PubMed]
27. Chen, J.; Huang, J.; Peng, J.; Yang, C.; Liao, Y.; Li, J.; Deng, Y.; Du, X. Effects of hypoxic stress on the digestion, energy metabolism, oxidative stress regulation, and immune function of the pearl oyster (*Pinctada fucata martensii*). *Aquac. Rep.* **2022**, *25*, 101246. [CrossRef]
28. Chen, Y.M.; Gao, P.L.; Tang, X.; Xu, C.G. Characterisation and bioactivities of an exopolysaccharide from an Antarctic bacterium *Shewanella frigidimarina* W32-2. *Aquaculture* **2020**, *530*, 735760. [CrossRef]
29. Zhao, H.Y.; Liu, T.J.; Zheng, F.R.; Lin, X.Z. Screening of exopolysaccharide-producing Antarctic bacteria and the preliminary study of effects of EPSs on nonspecific immunity of turbot (*Scophthalmus maximus*). *Adv. Mar. Sci.* **2013**, *31*, 237–246.
30. Balzaretti, S.; Taverniti, V.; Guglielmetti, S.; Fiore, W.; Minuzzo, M.; Ngo, H.N.; Ngere, J.N.; Sadiq, S.; Humphreys, P.N.; Laws, A.P. A novel rhamnose-rich hetero-exopolysaccharide isolated from *Lactobacillus paracasei* DG activates THP-1 human monocytic cells. *Appl. Environ. Microbiol.* **2017**, *83*, e02702-16. [CrossRef]
31. Zhang, Z.; Ran, C.; Ding, Q.-W.; Liu, H.-L.; Xie, M.-X.; Yang, Y.-L.; Xie, Y.-D.; Gao, C.-C.; Zhang, H.-L.; Zhou, Z.-G. Ability of prebiotic polysaccharides to activate a HIF1α-antimicrobial peptide axis determines liver injury risk in zebrafish. *Commun. Biol* **2019**, *2*, 274. [CrossRef]
32. Ivashkiv, L.B.; Donlin, L.T. Regulation of type I interferon responses. *Nat. Rev. Immunol.* **2014**, *114*, 36–49. [CrossRef]
33. Bansemir, A.; Blume, M.; Schroder, S.; Lindequist, U. Screening of cultivated seaweeds for antibacterial activity against fish pathogenic bacteria. *Aquaculture* **2006**, *252*, 79–84. [CrossRef]
34. Wang, Y.Q. The Study on Immunity of *Flavobacterium columnare* Exopolysaccharide in Grass Carp, *Ctenopharyngodon idellus*. Master's Thesis, Huazhong Agricultural University, Wuhan, China, 2018.

35. Luo, L.; Zhao, G.; Du, X.; Xu, Q.Y. Relationship Between Intestinal Flora and Immunity in Fish. *Chin. J. Fish.* **2018**, *31*, 51–54.
36. Wang, M.M.; Lv, C.J.; Yang, D.L.; Zhao, J.M. Effects of *Lactobacillus plantarum* (LP HMX-3) on growth, digestion, immunity and intestinal flora of *Apostichopus japonicus*. *J. Fish. China* 2022, 1–11.
37. Zhai, W.Y.; Guo, A.N. Research progress on the fish intestinal microbiota. *Henan Fish.* **2016**, *4*, 18–21.
38. Liu, L.; Liu, C.; Hua, H.; Zhao, W.; Zhu, H.; Cheng, Y.; Guo, Y.; Qian, H. Effect of polysaccharides from Tibetan turnip (*Brassica rapa* L.) on the gut microbiome after in vitro fermentation and in vivo metabolism. *Food Funct.* **2022**, *13*, 3063–3076. [CrossRef]
39. Wasmund, K.; Pelikan, C.; Schintlmeister, A.; Wagner, M.; Watzka, M.; Richter, A.; Bhatnagar, S.; Noel, A.; Hubert, C.R.J.; Rattei, T.; et al. Genomic insights into diverse bacterial taxa that degrade extracellular DNA in marine sediments. *Nat. Microbiol.* **2021**, *6*, 1102. [CrossRef]
40. Li, Y. Using Zebrafish Model to Study the Effect Mechanism of Three Natural Polysaccharides Against Common Carp Spring Viremia Virus. Ph.D. Thesis, Huazhong Agricultural University, Wuhan, China, 2021.
41. Caggianiello, G.; Kleerebezem, M.; Spano, G. Exopolysaccharides produced by lactic acid bacteria: From health-promoting benefits to stress tolerance mechanisms. *Appl. Microbiol. Biotechnol.* **2016**, *100*, 3877–3886. [CrossRef]
42. Schwaim, N.D.; Townsen, G.E.; Groisman, E.A. Prioritization of polysaccharide utilization and control of regulator activation in Bacteroides thetaiotaomicron. *Mol. Microbiol.* **2017**, *104*, 32–45. [CrossRef]
43. Zou, Y.Q.; Xiao, L.; Li, X.P.; Yu, J.; Liu, C. Megamonas Funiformis and Its Application. Guangdong Province. CN109504615A, 22 March 2019.
44. Bunker, J.J.; Flynn, M.; Koval, J.C.; Shaw, D.G.; Meisel, M.; McDonald, B.D.; Ishizuka, I.E.; Dent, A.L.; Wilson, P.C.; Jabri, B.; et al. Innate and adaptive humoral responses coat distinct commensal bacteria with immunoglobulin A. *Immunity* **2015**, *43*, 541–553. [CrossRef]
45. Peoples, L.M.; Kyaw, T.S.; Ugalde, J.A.; Mullane, K.K.; Chastain, R.A.; Yayanos, A.A.; Kusube, M.; Methe, B.A.; Bartlett, D.H. Distinctive gene and protein characteristics of extremely piezophilic Colwellia. *BMC Genom.* **2020**, *21*, 692. [CrossRef]
46. Ameen, Y. Bacterial warfare: Diversity, function, and role of Rhs polymorphic toxins within the Ralstonia solanacearum species complex. *Phytopathology* **2021**, *110*, 10.
47. Ma, C.C.; Guo, H.Y.; Chang, H.B.; Huang, S.; Jiang, S.; Huo, D.; Zhang, J.; Zhu, X. The effects of exopolysaccharides and exopolysaccharide-producing *Lactobacillus* on the intestinal microbiome of zebrafish (*Danio rerio*). *BMC Microbiol.* **2020**, *20*, 300. [CrossRef]
48. Aziz, T.; Naveed, M.; Makhdoom, S.I.; Ali, U.; Mughal, M.S.; Sarwar, A.; Khan, A.A.; Zhennai, Y.; Sameeh, M.Y.; Dablool, A.S.; et al. Genome Investigation and Functional Annotation of *Lactiplantibacillus plantarum* YW11 Revealing Streptin and Ruminococcin-A as Potent Nutritive Bacteriocins against Gut Symbiotic Pathogens. *Molecules* **2023**, *28*, 491. [CrossRef] [PubMed]
49. Aziz, T.; Naveed, M.; Sarwar, A.; Makhdoom, S.I.; Mughal, M.S.; Ali, U.; Yang, Z.; Shahzad, M.; Sameeh, M.Y.; Alruways, M.W.; et al. Functional Annotation of *Lactiplantibacillus plantarum* 13-3 as a Potential Starter Probiotic Involved in the Food Safety of Fermented Products. *Molecules* **2022**, *27*, 5399. [CrossRef] [PubMed]

foods MDPI

Article

Water-Insoluble Polysaccharide Extracted from *Poria cocos* Alleviates Antibiotic-Associated Diarrhea Based on Regulating the Gut Microbiota in Mice

Yong Lai [1], Huiling Deng [2,3], Qi Fang [1], Linhua Ma [1], Hui Lei [1], Xiurong Guo [1], Ya Chen [2] and Can Song [1,*]

[1] School of Pharmacy, Southwest Medical University, Luzhou 646000, China; yonglai3210@163.com (Y.L.); fq15510110160379@163.com (Q.F.); 15181428034@163.com (L.M.); huilei@swmu.edu.cn (H.L.); xiurongguo@swmu.edu.cn (X.G.)

[2] Chongqing Academy of Science and Technology, Chongqing 401121, China; 13618346171@163.com (H.D.); cq_chenya@163.com (Y.C.)

[3] Key Laboratory of Condiment Supervision Technology for State Market Regulation, Chongqing Institute for Food and Drug Administration, Chongqing 401121, China

* Correspondence: cansong@swmu.edu.cn; Tel./Fax: +86-0830-3162291

Abstract: Antibiotics are very effective in treating a variety of bacterial infections, while clinical overuse of antibiotics can lead to diseases such as antibiotic-associated diarrhea. Numerous studies have shown that natural polysaccharides can be used as prebiotics to alleviate antibiotic-associated diarrhea (AAD). *Poria cocos* is a medicinal and edible mushroom widely used for thousands of years in China, and our former study demonstrated that water-insoluble polysaccharide (PCY) has the potential prebiotic function. Therefore, we simulated the digestion and fermentation of PCY using feces from volunteers, and then administered it to C57BL/6 mice with AAD to study its effects on the gut microbiota and metabolites. The results indicated that PCY effectively alleviated the symptoms of AAD in mice, restored the intestinal barrier function, improved the content of short-chain fatty acids (SCFAs), decreased the level of inflammatory cytokines, and changed the structure of gut microbiota by increasing the relative abundance of norank_f__*Muribaculaceae* and unclassified_f__*Lachnospiraceae*, and decreasing that of *Escherichia-Shigella, Staphylococcus* and *Acinetobacter*. This study further demonstrated that PCY is an effective functional prebiotic for improving AAD disease, and provided a new avenue and insight for developing PCY as a functional food or prebiotic for alleviating gastrointestinal diseases.

Keywords: *Poria cocos*; water-insoluble polysaccharides; antibiotic-associated diarrhea; gut microbiota; short chain fatty acids

1. Introduction

The antibiotics are widely used as an effective drug to treat a wide range of bacterial infections in hospitals [1]. However, prolonged and inappropriate clinical use of antibiotics increases the risk of a number of complications, and one of the most common complications is antibiotic-associated diarrhea [2]. Normal gut microbiota plays a crucial role in human health [3], and many chronic diseases such as hypertension, diabetes, and inflammatory bowel disease are closely connected to the changes of gut microbiota [4]. Therefore, how to maintain the balance of intestinal microorganisms and reverse antibiotic-induced disorders of gut microbiota are inevitable hot topics in clinical practice.

Numerous studies have shown that the use of polysaccharide prebiotic can regulate gut microbiota and promote the structural integrity of intestinal tissues, thereby restoring the gut imbalance caused by AAD [5,6]. Polysaccharides are a class of long-chain carbohydrates that are widely found in plants and microorganisms [5,7]. Many studies have revealed that polysaccharides of fungal origin have anti-inflammatory, antioxidant, antitumor, antiviral, immunomodulatory and hypoglycemic abilities [8,9]. The intake of polysaccharides is crucial for maintaining intestinal health in human diets, and they are often divided into water-soluble polysaccharides and insoluble polysaccharides based on their water solubility [10,11]. *P. cocos* is a valuable and ancient traditional Chinese medicine, its main active ingredients contain water-soluble polysaccharides, water-insoluble polysaccharides, triterpenoids [12–14]. Modern pharmacological studies have shown that polysaccharides extracted from *P. cocos* have good biological activities such as anti-tumor, immunomodulation, hypoglycemia, lipid regulation and liver protection [15–17]. Our group previously studied the effect of three main active ingredients from *P. cocos* on mice and found that PCY can increase the number of *Lactobacillus* to regulate the gut microbiota, which is beneficial to the health of the organism [18]. Fungal polysaccharides from medical and edible mushroom sources have great prebiotic potential and are expected to be used to alleviate antibiotic-associated diarrhea, which deserves to be explored in depth.

In this study, we investigated prebiotic effects of water-insoluble polysaccharides of *P. cocos* origin. Firstly, we perform digestion and fermentation experiments of PCY to simulate its effects on healthy people. In addition, we evaluated the alterations and effects of PCY on mice with AAD, and explored its pharmacological mechanisms for regulating gut microbiota to alleviate antibiotic-associated diarrhea. We aimed to investigate the prebiotic properties of PCY for the alleviation of AAD, and to provide new ideas to address the clinical problem of antibiotic-induced intestinal diseases.

2. Materials and Methods

2.1. Reagents and Chemicals

P. cocos were purchased from Sichuan Xin RenTai Pharmaceutical Co., Ltd., Pengzhou, China. Standard monosaccharides were purchased from Shanghai Yuanye Bio-Technology Company, Shanghai, China. Standard SCFAs were purchased from SIGMA Company, Shanghai, China. Reducing sugar content (RSC) assay kit and inflammatory cytokine kit including TNF-α, IL-6, and IL-1β, were purchased from Beijing Solarbio Science Technology Co., Ltd., Beijing, China. Total Antioxidant Capacity Assay (TAOC) Kit with a rapid ABTS method was purchased from Beyotime biotechnology Co., Ltd., Nantong, China.

2.2. Extraction and Characteristics of PCY

The extraction and purification of PCY was performed according to a specific method that refers to our previous study [18]. The PMP (1-phenyl3-methyl-5-pyrazolone) pre-column derivatization-HPLC method was applied to analyze the monosaccharides of PCY, and the structure was applied scanning electron microscope (SEM) to observe.

2.3. In Vitro Simulated Digestion and Fermentation Experiments of PCY

In vitro simulated digestion and fecal fermentation experiments of PCY were carried out referenced to our team's previous study [19]. All the saliva, intestinal fluid, gastric fluid and fermentation media used in the experiment were prepared according to previous studies [20–22]. In the first part of the simulated digestion experiments, we determined data on residual ash weight, total antioxidant capacity and reducing sugar content of after 50 mg of PCY being treated with digestive solution. Then, after PCY was fermented by the feces of three healthy volunteers, we examined its pH, total antioxidant capacity and reducing sugar content, SCFAs concentration, and effect on the gut microbiota.

2.4. Animal Experiments

The Ethics Committee of the Animal Experimentation Center of Southwest Medical University implemented and approved all animal experimentation procedures (license number P20220707-01). All C57BL/6 mice were acclimatized for 1 week in a standard environment with 12-h light-dark cycles, a temperature of 22 °C, and 50% humidity. Twenty-eight C57BL/6 mice were randomly divided into four groups of seven mice each, namely normal control group (NC group), antibiotic-associated diarrhea group (AAD group), PCY treatment group (PCY group), and natural recovery group (NR group). The whole animal experiment was divided into two phases: pathological model phase (1–3d) and experimental recovery period (4–10d), except for the NC group for healthy blank control group only gavage saline per day, the rest of the group were pathological animal model group, during the pathological model period gavage 3 g/kg lincomycin hydrochloride twice a day [23]. In the experimental recovery period, mice of PCY group were gavaged with 300 mg/kg polysaccharides twice a day for 7 d, and we continuously recorded the scores of mouse weight, mental and diarrheal status, and finally collected mouse feces, blood, and cecum tissues and stored in −80 °C refrigerators for subsequent experimental determination.

2.5. Histological Examination of the Cecum

We fixed 4 groups of mouse cecum tissues in 10% paraformaldehyde, dehydrated them using anhydrous ethanol, and fixed them with paraffin embedding. The paraffin was cut into slice, and it was stained with hematoxylin for 10 min and washed with tap water. The above section was then differentiated using hydrochloric acid alcohol for 10 s and washed in warm water for 3 min until the color of the nuclei returned to blue. Next it was placed in 85% alcohol for 3 min, the sections were stained with eosin for 3 min and washed with tap water for 5 s, and finally the histological morphology and structure were observed under a light microscope.

2.6. 16S rRNA Sequencing

16S rRNA sequencing was carried in Meiji Biotechnology Co., Ltd., (Shanghai, China) using the Illumina platform. The detailed process and specific methods of high-throughput sequencing refer to our previous research [18]. Raw sequence data for fermentation and animal experiment microbiota are saved in the NCBI Sequence Read Archive (SRA) under registration number PRJNA932548 and PRJNA917814.

2.7. Detection of SCFAs

LC-MS (Liquid chromatography-mass spectrometry) was used to determine the concentration of SCFAs in mouse fecal samples. According to the SCFAs standard map, the library was retrieved by the data processing system of the chromatography workstation, the composition of SCFAs was confirmed with reference to relevant literature, and the concentration of SCFAs was calculated.

2.8. Cytokine Measurement

Enzyme-linked immunosorbent assay (ELISA) is an efficient detection method that uses specific binding of antigen and antibody to carry out immune reaction. The levels of inflammatory cytokines tumor necrosis factor alpha (TNF-α), interleukin 6 (IL-6), and interleukin 1β (IL-1β) were determined in mouse serum by using commercial ELISA kits.

2.9. Statistical Analysis

Significance of differences between groups was analyzed using GraphPad Prism 7.04 and SPSS statistical software 19.0. The hierarchical clustering analysis was performed with a Bray-Curtis distance matrix based on ASV level; PCoA was tested using ANOSIM with a Bray-Curtis distance matrix; the method used for the multi-group comparative analysis of microbiota was the Kruskal-Wallis H test with a FDR correction; the method used for the two-group comparative analysis of microbiota was the Wilcoxon rank- sum test with a two-tailed test and FDR correction; LEfSe analyses were performed using the multiple comparison strategy all-against-all (more strict).

3. Results

3.1. Characteristics of PCY

As seen in Figure S1, the monosaccharide composition of PCY determined is glucose. The results of scanning electron microscopy are in Figure S2, PCY exhibit a flaky or crumbly stacked morphology with smooth surface morphology, and it has a complete and uniform structure, which is uniformly smooth and tightly packed at both 30 μm and 100 μm magnifications.

3.2. In Vitro Simulated Digestion of the PCY

The whole digestive process includes three stages: saliva digestion, gastric juice digestion and intestinal juice digestion, and the environmental pH value of each stage is different. After 50 mg of water-insoluble polysaccharides entered the simulated digestion stage, the residual mass of polysaccharides gradually decreased, and the measured weight was 45.1 + 1.2 mg at 10 h (Figure 1B). As shown in Figure 1C, reducing sugars mainly include glucose, and a small part of water-insoluble polysaccharides is decomposed into reducing sugars under the action of saliva, gastric juice and intestinal juice. The results show that the RSC gradually increases with time, and the RSC is 0.295 + 0.018 mg/mL when the simulated digestion is completed. When using Trolox as a standard, the TAOC of the sample is expressed as TEAC. As shown in Figure 1D, the TAOC of PCY gradually increased during the simulated digestion, and the TAOC of polysaccharide digestion was 0.829 + 0.08 mM at 10 h.

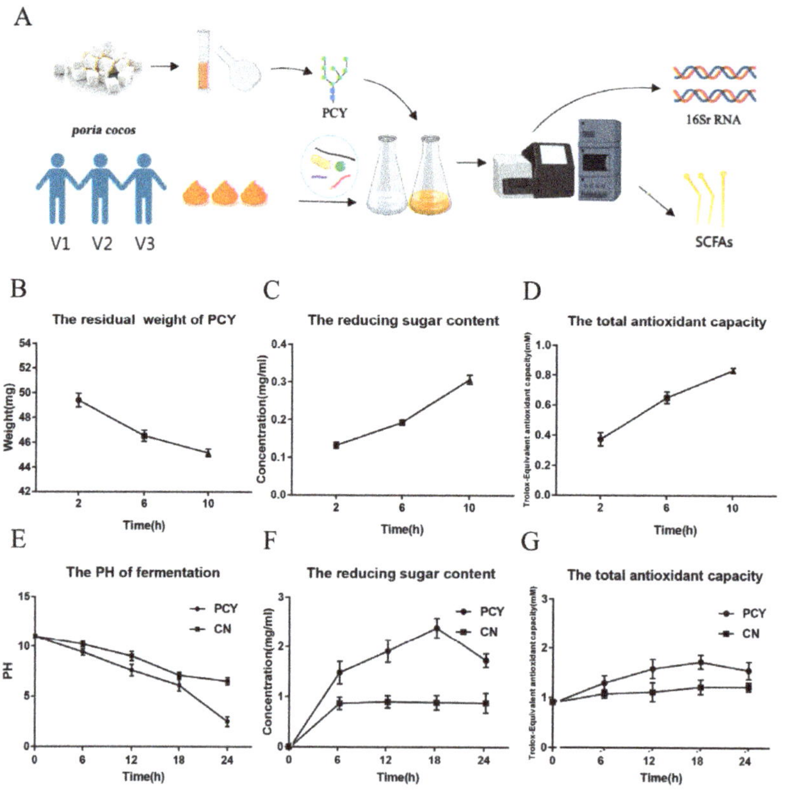

Figure 1. (**A**) Schematic diagram of simulated PCY fermentation in vitro. (**B**) The residual ash weight of PCY in digestion. (**C**) The RSC of PCY in digestion. (**D**) The TAOC of PCY in digestion. (**E**) The pH of PCY in fermentation. (**F**) The RSC of PCY in fermentation. (**G**) The TAOC of PCY in fermentation.

3.3. In Vitro Fermentation of PCY

A schematic diagram of healthy human fecal fermentation for PCY is shown in Figure 1A. The changes of pH, RSC, and TAOC during the fermentation of human feces were determined (Figure 1E–G). The pH of the fermentation broth decreased with time, while the RSC and TAOC increased with time. The results of microbiota analysis of PCY after 24 h fermentation in healthy human feces and the results of its metabolites are shown in Figure 2. The results of microbial composition of fresh feces from healthy volunteers are shown in the supplementary material (Figure S3), and the normal human gut microbes at the phylum level contain mainly Firmicutes, Bacteroidota, Proteobacteria, Actinobacteriota, and Fusobacteriota. At the end of fermentation, in the fisher's exact test bar plot of the PCY and NC groups, the relative abundance of Firmicutes and Bacteroidota significantly increased, while the relative abundance of Actinobacteriota significantly decreased on phylum level (Figure 2A–C). The results of SCFAs in the PCY and NC groups are shown in the Figure 2D, and the concentration of propionic and acetic acid in the PCY group was higher, while the concentration of acetic, propionic, butyric, valeric, and caproic acid have no significant difference.

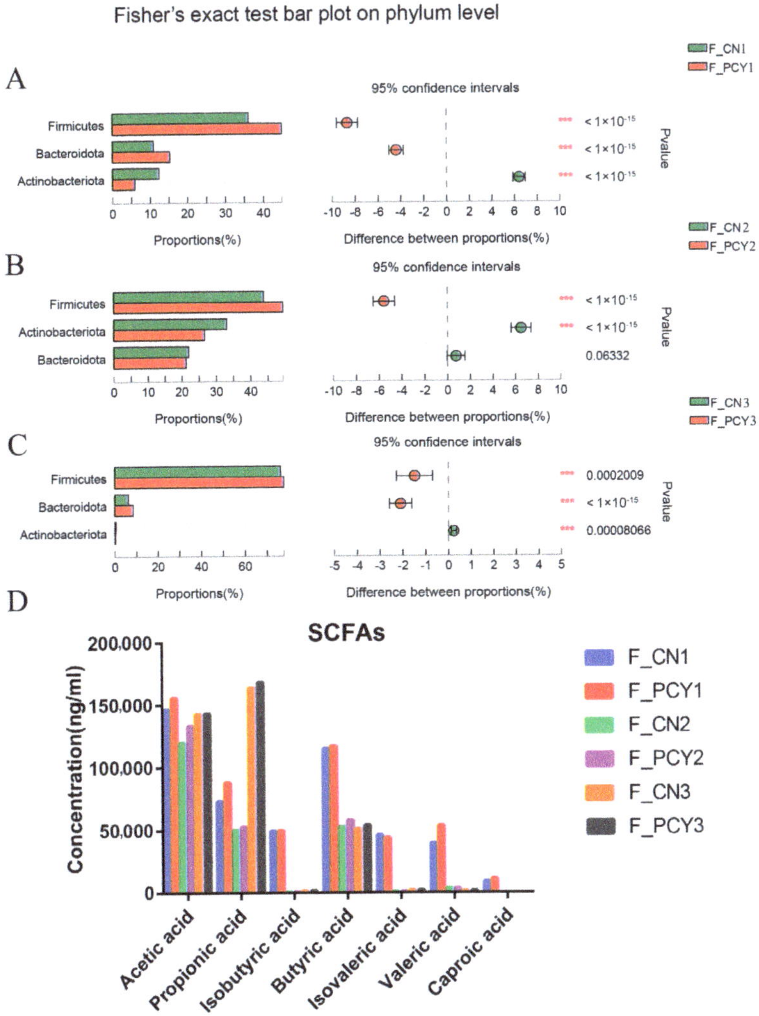

Figure 2. (**A**) Difference between F_CN1 and F_PCY1 at the phylum level. (**B**) Difference between F_CN2 and F_PCY2 at the phylum level. (**C**) Difference between F_CN3 and F_PCY3 at the phylum level. (**D**) The changes of SCFAs in all groups. *** $p < 0.001$.

3.4. General Data of C57BL/6 Mice

The whole animal experiment lasted 10 days, which was divided into a 3 daypathological model modeling period and a 7-day experimental treatment period (Figure 3A). As shown in Figure 3B, mice's weight dropped rapidly after antibiotics during the model period, and then the PCY group of mice received polysaccharide treatment and recovered their weight better compared to the NR group. Moreover, the status scale for antibiotic-associated diarrhea in C57BL/6 mice is shown in Figure 3C, the diarrhea status score of mice in the PCY group significantly decreased than that in the NR group during the monitoring of feeding (Figure 3D).

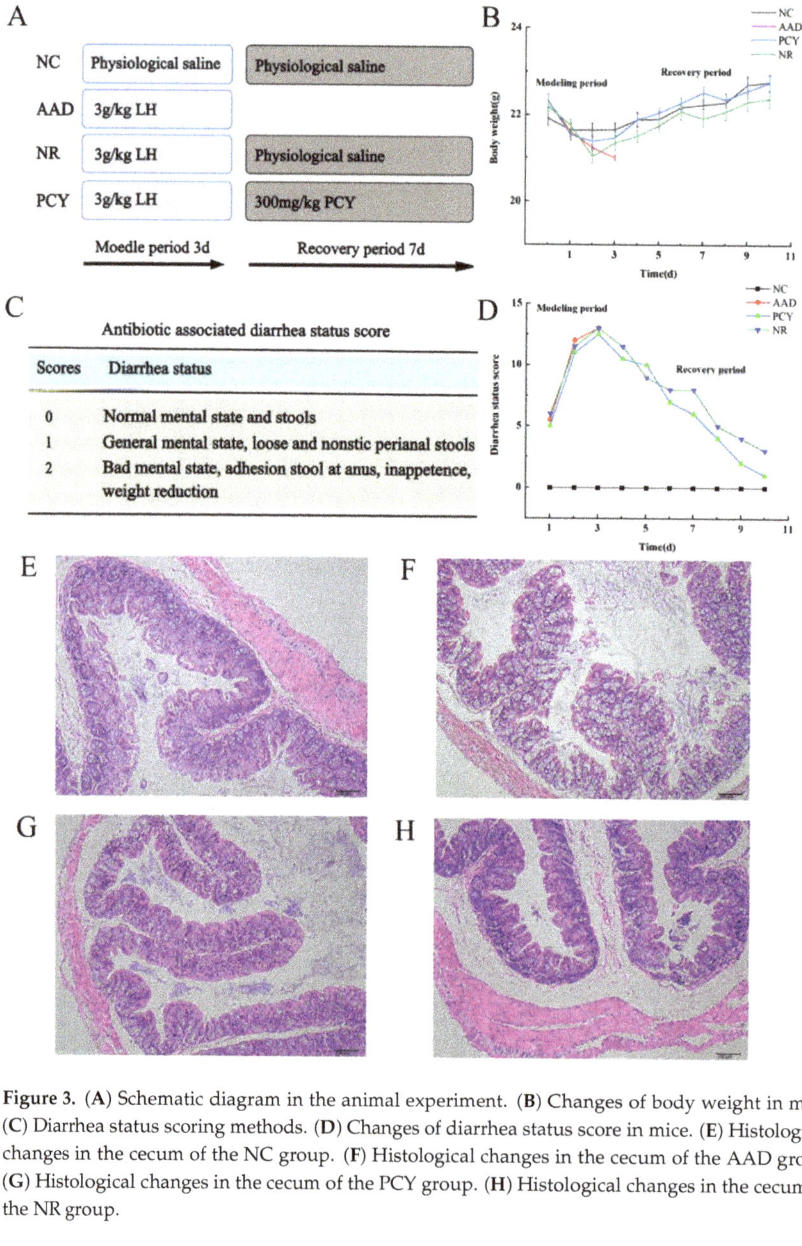

Figure 3. (**A**) Schematic diagram in the animal experiment. (**B**) Changes of body weight in mice. (**C**) Diarrhea status scoring methods. (**D**) Changes of diarrhea status score in mice. (**E**) Histological changes in the cecum of the NC group. (**F**) Histological changes in the cecum of the AAD group. (**G**) Histological changes in the cecum of the PCY group. (**H**) Histological changes in the cecum of the NR group.

3.5. Effect of PCY on Cecal Tissues

After HE staining and photographing, Figure 3 exhibited the histopathological changes of the cecum in C57BL/6 mice. The histological features of the blank control group are shown in Figure 3E, with intact intestinal mucosa and regular arrangement of epithelial cells. The cecal tissues of mice in the AAD group showed severe histopathological changes, such as mucosal defects, massive inflammatory cells infiltration, edema, and incomplete goblet cells (Figure 3F). After a one-week recovery period, the mice with the administration of PCY has the best effect in the histological characteristics of the cecum (Figure 3G). Compared with mice in the NR group, PCY significantly improved the defects of the intestinal mucosa, reduced the number of inflammatory cells, and reduced the edema of the submucosa (Figure 3H). It has been shown that the intestinal mucosa is defective in the intestinal tissue of AAD hosts, and there are pathological features such as inflammatory cell infiltration and edema, which are consistent with the characteristics of the mice in the AAD group in this study.

3.6. Composition and Diversity of the Gut Microbiota

The feces of C57BL/6 mice were used for 16S rRNA in the experiment. The PCR amplification results of feces in the AAD group are shown in the supplementary material (Figure S4), and the serious damage caused by antibiotics to the gut microbiota fully demonstrates that our experiments have successfully created a pathological model, which is consistent with the previous study [24]. The Rarefaction curve is used to determine whether the amount of sequencing data for the gut microbiota of each sample is reasonable, and the results show that at the current sequencing depth, each sample can be used for further analysis (Figure S5).

α-diversity contains the Simpson index, Ace index, and Shannon index (Table S1). As shown in Figure 4A–C, the types and numbers of gut microorganisms in the AAD and NR groups were significantly lower than those in the NC group, and the differences were statistically significant. Results of α-diversity demonstrated that the microbial communities of the experimental mice were severely damaged by lincomycin hydrochloride, and the application of PCY improved the microbial diversity of the mouse gut microbiota compared to naturally recovered mice. Firmicutes and Bacteroides are the two dominant phylum for maintaining gut microbiota balance at the phylum level. As demonstrated in Figure 4D, there are Bacteroidota, Proteobacteria, Firmicutes, Campilobacterota, and Verrucomicrobiota on mice gut microbiota in NC group. The relative abundance of Proteobacteria in the NR group surprisingly accounted for most of the bacterial phylum level, which was significantly higher than that in the NC group, while the abundance of Proteobacteria was significantly reduced after the administration of PCY. The number of *Escherichia-Shigella* accounts for half of the level of gut microbiota genus in the NR group, and it is the most common pathogen in humans (Figure 4E). As shown in Figure S6, the heat map detected the microbial community changes at the genus level, PCY has obviously regulated the genus level of the mouse gut microbial community.

β-Diversity was used to investigate similarities between groups. PCoA (principal co-ordinates analysis) is a visualization method to study data similarities or differences, showing differences between and within groups by analyzing structural changes in the gut microbiota of mice. As shown in Figure 4G, there are significant differences between the three groups (p = 0.001). The intestinal microbiota between the experimental group and the control group showed a clear separation, and the inconsistency with the blank control group without any treatment indicated the successful establishment of the AAD model. In Figure 4F, Hierarchical clustering tree on ASV level demonstrated that the clustering tree branches of the NC and PCY groups are clearly similar, while the NR group differs from them. The above results show that in addition to being able to reverse the damage caused by antibiotics, PCY can also cause better probiotic changes by regulating the microbiota of AAD mice.

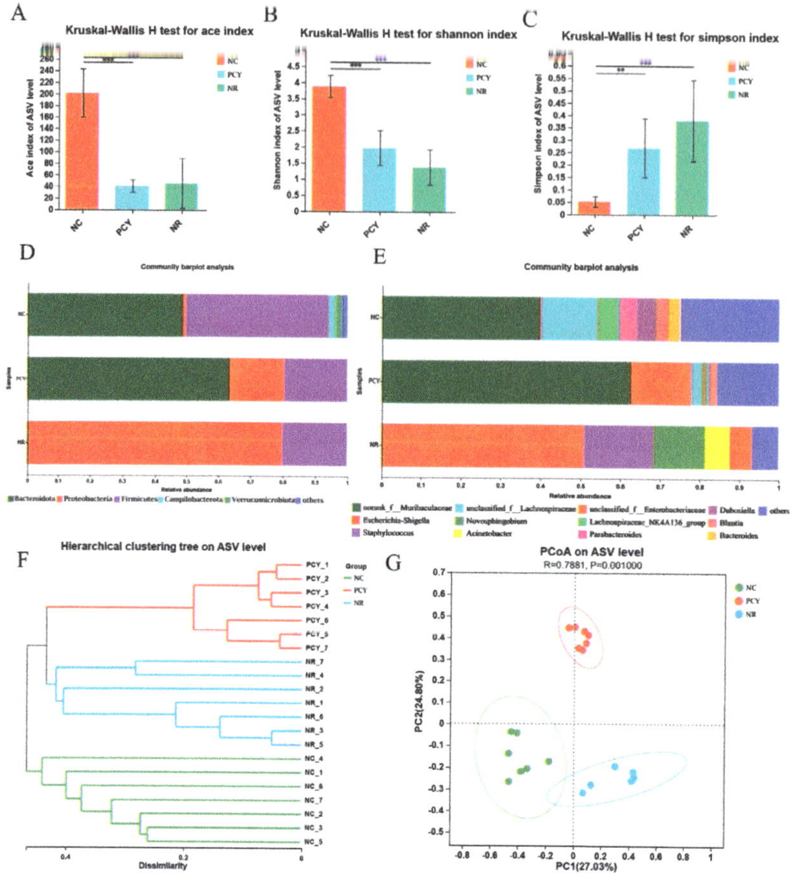

Figure 4. (**A**) Ace index comparation. (**B**) Shannon index comparation. (**C**) Simpson index comparation. (**D**) Difference of three groups at the phylum level. (**E**) Difference of three groups at the genus level. (**F**) Hierarchical clustering tree on ASV level of mice in three groups. (**G**) PCoA on ASV level of mice in three groups. ** $p < 0.01$, *** $p < 0.001$.

3.7. Composition Analysis of the Gut Microbiota

The comparative analysis of the three groups of gut microbiota at the genus level are presented by Kruskal-Wallis H test bar plot and Wilcoxon rank-sum test bar plot. As demonstrated in Figure 5A, PCY increased the relative abundance of norank_f__*Muribaculaceae, Erysipelatoclostridium* and *Clostridium_innocuum_*group and decreased the relative abundance of *Staphylococcus* and *Acinetobacter* compared to the NR group. The Kruskal-Wallis H test box plot of norank_f__*Muribaculaceae* and *Staphylococcus* in three groups were seen in Figure 5B,C, there are significant difference on norank_f__*Muribaculaceae* and *Staphylococcus* between groups. In the comparison of NC and NR groups, the relative abundance of *Escherichia-Shigella, Staphylococcus* and *Acinetobacter* increased in the NR group, while the relative abundance of norank_f__*Muribaculaceae* and unclassified_f__*Lachnospiraceae* decreased (Figure 5D). As seen in Figure 5E, the use of antibiotics increased the relative abundance of *Escherichia-Shigella* and decreased the relative abundance of unclassified_f__*Lachnospiraceae, Lachnospiraceae*_NK4A136_group, *Dubosiella* and *Bacteroides* compared to the NC group. As shown in Figure 5F, PCY significantly increased the relative abundance of norank_f__*Muribaculaceae* compared to the NR group ($p < 0.01$), while the beneficial bacteria were almost absent in the naturally recovery group. The results of linear discriminant analysis effect size (LEfSe) are shown in histogram of LDA value distribution and branching of species evolution (Figure 5G,H). We compared high-dimensional categories and explored differences in bacterial community dominance. There were 14 significantly different species in the NC group, the PCY group had 15 significantly different species in abundance, significant bacteria were *Muribaculaceae* and *Lachnoclostridium*, while the NR group had only 10 significantly different species with many harmful bacteria.

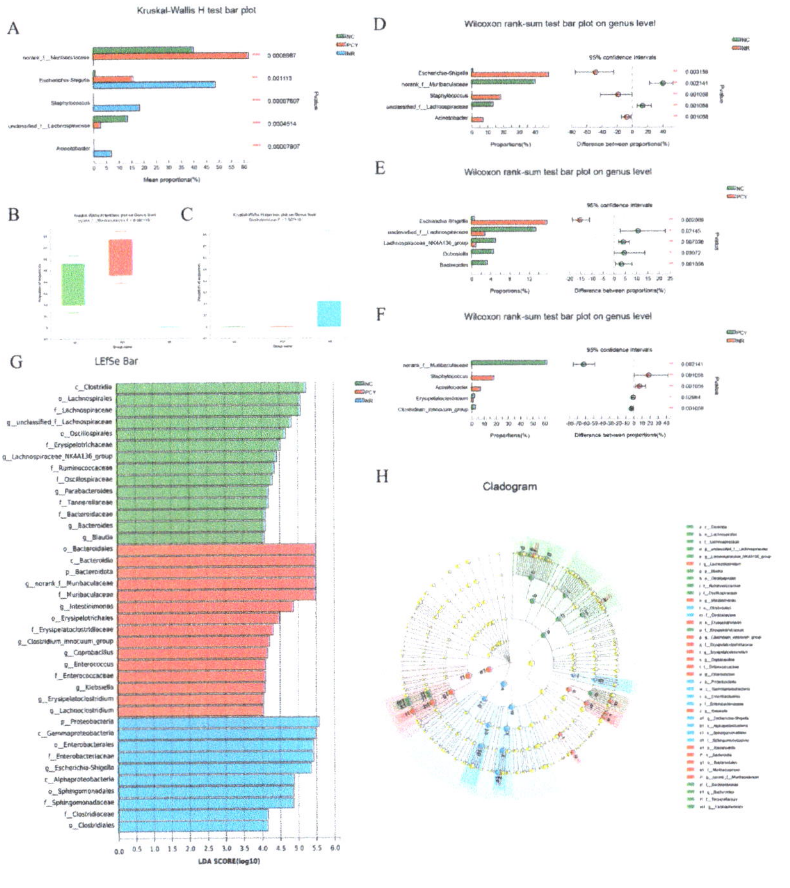

Figure 5. (**A**) Difference of three groups at the genus level. (**B**) Kruskal−Wallis H test box plot of norank_f__*Muribaculaceae*. (**C**) Kruskal−Wallis H test box plot of *Staphylococcus*. (**D**) Difference between NC and NR groups at the genus level. (**E**) Difference between NC and PCY groups at the genus level. (**F**) Difference between PCY and NR groups at the genus level. (**G**) LEfSe analysis. (**H**) Taxonomic cladogram of LEfSe. * $p < 0.05$, ** $p < 0.01$, *** $p < 0.001$.

The biofilm and potential pathogenicity predicting of the three groups microbiota were evaluated using the Kruskal-Wallis H test method, as shown in Figure 6A–D, microbiota in the NC and PCY groups had less biofilm and potentially pathogenic bacteria compared to the NR group. COG (Clusters of Orthologous Groups) is a method of differential gene function annotation and has a prominent role in predicting the function of proteins [25]. The COG function classifications are shown in Figure 6E and Figure S7. In addition, the correlation heat map of protein function prediction for the three groups is shown in Figure S8.

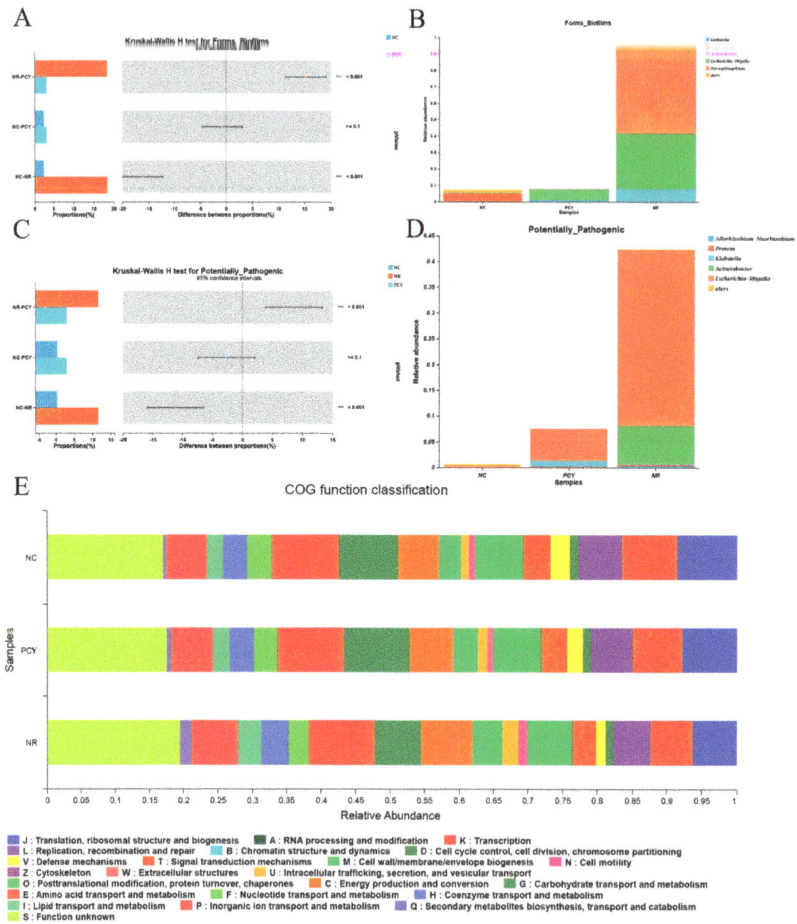

Figure 6. (**A**) Forms of Biofilms comparation. (**B**) The relative abundance of Biofilms in three groups. (**C**) Potentially Pathogenic comparation. (**D**) The relative abundance of Potentially Pathogenic in three groups. (**E**) COG function classification. *** $p < 0.001$.

3.8. Effect of PCY on SCFAs

The results of SCFAs content in the mouse feces in each group are shown in Figure 7A. The concentration of acetic acid and propionic acid significantly decreased in the AAD group than the NC group ($p < 0.001$). The use of lincomycin hydrochloride decreased SCFAs in mice compared to healthy mice, probably because the antibiotic suppressed SCFA-producing bacteria. The level of acetic acid in the PCY group (1.63 $\pm$ 0.48 μg/mg) was significantly higher than that in the NR group (0.54 $\pm$ 0.22 μg/mg) and the AAD group (0.15 $\pm$ 0.01 μg/mg) ($p < 0.001$). butyric acid in the PCY group was 0.42 $\pm$ 0.2 μg/mg, which was statistically higher than that in the NR group ($p < 0.001$). In addition, the concentrations of butyric, valeric, iso-butyric, isovaleric, hexanoic, and iso-hexanoic acid in the PCY group did not change significantly between groups in our experiment.

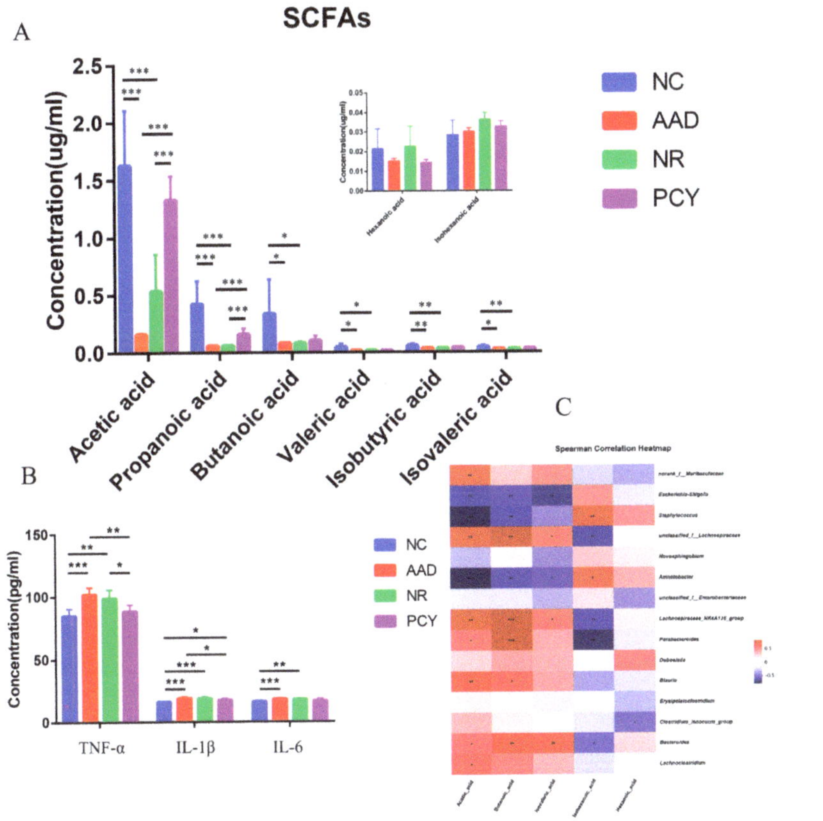

Figure 7. (**A**) Changes in concentrations of SCFAs. (**B**) Changes in the secretion of inflammatory cytokines. (**C**) Spearman Correlation Heatmap of bacteria and SCFAs. * $p < 0.05$, ** $p < 0.01$, *** $p < 0.001$.

The value of environmental factor VIF is shown in Table S2, the strength of the correlation was assessed using the color depth on the heatmap, and the results indicated that acetic, butyric, isovaleric, hexanoic and iso-hexanoic acid are strongly correlated with some bacteria after our filtering (Figure 7C). For example, acetic acid was positively correlated with norank_f__*Muribaculaceae* and unclassified_f__*Lachnospiraceae*; acetic acid was negatively correlated with *Staphylococcus* and *Acinetobacter*.

3.9. Effect of PCY on Cytokines

As demonstrated in Figure 7B, the concentrations of TNF-α, IL-6 and IL-1β were significantly higher in the AAD group than they were in the NC group ($p < 0.001$), and the concentrations of TNF-α, IL-6 and IL-1β were also higher in the NR group than they were in the NC group ($p < 0.01$), indicating the success of our pathological model of AAD. PCY significantly decreased the levels of TNF-α and IL-1β compared with the AAD group ($p < 0.05$) and decreased the levels of TNF-α compared with the NR group ($p < 0.05$). The inflammatory cytokine concentration of PCY group mice is more similar to that of healthy mice in NC group, while the inflammatory cytokines in the serum of naturally recovered mice tended to be closer to those of mice in the AAD model group, indicating that the intake of PCY accelerated the recovery of mice to a healthy state.

4. Discussion

In recent years, numerous studies have revealed a complex and close connection between various target organs and gut microbiota, including the gut-brain axis, the gut-lung axis, the gut-skin axis, and the gut-liver axis [26,27]. Gut microbiota is in dynamic equilibrium, but is susceptible to a variety of factors such as diet, disease, and antibacterial drugs. The occurrence and development of a series of diseases are inextricably linked to the changes of the gut microbiome, among them, antibiotic-related diarrhea, as the most common type of disease in clinical practice, is directly related to the gut microbiota [28]. A healthy gut contains thousands of microorganisms, among which the balance of commensal bacteria, probiotics and pathogens is very important, and together they play a unique role in regulating the gut microbiota, such as inflammatory signaling, short-chain fatty acids, bile acids and lipid metabolism [29,30]. Some polysaccharides have been reported to be used as prebiotics to improve AAD, cereus sinensis polysaccharide improves AAD symptoms by reversing microbial dysfunction, and Panax ginseng polysaccharides altered the composition and diversity of gut microbiota in mice with AAD to promote recovery of health [24,28]. In this project, mice were able to repair intestinal mucosal integrity and intestinal barrier function better and faster by ingesting *P. cocos* water-insoluble polysaccharides, suggesting that PCY has the potential to be a prebiotic for improving intestinal function.

In our primary simulated digestion, polysaccharides readily form polymers in solution, and the disruption of aggregates of polymer chains and cleavage of glycosidic bonds lead to changes in polysaccharide body weight according to previous study [22,31]. During the experiment, the residual weight of PCY decreased, the RSC gradually increased, and the TAOC gradually increased, suggesting that the theory is consistent with our experimental results [32]. Together with the results of RSC and TAOC, we hypothesized that the disruption of aggregates in water-insoluble polysaccharides and the breakage of some glycosidic bonds led to a slight decrease in the weight of PCY. The results of microbiota analysis in simulated fermentation told us that PCY has the ability to enhance Firmicutes and Bacteroidota, two major bacterial phyla of the gut microbiota contain many probiotic bacteria, such as *Lactobacillus* [33]. Studies have demonstrated that the ratio of Firmicutes to Bacteroidetes abundance is closely connected to the development of obesity, with higher relative abundance of Bacteroidetes being less likely to be obese [34]. In addition, acetic acid is the main metabolite of the fermentation of indigestible polysaccharides and is vital source of energy [35]. Propionic acid is the main product of the phylum Bacteroidetes, which is involved in the regulation of immunity and can inhibit hepatic cholesterol synthesis [36]. The results indicated that the gut microbiota of each volunteer is individual and the results obtained with or without the addition of water insoluble polysaccharides to the samples are different. Consequently, the administration of PCY had effect on the gut microbiota and increased the concentration of SCFAs after the fermentation.

The results of 16Sr RNA high sequencing in animal experiment have demonstrated the strong prebiotic potential of PCY, as evidenced by its ability to increase probiotics and decrease harmful bacteria in mice gut microbiota, alleviating the damage to the intestinal microbiota from antibiotic-associated diarrhea. Researches have indicated that Proteobacteria contains many pathogenic bacteria, such as *Escherichia coli, Helicobacter pylori* and *Salmonella*, which can cause dysbiosis in gut microbiota [37]. Proteobacteria are considered to be one of the pathogenic endotoxin-producing bacteria, and the accumulation of large amounts of endotoxin in the intestine can cause many pathological changes, such as intestinal inflammation, immune disorders and functional damage to the intestinal mucosa [38]. Recent studies have shown that expansion of *Escherichia-Shigella* in gut is strongly associated with the occurrence and development of the disease [39]. Comparison of the three groups at the fecal bacterial phylum and genus level showed that antibiotics cause disturbances in the gut microbiota and that the naturally recovered gut microbiota leads to an undesirable multiplication of Proteobacteria. The administration of PCY can alleviate this process and reduce the number of Proteobacteria bacteria to roughly normal levels. *Muribaculaceae* belong to the classification of the phylum Bacteroidetes and are the main utilizers of mucin monosaccharides [40]. Numerous studies have shown that *Muribaculaceae* play an important role in the energy metabolism of the intestine, are functionally diverse in degrading complex carbohydrates, and are considered to exist as beneficial bacteria [41,42]. Treatment with lincomycin hydrochloride can deplete *Muribaculaceae*, and the number of *Muribaculaceae* cannot be restored to the original number even after stopping antibiotic treatment and waiting for the organism to recover naturally, while PCY can facilitate this process. PCY intake increased the relative abundance of unclassified_f__*Lachnospiraceae* among the three groups ($p < 0.01$). unclassified_f__ *Lachnospiracea* is a potential beneficial bacterium, which has considerable capacity in the utilization of dietary fiber such as polysaccharides, and ferments carbohydrates to produce butyric and acetic acid to provide energy [43]. The increase in the abundance of *Lachnospiraceae* is closely related to the increase of plant diet rich in fiber and short-chain fatty acids. Thus, the upregulation of *Lachnospiraceae* and *Muribaculaceae* may indicated that PCY has the potential to promote the treatment of gastrointestinal diseases. *Escherichia-Shigella, Staphylococcus* and *Acinetobacter* were abundantly present in the NR group, while they were little in the NC and PCY groups with statistically significant differences. This result clearly indicates that PCY has the ability to reduce harmful bacteria, thus creating a near normal intestinal microbial environment. *Acinetobacter* are a group of non-fermentative gram-negative bacilli and *Staphylococcus* are common purulent cocci, both are widely found in nature and are important conditional pathogens and common bacteria causing hospital infections today [44,45]. *Escherichia-Shigella* is normally present in the healthy intestine, but its excess can cause disease, and studies have shown a strong correlation between *Escherichia-Shigella* and the severity of non-alcoholic steatohepatitis disease [46]. Naturally recovered mice produce numerous pathogens in the intestine that hinder the process of restoring health. Interestingly, the PCY treated mice, like the normal group of mice, do not produce these pathogens and increase the number of beneficial bacteria such as norank_f__*Muribaculaceae* and unclassified_f__*Lachnospiraceae*.

SCFAs, as carboxylic acids with less than 6 carbon atoms, are important metabolites of intestinal microorganisms [47]. SCFAs are a key source of energy for intestinal cells and can regulate energy homeostasis and metabolism, and SCFAs can regulate immune response and participate in inflammatory response [48]. SCFAs have also been found to inhibit tumor cell proliferation and promote apoptosis [49]. In our study, antibiotics significantly disrupted the intestinal barrier and affected the structure and function of the intestinal microbial community in AAD mice, leading to a substantial reduction in SCFAs. The damage caused by the drug could not be reversed by the natural recovery of the mouse organism even within one week after the cessation of antibiotics. However, the use of PCY was effective in ameliorating this situation, with significantly higher concentrations of butyric acetate in the major SCFAs compared to mice in the AAD group. PCY, a water-insoluble polysaccharide that is difficult to digest, can be used by bacteria in the intestine to produce SCFAs, so the concentration of SCFAs were increased simultaneously, which indicates that the results of our metabolome are in general agreement with the previous microbiota analysis.

The pathophysiological features of antibiotic-associated diarrhea include an inflammatory response and the release of multiple inflammatory cytokines, and excessive inflammatory cytokines release leads to severe intestinal mucosal damage and dysfunction [50,51]. TNF-α, IL-1β, and IL-6 are the major inflammatory cytokines in the body and play an important role in maintaining gastrointestinal homeostasis [52]. TNF-α is a key promoter of the inflammatory response following intestinal disruption and be able to promote the production and gene expression of IL-1β, and IL-6 [53]. AAD is strongly associated with disruption of intestinal structure, dysbiosis of gut microbiota, inflammatory response and SCFA reduction, as shown in our study, in which mice in the AAD model group had defective and incomplete intestinal mucosa, altered intestinal microbial community structure and composition, reduced beneficial bacteria, increased harmful bacteria, reduced production of SCFAs, and overproduction of inflammatory cytokines. Naturally recovered pathological mice did not perform well in the recovery process, accompanied by the proliferation of some harmful bacteria such as unclassified_f__*Lachnospiraceae* and norank_f__*Muribaculaceae*, which can lead to a large accumulation of harmful substances in the intestine and have a bad effect on improving the condition. Surprisingly, after one week administration of PCY, the mice showed great improvement and recovery, as evidenced by homogeneous and intact intestinal mucosal structure, reduced inflammatory cell infiltration and an overall approach to the intestine of healthy mice, normalized gut microbiota structure and composition, increased relative abundance of probiotics, increased concentration of functionally important SCFAs, and decreased inflammatory cytokines to normal levels. Schematic representation of PCY to alleviate AAD in C57BL/6 mice are in Figure 8. Although our study is not exhaustive and we have not explored the restorative mechanism of a specific protein for AAD, the alteration of gut microbiota, the increase of SCFAs and the decrease of inflammatory cytokines are sufficient to demonstrate the effective alleviating effect of PCY on AAD. Particularly, previous studies on water-soluble polysaccharides have been very detailed, and it is valuable that PCY, as a special water-insoluble polysaccharide, can exert such an excellent effect, which is very important for the in-depth development of Chinese traditional medicine *Poria cocos*. Discovering the new function of water insoluble polysaccharide can realize the secondary utilization of common Chinese medicine, and it is also another novel strategy to find prebiotics to improve gastrointestinal diseases.

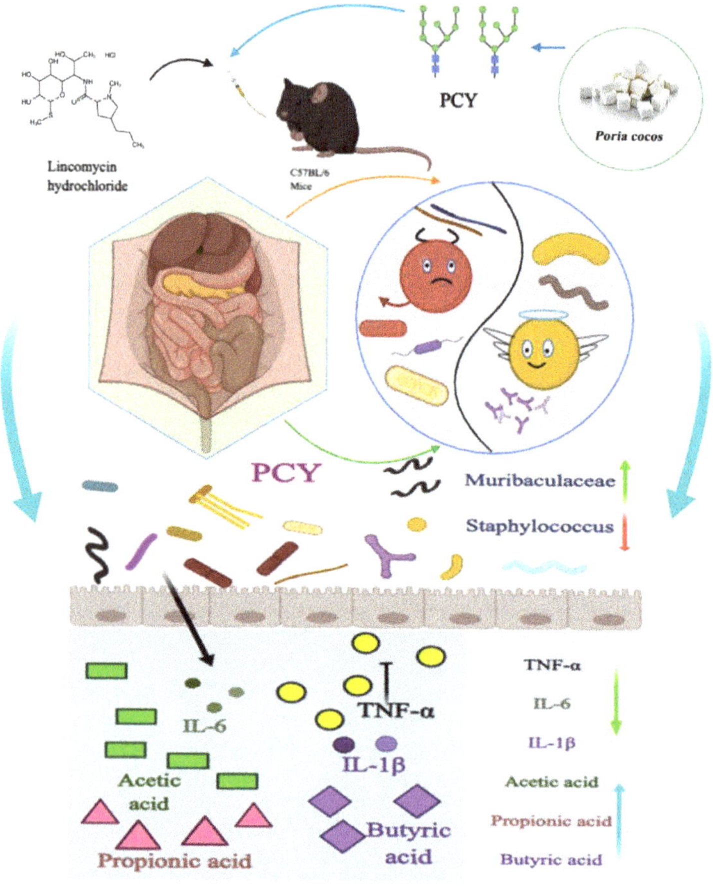

Figure 8. Schematic representation of PCY to alleviate AAD in C57BL/6 mice.

5. Conclusions

In summary, we first performed digestion simulations and in vitro fermentation experiments to study the properties of PCY. And then we acted PCY on mice with AAD to investigate its effects on the mice's gut microbiota and metabolites. The results suggest that PCY have beneficial effects on AAD mice. Specifically, PCY facilitated the recovery of intestinal mucosa and barrier function, altered the composition and structure of the gut microbiota, increasing the relative abundance of beneficial bacteria, such as norank_f__*Muribaculaceae* and unclassified_f__*Lachnospiraceae*, and decreasing the relative abundance of harmful bacteria, such as *Escherichia-Shigella*, *Staphylococcus* and *Acinetobacter*, and increasing the production of important SCFAs, such as acetic acid and propionic acid. In addition, PCY decreased the levels of inflammatory cytokines and inhibited the inflammatory response, thereby better facilitating the recovery of AAD in mice. This study further demonstrates the potential of PCY as a prebiotic, and the results provide a basis for the development of a prebiotic that is effective in alleviating AAD.

Supplementary Materials: The following supporting information can be downloaded at: https://www.mdpi.com/article/10.3390/foods12163080/s1, Figure S1: The high-performance liquid chromatograms of PCY; Figure S2: The scanning electron micrograph of PCY in 100 μm and 30 μm; Figure S3: The results of microbial composition of fresh feces from healthy volunteers; Figure S4: The PCR amplification results of AAD group mice's feces; Figure S5: The Rarefaction curve; Table S1: The α-diversity of microbial communities in mice; Figure S6: The heat map of microbial community at the genus level; Figure S7: The COG function classification in three groups; Figure S8: The correlation heat map of protein function prediction for the three groups; Table S2: The value of environmental factor VIF.

Author Contributions: Investigation, Y.L.; Data curation, Y.L.; Writing-original draft, Y.L.; Methodology, H.D.; Validation, H.D.; Data curation, Q.F.; Data curation, L.M.; Software, H.L.; Validation, X.G.; Software, Y.C.; Conception, C.S.; Methodology, C.S.; Resources, C.S.; Funding acquisition, C.S.; Writing—review & editing, C.S. All authors have read and agreed to the published version of the manuscript.

Funding: This work supported by Special Incentives and Guidance Project of Chongqing Academy of Science and Technology (cast2022jxj140006).

Institutional Review Board Statement: All animal experimental procedures were performed and approved by the Ethics Committee of the Animal Experimentation Center of Southwest Medical University (license number P20220707-01).

Data Availability Statement: The data presented in this study are available on request from the corresponding author.

Conflicts of Interest: The authors declare no conflict of interest.

References

1. Mantegazza, C.; Molinari, P.; D'Auria, E.; Sonnino, M.; Morelli, L.; Zuccotti, G.V. Probiotics and antibiotic-associated diarrhea in children: A review and new evidence on Lactobacillus rhamnosus GG during and after antibiotic treatment. *Pharmacol. Res.* **2018**, *128*, 63–72. [CrossRef] [PubMed]
2. Harlow, B.E.; Lawrence, L.M.; Flythe, M.D. Diarrhea-associated pathogens, lactobacilli and cellulolytic bacteria in equine feces: Responses to antibiotic challenge. *Vet. Microbiol.* **2013**, *166*, 225–232. [CrossRef] [PubMed]
3. Vujkovic-Cvijin, I.; Sklar, J.; Jiang, L.; Natarajan, L.; Knight, R.; Belkaid, Y. Host variables confound gut microbiota studies of human disease. *Nature* **2020**, *587*, 448–454. [CrossRef]
4. Khan, S.; Luck, H.; Winer, S.; Winer, D.A. Emerging concepts in intestinal immune control of obesity-related metabolic disease. *Nat. Commun.* **2021**, *12*, 2598. [CrossRef] [PubMed]
5. Nowakowski, P.; Markiewicz-Zukowska, R.; Bielecka, J.; Mielcarek, K.; Grabia, M.; Socha, K. Treasures from the forest: Evaluation of mushroom extracts as anti-cancer agents. *Biomed. Pharmacother.* **2021**, *143*, 112106. [CrossRef] [PubMed]
6. Li, C.; Zhou, K.; Xiao, N.; Peng, M.; Tan, Z. The Effect of Qiweibaizhu Powder Crude Polysaccharide on Antibiotic-Associated Diarrhea Mice Is Associated With Restoring Intestinal Mucosal Bacteria. *Front. Nutr.* **2022**, *9*, 952647. [CrossRef]
7. Yu, Y.; Shen, M.; Song, Q.; Xie, J. Biological activities and pharmaceutical applications of polysaccharide from natural resources: A review. *Carbohydr. Polym.* **2018**, *183*, 91–101. [CrossRef]

8. Yin, J.; Ren, W.; Wei, B.; Huang, H.; Li, M.; Wu, X.; Wang, A.; Xiao, Z.; Shen, J.; Zhao, Y.; et al. Characterization of chemical composition and prebiotic effect of a dietary medicinal plant Penthorum chinense Pursh. *Food Chem.* **2020**, *319*, 126568. [CrossRef]
9. Li, X.; Guo, R.; Wu, X.; Liu, X.; Ai, L.; Sheng, Y.; Song, Z.; Wu, Y. Dynamic digestion of tamarind seed polysaccharide: Indigestibility in gastrointestinal simulations and gut microbiota changes in vitro. *Carbohydr. Polym.* **2020**, *239*, 116194. [CrossRef]
10. RuiDian, K.; ShunFa, L.; Yi, C.; ChuRong, J.; QiaGuang, S. Analysis of chemical composition of polysaccharides from Poria cocos Wolf and its anti-tumor activity by NMR spectroscopy. *Carbohydr. Polym.* **2010**, *80*, 31–34. [CrossRef]
11. Meng, X.; Liang, H.; Luo, L. Antitumor polysaccharides from mushrooms: A review on the structural characteristics, antitumor mechanisms and immunomodulating activities. *Carbohydr. Res.* **2016**, *424*, 30–41. [CrossRef] [PubMed]
12. Sun, S.-S.; Wang, K.; Ma, K.; Bao, L.; Liu, H.-W. An insoluble polysaccharide from the sclerotium of Poria cocos improves hyperglycemia, hyperlipidemia and hepatic steatosis in ob/ob mice via modulation of gut microbiota. *Chin. J. Nat. Med.* **2019**, *17*, 3–14. [CrossRef] [PubMed]
13. Zou, Y.T.; Zhou, J.; Wu, C.Y.; Zhang, W.; Shen, H.; Xu, J.D.; Zhang, Y.Q.; Long, F.; Li, S.L. Protective effects of Poria cocos and its components against cisplatin-induced intestinal injury. *J. Ethnopharmacol.* **2021**, *269*, 113722. [CrossRef]
14. Khan, I.; Huang, G.; Li, X.; Leong, W.; Xia, W.; Hsiao, W.L.W. Mushroom polysaccharides from Ganoderma lucidum and Poria cocos reveal prebiotic functions. *J. Funct. Foods* **2018**, *41*, 191–201. [CrossRef]
15. Feng, Y.L.; Cao, G.; Chen, D.Q.; Vaziri, N.D.; Chen, L.; Zhang, J.; Wang, M.; Guo, Y.; Zhao, Y.Y. Microbiome-metabolomics reveals gut microbiota associated with glycine-conjugated metabolites and polyamine metabolism in chronic kidney disease. *Cell Mol. Life Sci.* **2019**, *76*, 4961–4978. [CrossRef] [PubMed]
16. Sun, Y. Biological activities and potential health benefits of polysaccharides from Poria cocos and their derivatives. *Int. J. Biol. Macromol.* **2014**, *68*, 131–134. [CrossRef]
17. Cheng, Y.; Xie, Y.; Ge, J.C.; Wang, L.; Peng, D.Y.; Yu, N.J.; Zhang, Y.; Jiang, Y.H.; Luo, J.P.; Chen, W.D. Structural characterization and hepatoprotective activity of a galactoglucan from Poria cocos. *Carbohydr. Polym.* **2021**, *263*, 117979. [CrossRef]
18. Lai, Y.; Yu, H.; Deng, H.; Fang, Q.; Lei, H.; Liu, L.; Wu, N.; Guo, X.; Song, C. Three main metabolites from Wolfiporia cocos (F. A. Wolf) Ryvarden & Gilb regulate the gut microbiota in mice: A comparative study using microbiome-metabolomics. *Front. Pharmacol.* **2022**, *13*, 911140. [CrossRef]
19. Lai, Y.; Deng, H.; Chen, M.; Fan, C.; Chen, Y.; Wang, F.; Zhou, Q.; Song, C. In vitro fermentation properties of grape seed polysaccharides and the effect on regulating gut microbiota in mice. *Food Measure* **2023**. [CrossRef]
20. Chen, P.; Chen, X.; Hao, L.; Du, P.; Li, C.; Han, H.; Xu, H.; Liu, L. The bioavailability of soybean polysaccharides and their metabolites on gut microbiota in the simulator of the human intestinal microbial ecosystem (SHIME). *Food. Chem.* **2021**, *362*, 130233. [CrossRef]
21. Pham, V.T.; Mohajeri, M.H. The application of in vitro human intestinal models on the screening and development of pre- and probiotics. *Benef. Microbes* **2018**, *9*, 725–742. [CrossRef] [PubMed]
22. Ma, Y.; Jiang, S.; Zeng, M. In vitro simulated digestion and fermentation characteristics of polysaccharide from oyster (*Crassostrea gigas*), and its effects on the gut microbiota. *Food Res. Int.* **2021**, *149*, 110646. [CrossRef] [PubMed]
23. Bie, N.; Duan, S.; Meng, M.; Guo, M.; Wang, C. Regulatory effect of non-starch polysaccharides from purple sweet potato on intestinal microbiota of mice with antibiotic-associated diarrhea. *Food Funct.* **2021**, *12*, 5563–5575. [CrossRef]
24. Cui, M.; Wang, Y.; Elango, J.; Wu, J.; Liu, K.; Jin, Y. Cereus sinensis Polysaccharide Alleviates Antibiotic-Associated Diarrhea Based on Modulating the Gut Microbiota in C57BL/6 Mice. *Front. Nutr.* **2021**, *8*, 751992. [CrossRef]
25. Galperin, M.Y.; Kristensen, D.M.; Makarova, K.S.; Wolf, Y.I.; Koonin, E.V. Microbial genome analysis: The COG approach. *Brief. Bioinform.* **2019**, *20*, 1063–1070. [CrossRef] [PubMed]
26. Morais, L.H.; Schreiber, H.L.T.; Mazmanian, S.K. The gut microbiota-brain axis in behaviour and brain disorders. *Nat. Rev. Microbiol.* **2021**, *19*, 241–255. [CrossRef]
27. Chang, C.J.; Lin, C.S.; Lu, C.C.; Martel, J.; Ko, Y.F.; Ojcius, D.M.; Tseng, S.F.; Wu, T.R.; Chen, Y.Y.; Young, J.D.; et al. Ganoderma lucidum reduces obesity in mice by modulating the composition of the gut microbiota. *Nat. Commun.* **2015**, *6*, 7489. [CrossRef]
28. Li, S.; Qi, Y.; Chen, L.; Qu, D.; Li, Z.; Gao, K.; Chen, J.; Sun, Y. Effects of Panax ginseng polysaccharides on the gut microbiota in mice with antibiotic-associated diarrhea. *Int. J. Biol. Macromol.* **2019**, *124*, 931–937. [CrossRef]
29. Reichardt, N.; Vollmer, M.; Holtrop, G.; Farquharson, F.M.; Wefers, D.; Bunzel, M.; Duncan, S.H.; Drew, J.E.; Williams, L.M.; Milligan, G.; et al. Specific substrate-driven changes in human faecal microbiota composition contrast with functional redundancy in short-chain fatty acid production. *ISME J.* **2018**, *12*, 610–622. [CrossRef]
30. Luo, Y.; Fang, Q.; Lai, Y.; Lei, H.; Zhang, D.; Niu, H.; Wang, R.; Song, C. Polysaccharides from the leaves of Polygonatum sibiricum Red. regulate the gut microbiota and affect the production of short-chain fatty acids in mice. *AMB Express* **2022**, *12*, 35. [CrossRef]
31. Liu, Y.; Li, Y.; Ke, Y.; Li, C.; Zhang, Z.; Wu, Y.; Hu, B.; Liu, A.; Luo, Q.; Wu, W. In vitro saliva-gastrointestinal digestion and fecal fermentation of Oudemansiella radicata polysaccharides reveal its digestion profile and effect on the modulation of the gut microbiota. *Carbohydr. Polym.* **2021**, *251*, 117041. [CrossRef] [PubMed]
32. Fu, Z.; Han, L.; Zhang, P.; Mao, H.; Zhang, H.; Wang, Y.; Gao, X.; Liu, E. Cistanche polysaccharides enhance echinacoside absorption in vivo and affect the gut microbiota. *Int. J. Biol. Macromol.* **2020**, *149*, 732–740. [CrossRef] [PubMed]
33. Song, C.; Huang, F.; Liu, L.; Zhou, Q.; Zhang, D.; Fang, Q.; Lei, H.; Niu, H. Characterization and prebiotic properties of pectin polysaccharide from *Clausena lansium* (Lour.) Skeels fruit. *Int. J. Biol. Macromol.* **2022**, *194*, 412–421. [CrossRef] [PubMed]

34. Su, L.; Xin, C.; Yang, J.; Dong, L.; Mei, H.; Dai, X.; Wang, Q. A polysaccharide from Inonotus obliquus ameliorates intestinal barrier dysfunction in mice with type 2 diabetes mellitus. *Int. J. Biol. Macromol.* **2022**, *214*, 312–323. [CrossRef]
35. Zhu, Z.; Huang, R.; Huang, A.; Wang, J.; Liu, W.; Wu, S.; Chen, M.; Chen, M.; Xie, Y.; Jiao, C.; et al. Polysaccharide from Agrocybe cylindracea prevents diet-induced obesity through inhibiting inflammation mediated by gut microbiota and associated metabolites. *Int. J. Biol. Macromol.* **2022**, *209*, 1430–1438. [CrossRef] [PubMed]
36. Nie, Y.; Liu, Q.; Zhang, W.; Wan, Y.; Huang, C.; Zhu, X. Ursolic acid reverses liver fibrosis by inhibiting NOX4/NLRP3 inflammasome pathways and bacterial dysbiosis. *Gut. Microbes* **2021**, *13*, 1972746. [CrossRef] [PubMed]
37. Shin, N.R.; Whon, T.W.; Bae, J.W. Proteobacteria: Microbial signature of dysbiosis in gut microbiota. *Trends Biotechnol.* **2015**, *33*, 496–503. [CrossRef]
38. Litvak, Y.; Byndloss, M.X.; Tsolis, R.M.; Baumler, A.J. Dysbiotic Proteobacteria expansion: A microbial signature of epithelial dysfunction. *Curr. Opin Microbiol.* **2017**, *39*, 1–6. [CrossRef]
39. Xin, F.Z.; Zhao, Z.H.; Liu, X.L.; Pan, Q.; Wang, Z.X.; Zeng, L.; Zhang, Q.R.; Ye, L.; Wang, M.Y.; Zhang, R.N.; et al. Escherichia fergusonii Promotes Nonobese Nonalcoholic Fatty Liver Disease by Interfering With Host Hepatic Lipid Metabolism Through Its Own msRNA 23487. *Cell Mol. Gastroenterol. Hepatol.* **2022**, *13*, 827–841. [CrossRef]
40. Lagkouvardos, I.; Lesker, T.R.; Hitch, T.C.A.; Galvez, E.J.C.; Smit, N.; Neuhaus, K.; Wang, J.; Baines, J.F.; Abt, B.; Stecher, B.; et al. Sequence and cultivation study of Muribaculaceae reveals novel species, host preference, and functional potential of this yet undescribed family. *Microbiome* **2019**, *7*, 28. [CrossRef]
41. Pereira, F.C.; Wasmund, K.; Cobankovic, I.; Jehmlich, N.; Herbold, C.W.; Lee, K.S.; Sziranyi, B.; Vesely, C.; Decker, T.; Stocker, R.; et al. Rational design of a microbial consortium of mucosal sugar utilizers reduces Clostridiodes difficile colonization. *Nat. Commun.* **2020**, *11*, 5104. [CrossRef] [PubMed]
42. Li, L.; Qiu, N.; Meng, Y.; Wang, C.; Mine, Y.; Keast, R.; Guyonnet, V. Preserved egg white alleviates DSS-induced colitis in mice through the reduction of oxidative stress, modulation of infl ammatory cytokines, NF-κB, MAPK and gut microbiota composition. *Food Sci. Hum. Wellness* **2023**, *12*, 312–323. [CrossRef]
43. Lai, Y.; Fang, Q.; Guo, X.; Lei, H.; Zhou, Q.; Wu, N.; Song, C. Effect of polysaccharides from Dictyophora indusiata on regulating gut microbiota and short-chain fatty acids in mice. *J. Food Meas. Charact.* **2022**, *17*, 1–11. [CrossRef]
44. Biclot, A.; Huys, G.R.B.; Bacigalupe, R.; D'Hoe, K.; Vandeputte, D.; Falony, G.; Tito, R.Y.; Raes, J. Effect of cryopreservation medium conditions on growth and isolation of gut anaerobes from human faecal samples. *Microbiome* **2022**, *10*, 80. [CrossRef] [PubMed]
45. Dong, R.; Bai, M.; Zhao, J.; Wang, D.; Ning, X.; Sun, S. A Comparative Study of the Gut Microbiota Associated With Immunoglobulin a Nephropathy and Membranous Nephropathy. *Front. Cell. Infect. Microbiol.* **2020**, *10*, 557368. [CrossRef]
46. Zhang, Z.; Liang, X.; Tong, L.; Lv, Y.; Yi, H.; Gong, P.; Tian, X.; Cui, Q.; Liu, T.; Zhang, L. Effect of *Inonotus obliquus* (Fr.) Pilat extract on the regulation of glycolipid metabolism via PI3K/Akt and AMPK/ACC pathways in mice. *J. Ethnopharmacol.* **2021**, *273*, 113963. [CrossRef]
47. Tan, J.; McKenzie, C.; Potamitis, M.; Thorburn, A.N.; Mackay, C.R.; Macia, L. The role of short-chain fatty acids in health and disease. *Adv. Immunol.* **2014**, *121*, 91–119. [CrossRef]
48. Liu, C.; Du, P.; Guo, Y.; Xie, Y.; Yu, H.; Yao, W.; Cheng, Y.; Qian, H. Extraction, characterization of aloe polysaccharides and the in-depth analysis of its prebiotic effects on mice gut microbiota. *Carbohydr. Polym.* **2021**, *261*, 117874. [CrossRef]
49. Yao, Y.; Cai, X.; Fei, W.; Ye, Y.; Zhao, M.; Zheng, C. The role of short-chain fatty acids in immunity, inflammation and metabolism. *Crit. Rev. Food Sci. Nutr.* **2022**, *62*, 1–12. [CrossRef]
50. Yuan, J.; Qin, S.; Hu, S.; Liu, Z.; Song, Y.; Li, L. Restoration of cefixime-induced gut microbiota changes by a prebiotic blend in a mouse model. *Appl. Microbiol. Biotechnol.* **2022**, *106*, 5197–5209. [CrossRef]
51. Mekonnen, S.A.; Merenstein, D.; Fraser, C.M.; Marco, M.L. Molecular mechanisms of probiotic prevention of antibiotic-associated diarrhea. *Curr. Opin Biotechnol.* **2020**, *61*, 226–234. [CrossRef] [PubMed]
52. Silverman, M.A.; Konnikova, L.; Gerber, J.S. Impact of Antibiotics on Necrotizing Enterocolitis and Antibiotic-Associated Diarrhea. *Gastroenterol. Clin. North Am.* **2017**, *46*, 61–76. [CrossRef] [PubMed]
53. Li, Q.; Cui, Y.; Xu, B.; Wang, Y.; Lv, F.; Li, Z.; Li, H.; Chen, X.; Peng, X.; Chen, Y.; et al. Main active components of Jiawei Gegen Qinlian decoction protects against ulcerative colitis under different dietary environments in a gut microbiota-dependent manner. *Pharmacol. Res.* **2021**, *170*, 105694. [CrossRef] [PubMed]

foods

MDPI

Review

An Update on the Pivotal Roles of Probiotics, Their Components, and Metabolites in Preventing Colon Cancer

Xue Deng [1,†], Jing Yang [2,†], Yu Zhang [1], Xiaoyong Chen [1], Chen Wang [1], Huayi Suo [1,3] and Jiajia Song [1,*]

1 College of Food Science, Southwest University, Chongqing 400715, China; dx15708131969@email.swu.edu.cn (X.D.); y000063@swu.edu.cn (Y.Z.); cxy0520@swu.edu.cn (X.C.); wangchen20220307@swu.edu.cn (C.W.); birget@swu.edu.cn (H.S.)
2 Chongqing Engineering Research Center for Processing & Storage of Distinct Agricultural Products, Chongqing Technology and Business University, Chongqing 400067, China; jyang@ctbu.edu.cn
3 National Citrus Engineering Research Center, Southwest University, Chongqing 400712, China
* Correspondence: jiajias@swu.edu.cn
† These authors contributed equally to this work.

Abstract: Diet, lifestyle, and gut microbiota composition are key risk factors for the progression of colon cancer. Probiotics are living microorganisms that can offer health benefits to the parasitifer when ingested in competent quantities. Several in vivo, in vitro, and clinical studies have demonstrated that probiotics can prevent and mitigate the development of colon cancer. The anti-colon cancer mechanisms of probiotics include the suppression of cell proliferation and the promotion of cancer cell apoptosis, immunomodulation, the modulation of intestinal microorganisms and their metabolism, strengthening the intestinal barrier, and antioxidant effects. This article describes the pathogenesis of colon cancer and the available therapeutic options. In addition, this paper reviews the mechanisms by which probiotics mitigate colon cancer as well as the mitigating effects of probiotic components and metabolites on colon cancer.

Keywords: probiotics; colon cancer; colon cancer treatment; anti-cancer mechanisms

1. Introduction

Colon cancer is the third most common cancer worldwide. Being a malignant tumor, the incidence of and mortality rate associated with colon cancer are currently high worldwide [1]. According to statistics, the incidence of colon cancer in 2020 was 10% and the associated mortality rate was 9.4% [2]. Most colon cancers could be attributed to lifestyle and aging, whereas a few cases are due to underlying genetic disorders. Other factors, such as dietary habits and gut microbiota, are also related to colon cancer development [3,4].

Conventional therapeutic strategies for colon cancer comprise surgical removal, chemotherapy, and radiotherapy. Surgery is usually indicated in patients whose colonic lesions are early and localized. Recovery is expected after surgical removal, which is one of the efficacious treatments for localized cancer [5]. On the contrary, those with metastatic colon cancer tend to opt for chemotherapy as their primary treatment. Radiotherapy, which is yet another treatment modality, uses ionizing radiation to inhibit the multiplication of malignant cells and can effectively address prevalent forms of cancer, including skin cancers, lung carcinomas, lymphomas, head and neck carcinomas, among others [5,6]. Immunotherapy is a new therapeutic strategy for colon cancer, and its efficacy has been proven in clinical trials for patients with colon cancer, especially those with high microsatellite instability (MSI-H) [7]. However, some of the current treatments exert several adverse effects on the body; hence, exploring new alternatives for the prevention and treatment of colon cancer becomes necessary.

The World Food and Agriculture Organization has provided a definition of probiotics as "living microorganisms that, when administered in adequate quantities, provide health benefits to the host". There is growing evidence to support the potential role of probiotics in colon cancer prevention. In vitro experiments have demonstrated the proliferation suppression of colon cancer cells by various probiotics. *Pediococcus pentosaceus* GS4, *Pediococcus pentosaceus* SP2, and *Lacticaseibacillus paracasei* SP5 are a few examples [8,9]. Administration of *Lacticaseibacillus rhamnosus* GG and *Lactobacillus acidophilus* to mice with colon cancer has been shown to reduce tumor incidence [10]. *Lacticaseibacillus rhamnosus* and *Lactobacillus acidophilus* have been reported to alleviate bowel symptoms and enhance the quality of life of patients with colon cancer [11]. In addition, probiotic metabolites and bacterial fractions can also offer resistance to colon cancer. Cell-free supernatant of *Lacticaseibacillus rhamnosus* has been documented to inhibit the metastasis and proliferation of HT-29 cells and promote their apoptosis [12]. Furthermore, probiotic extracellular polysaccharides (EPSs) and S-layer protein (Slp) can inhibit colon cancer [13,14].

This review summarizes the pathogenesis and treatment of colon cancer and also examines the recent advances in probiotics as well as the mechanisms by which they alleviate colon cancer. In addition, the progress in studies pertaining to the inhibition of colon cancer by probiotic components and related metabolites is presented.

2. Risk Factors of Colon Cancer

2.1. Diet

Epidemiologic and experimental findings often link food and nutrient intake to colon cancer risk. Calcium occurs in all types of foods, and increasing its intake may reduce colon cancer risk [15]. Moreover, according to epidemiological studies, vitamin D deficiency is yet another factor associated with an increased incidence of colon cancer that also adversely affects the survival of patients with colon cancer [16]. A meta-analysis revealed that vitamin D intake has been clinically associated with improved long-term survival in patients with colon cancer [17]. Moreover, dietary fiber and whole grain intake exerts a positive effect on cancer prevention [18]. On the contrary, excessive intake of red and processed meats significantly increases the risk of colon cancer [19]. Heme iron in red meat leads to the formation of endogenous N-nitroso compounds as well as heterocyclic amino acids and polycyclic aromatic hydrocarbons when cooked at high temperatures. The high intake of all these foods has been linked to colon tumors [20,21]. Sweets typically contain higher amounts of carbohydrates and fats. Furthermore, research indicates a significant correlation between a high consumption of sweets and an elevated risk of colon cancer [22]. Notably, Western dietary patterns characterized by increased intake of processed and red meat, refined grains, and sweets have been associated with the development of colon cancer [23–26]. Conversely, adopting a diet abundant in fruits, vegetables, dairy products, and whole-grain products has been linked to a reduced likelihood of colon cancer development [26,27].

2.2. Obesity

Obesity is characterized by chronic inflammation and leads to an increased risk of colon cancer as well as increased mortality and complications after colon cancer surgery [28]. Elevated levels of inflammatory factors, such as tumor necrosis factor-α (TNF-α), interleukin-1 (IL-1), and IL-6, have the potential to initiate colon carcinogenesis [29,30]. Obesity can trigger an increase in the levels of insulin and insulin-like growth factor 1 (IGF-1) in the blood [31]. Insulin and IGF-1 have significant implications in colon cancer development, as they activate the phosphatidylinositol 3-kinase (PI3K)/Akt signaling pathway, stimulating cell growth and cell cycle signaling [31]. Moreover, studies indicated that insulin and IGF-1 can enhance the proliferation of colon cancer cells and facilitate the growth of colon cancer allografts in obese mice [32,33]. Obesity has been correlated with higher levels of leptin, a hormone originating from adipose tissue. Leptin plays a crucial role in regulating the homeostatic balance of adipose tissue mass. Some studies have demonstrated that leptin can stimulate the growth of colon cancer cells [34,35]. Conversely, adiponectin, which enhances energy expenditure, can mitigate colon cancer related to obesity by inhibiting the mammalian target of the rapamycin (mTOR) pathway [36].

2.3. Gut Microorganisms

Colon cancer progression is intricately linked to ecological disturbances in the gut microbiota. Colon cancer is induced by oncogenic bacteria, such as enterotoxigenic *Bacteroides fragilis* (ETBF), via direct interactions with colonic epithelial cells and by altering the microbiota composition at the colonic site [37]. The toxin fragilysin produced by ETBF binds to colonic epithelial cells and in turn activates the proto-oncogene MYC and promotes the expression of the Wnt/β-linker proteins in the nucleus. These alterations induce the hyperproliferation of colonic epithelial cells [38,39]. ETBF may induce changes in the bacterial communities in the local environment of tumors and may permit the establishment of transient bacteria, such as *Fusobacterium nucleatum* [40]. *Fusobacterium nucleatum* stimulates the growth of colon tumor cells by triggering autophagy through the Toll-like receptor-4 pathway [41]. Escherichia coli inhabits the human gut and specific pathogenic strains have been linked to colon cancer. The production of colicin by *Escherichia coli* can lead to DNA damage and cell mutations, a phenomenon closely associated with the development of colon cancer [42,43]. Additionally, *Enterococcus faecalis* is capable of generating reactive oxygen species (ROS) and superoxide anions, which can induce DNA damage, potentially contributing to the development of cancer [38]. An increased abundance of *Bacteroides fragilis*, *Escherichia coli*, *Enterococcus faecalis*, *Fusobacterium nucleatum*, and *Streptococcus gallolyticus* and a decreased abundance of *Bifidobacterium*, *Clostridium*, *Faecalibacterium*, and *Roseburia* have been observed in patients with colon cancer [44]. A substantial reduction in butyrate-producing bacteria, such as *Lachnospiraceae*, among individuals with colon cancer, may result in an upsurge in pathogenic microorganisms. This, in turn, contributes to the progression of colon cancer [45]. Sulfidogenic bacteria, such as *Fusobacterium*, *Desulfovibrio*, and *Bilophila wadsworthia*, have been shown to induce genomic instability, which has been linked to colon cancer via the generation of hydrogen sulfide [46]. The intestinal microbiota participates in the metabolic transformation of the tumor and tumor-suppressor metabolites, which influences the development of colon cancer [47]. Secondary bile acids serve as predisposing agents for colon cancer. Gut microbes participate in the synthesis and transformation of these secondary bile acids, including lithocholic acid and deoxycholic acid [48]. Additionally, carnitine found in red meat produces trimethylamine, which, in collaboration with gut microbes, generates trimethylamine-n-oxide, a metabolite associated with the development of colon cancer [49].

2.4. Genetic Instability

The pathogenesis of colon cancer can be attributed to genetic or epigenetic changes, with genomic instability serving as the underlying process. In the pathogenesis of colon cancer, at least three pathways of genomic instability exist: chromosomal instability, microsatellite instability, and aberrant DNA methylation [50]. Colon cancer progresses through several stages that involve the gradual accumulation of mutations in many tumor suppressor genes and proto-oncogenes. Along with the activation of oncogenes, some tumor suppressor genes such as adenomatous polyposis coli (APC) and p53 are mutated and inactivated [51,52]. APC is recognized as a tumor suppressor protein that opposes the Wnt signaling pathway [53]. The inhibition of the Wnt signaling pathway has been linked to the effectiveness of certain dietary agents in preventing colon cancer [54]. Additionally, mutations in APC are closely associated with the initiation of colon tumorigenesis [55].

3. Treatment Options for Colon Cancer

3.1. Surgical Treatment

To date, the standard therapeutic option for colon cancer, which is a malignant tumor, is surgical removal. Surgical treatment can be classified as radical intervention and palliative. The former includes right or left hemicolectomy, segmental colectomy on transverse or sigmoid colon, and subtotal or total colectomy. The latter is categorized as colostomy and colonic bypass and is applied in advanced nonresectable stages [56].

3.2. Adjuvant Chemotherapy

Fluorouracil operates by inhibiting thymidylate synthase, an essential enzyme in DNA synthesis. Leucovorin, on the other hand, acts as a reductive folate, augmenting the inhibitory impact of fluorouracil on thymidylate synthase. For patients with colon cancer, fluorouracil combined with leucovorin (FL) was the earliest standard adjuvant therapy, and it reduced the likelihood of tumor recurrence [57]. Oxaliplatin, a third-generation platinum derivative, combats cancer by antagonizing DNA replication and transcription. Patients were treated with FL alone or in combination with oxaliplatin after radical resection for stage II or III colon cancer. The rate of disease-free survival at 3 years was 78.2% in the oxaliplatin-added group and 72.9% in the no-addition group. The occurrence rate of febrile neutropenia was 1.8% and that of adverse reactions in the gastrointestinal tract was low in the group to which oxaliplatin was added [58]. The use of oxaliplatin in combination with fluoropyrimidine significantly improved disease-free and overall survival in patients with stage III colon cancer with MSI [59]. Capecitabine, an oral prodrug of 5-fluorouracil (5-FU), undergoes conversion into 5-FU within the body, where it effectively hinders DNA synthesis. Postoperative therapy with capecitabine substantially decreased the likelihood of death and recurrence compared with patients treated using surgery only. Furthermore, a 16-week capecitabine chemotherapeutic regimen improved the survival in elderly patients with stage III colon cancer [60]. Patients receiving capecitabine and oxaliplatin had elevated disease-free survival rates and overall survival rates of 62% and 80%, respectively, at 2 years. In addition, the efficacy of the combination of capecitabine and oxaliplatin was comparable to that of 5-FU [61]. This finding implies that adjuvant therapy with capecitabine in combination with oxaliplatin is feasible in patients with early stage colon cancer. Irinotecan functions as an inhibitor of topoisomerase I, preventing DNA untangling and, in turn, disrupting DNA replication and cell division. In a meta-analysis of randomized controlled studies, the supplementation with irinotecan considerably enhanced the progression-free survival rate and median progression-free survival and reduced disease progression compared with FL. However, an increased incidence of grade $\geq$3 adverse events was observed with irinotecan treatment. These adverse reactions, which mainly involved diarrhea and neutropenia, were generally acceptable [62]. Bevacizumab is a targeted drug that obstructs the vascular endothelial growth factor, thereby decelerating tumor growth and enhancing the efficacy of chemotherapy. When bevacizumab was added to first-line chemotherapy in patients with metastatic colon cancer, progression-free survival was significantly improved. Bevacizumab monotherapy may be considered in certain circumstances, such as cumulative toxicity or patient refusal, especially in RAS wild-type patients [63].

3.3. Immunotherapy

In a study, patients with stage II and III colon cancer were treated with active specific immunotherapy (ASI) using autologous tumor cells and an immunomodulating adjuvant bacillus Calmette–Guérin vaccine. For patients with stage II disease, ASI not only significantly prolonged the recurrence-free interval but also improved the 5-year overall survival and recurrence-free survival. Nonetheless, a statistically significant prognostic effect was not achieved in patients with stage III disease [64]. Pembrolizumab, an antibody targeting programmed death receptor I, achieves its anti-tumor effects by bolstering autoimmunity. Moreover, 307 treatment-naïve patients with MSI-H or mismatch repair-deficient (dMMR) colon cancer were treated with pembrolizumab or chemotherapy. The former was better than the latter in terms of progression-free survival and overall remission rate. Furthermore, there were fewer treatment-related adverse events with pembrolizumab. This result signifies that pembrolizumab can be used as first-line treatment for MSI-H-dMMR metastatic colon cancer [7]. Edrecolomab is a mouse-derived monoclonal IgG2a antibody that specifically binds to antigens associated with human tumors. In studies involving patients with resected stage III colorectal cancer, edrecolomab treatment resulted in a 32% increase in overall survival and a 23% reduction in the rate of tumor recurrence compared with the untreated group [65]. The administration of edrecolomab is a novel immunotherapeutic strategy that holds promise as the preferred monotherapy for stage II colon cancer and the combination chemotherapeutic regimen for stage III colon cancer.

3.4. Radiotherapy

Although adjuvant radiotherapy is not routinely recommended for patients with colon cancer, it is indicated in those with a high risk of local recurrence, such as advanced local disease and/or positive margins. Patients with localized lesions in advanced stages and those with positive resection margins exhibited improved overall survival when radiotherapy was administered [66]. Randomly assigned patients with colon cancer were treated with fluorouracil and levamisole (levamisole, the levorotatory isomer of tetramisole, restores the function of suppressed macrophages and T cells to their normal state), followed by radiotherapy in one group and no radiotherapy in the other group. The results revealed that the 5-year overall survival rates for patients treated with chemotherapy alone and those treated with a combination of chemotherapy and radiotherapy were 62% and 58%, respectively. The 5-year disease-free survival rate was 51% in both groups, which was not significantly different. However, radiation therapy resulted in increased toxicity [67]. Risk factors and the treatment of colon cancer are shown in Figure 1.

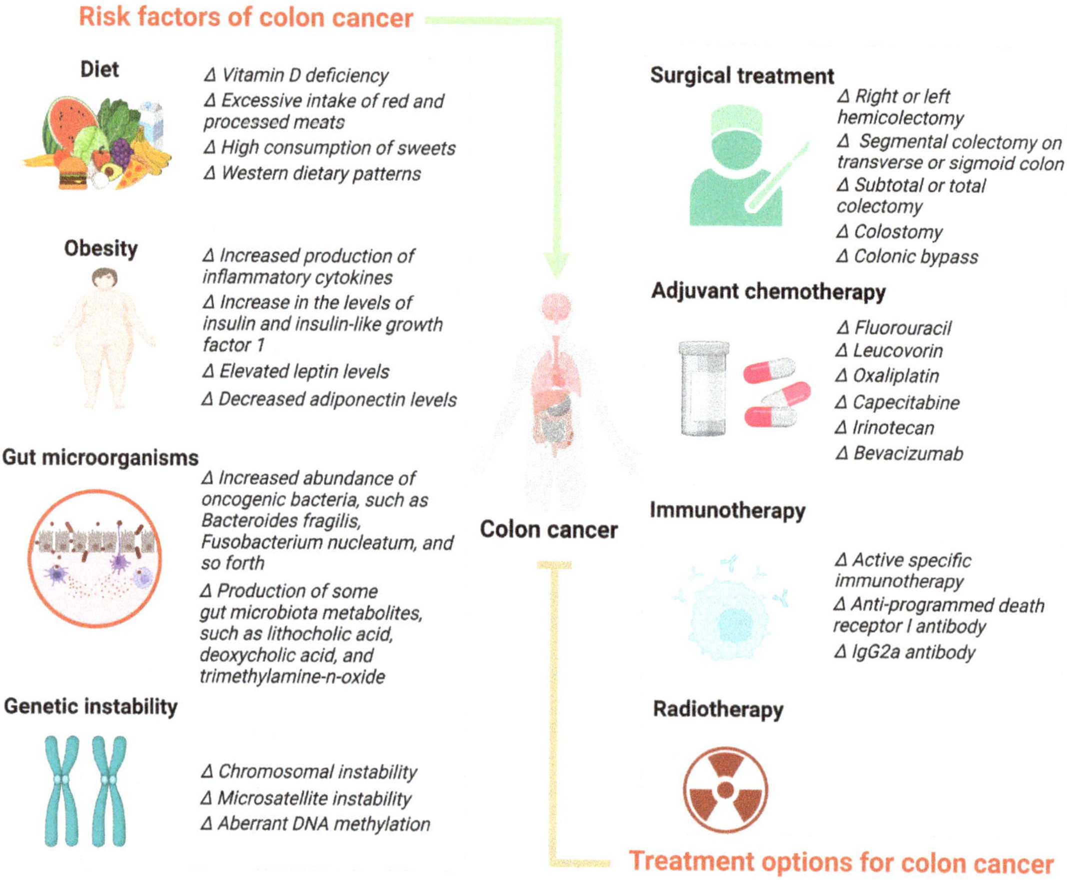

Figure 1. Risk factors and treatments of colon cancer.

4. Curative Effects and Mechanisms of Action of Probiotics on Colon Cancer

4.1. Inhibition of Proliferation and Induction of Apoptosis in Cancer Cells

Cytotoxicity assay revealed that colon cancer cell growth was suppressed following the application of probiotics. *Lacticaseibacillus casei* ATCC393 inhibited the growth of CT26 and HT-29 cells by 52% and 78%, respectively, after 24 h of treatment at a concentration of 10^9 CFU/mL [68]. The sulforhodamine B colorimetric assay indicated that treatment with *Pediococcus pentosaceus* SP2 at 10^8 CFU/mL for 48 h reduced the viability of HT-29 cells by 80% [8]. The methylthiazolyldiphenyl-tetrazolium (MTT) assay (a method for measuring cell viability) showed that when the probiotic *Pediococcus pentosaceus* GS4 acted on HCT-116 cells, cell viability was only 32.66% [9]. Following the treatment of colon cancer cells with *Lacticaseibacillus casei* JY300-8, the impact of Lacticaseibacillus casei JY300-8 on cell proliferation activity was assessed using the MTT assay. This suppressed the growth of Caco-2 cells by up to 65.27% and that of HT-29 and HCT-116 cells by >80% [69]. The treatment of HT-29 cells with the probiotic mixture of *Lactobacillus sporogenes* and *Clostridium butyricum* and the drug 5-FU resulted in 57% and 68% cell survival, respectively, after 72 h, whereas the combination of probiotics and drugs resulted in a cell survival rate of only 38% [70]. *Lactobacillus acidophilus* KLDS1.0901 was used to treat HT-29 cells at different multiplicities of infection (MOI, rate of *Lactobacillus* to cells), i.e., 1, 10, 50, and 100, and the treatment durations were 12, 24, and 48 h. The maximum inhibition of HT-29 cells at MOI = 100 was 40.51% at 48 h [71]. Moreover, these live probiotics exhibited differing inhibitory effects on colon cancer cells at varying concentrations and time periods. Also, the inhibitory effect of inactivated probiotics on colon cancer has been experimentally examined. The growth of HT-29 cells was considerably suppressed after treatment with heat-inactivated *Levilactobacillus brevis* and *Lacticaseibacillus paracasei* (heated at 80 °C for 30 min). The inhibitory effect of *Levilactobacillus brevis* was stronger than that of *Lacticaseibacillus paracasei*, and there was no significant toxicity to normal HEK-293 cells [72]. After the heat inactivation of *Saccharomyces cerevisiae*, the survival rate of SW480 cells decreased to 28.49%, and the inhibitory effect was higher than that of the anticancer drug 5-FU [73]. Hence, microbial viability may not have an effect on the probiotic's ability to suppress the growth of cancer cells.

Flow cytometry analysis revealed that the colon cancer cell cycle was altered after probiotic treatment and that the proportion of cells in the G_1 phase increased. According to the findings, 51.69% of the cells were blocked in G_1 phase after *Lacticaseibacillus paracasei* X12 treatment. The treatment further upregulated p27, a cell cycle protein-dependent kinase inhibitor involved in the mTOR/4E-BP1 pathway, which blocked the cell cycle [74]. In addition, the expression levels of certain cell cycle-related genes were altered during the process. The expression levels of the cell cycle proteins cyclin A, cyclin B_1, cyclin B_2, cyclin D_1, and cyclin E were significantly downregulated [68,75]. In other studies, chromatin condensation and nuclear fragmentation were observed upon treating colon cancer cells with probiotics (*Lactobacillus sporogenes*, *Clostridium butyricum*, and *Saccharomyces cerevisiae*), which implies the occurrence of apoptosis [70,73]. Apoptosis was further confirmed by staining the cells with Annexin V-FITC/PI. A significant increase in the proportion of early and late apoptotic cells was noted, with >90% of cells undergoing apoptosis [68]. The treatment of HT-29 cells with *Lactobacillus acidophilus* KLDS1.0901 dramatically increased the number of apoptotic cells compared with untreated cells, which reduced the mitochondrial membrane potential and increased the ROS levels [71]. The apoptosis-promoting genes Bax, Bid, Bad, and Bak and the apoptosis inhibitor genes Bcl-2 and Bcl-XL were altered during apoptosis [76,77]. *Levilactobacillus brevis* and *Lacticaseibacillus paracasei* enhanced the mRNA expression levels of Bax, caspase-3, and caspase-9 and inhibited the expression of Bcl-2 in HT-29 cells by initiating the mitochondrial pathway, thereby promoting apoptosis [72]. The administration of *Lacticaseibacillus rhamnosus* GG and *Lactobacillus acidophilus* decreased the tumor incidence, tumor load, and tumor recurrence in a 1,2-dimethylhydrazine (DMH)-induced colon tumor model. These two strains induced apoptosis by downregulating the expression of the proto-oncogene K-ras and upregulating the expression of p53 [10]. After the induction of colon cancer in mice with the drug azoxymethane (AOM), *Pediococcus pentosaceus* GS4 reduced disease severity and tumor characteristics and also lowered the ROS levels to alleviate oxidative stress. Furthermore, GS4 inhibited the expression of nuclear factor κB (NF-κB) and p-Akt, which promoted apoptosis via the Akt/NF-κB signaling pathway [9].

Taken together, probiotics suppress the development of colon cancer by downregulating the genes encoding cell cycle proteins to block the cell cycle and upregulating the expressions of genes and proteins involved in the apoptotic process.

4.2. Immunomodulation

Probiotics prevent disease via immunomodulation, which is a common mechanism. Probiotics can relieve colon cancer by modulating inflammatory factors and chemokines, increasing the number of T-cells, producing immunoglobulins, and activating phagocytosis.

Colon cancer requires the involvement of certain inflammatory cytokines, such as TNF-α, IL-1β, IL-6, and IL-17. On the contrary, some anti-inflammatory factors, such as interferon (IFN)-γ and IL-10, can mitigate colon cancer. Oral treatment with *Lacticaseibacillus casei* significantly reduced the tumor volume in mice. Cytokine production, as determined using an enzyme-linked immunosorbent assay, revealed considerable elevation in the levels of IFN-γ, IL-12, and IL-10 [78]. The probiotics *Lactobacillus coryniformis* MXJ32 and *Clostridium butyricum* reduced the occurrence of colonic lesions and attenuated the severity of the disease in colon cancer mouse models by enhancing the expression of anti-inflammatory cytokines and reducing the expression of proinflammatory cytokines [77,79]. DMH induced colon cancer in rats, resulting in a significant increase in the expression of cyclooxygenase-2 (COX-2). COX-2 catalyzes the production of prostaglandins from arachidonic acid, thereby stimulating cell proliferation and proinflammatory processes. However, the administration of probiotics *Lactiplantibacillus plantarum* and *Lacticaseibacillus rhamnosus* GG resulted in reduced COX-2 expression [80].

Colon cancer was induced in mice via the subcutaneous injection of CT26 cells, and the animals were treated with the oral administration of *Lacticaseibacillus casei*. The findings indicated a distinct increase in tumor-infiltrating lymphocytes and an increase in the ligands for the CC chemokine receptor 5 (CCR5) and CXC chemokine receptor 3 (CXCR3). This observation could be ascribed to the generation of Th1 immunostimulatory cytokines and chemokines in tumor tissues as well as the transference of T cells in tumors [78]. In addition, treatment with *Lactiplantibacillus pentosus* B281 and *Lactiplantibacillus plantarum* B282 caused a significant infiltration of leukocytes in mice with colon cancer, and most of the infiltrating leukocytes were neutrophils. Both strains increased the expressions of IL-1α, IL-1β, IL-6, chemokine (C-X-C motif) ligand (CXCL)-1, chemokine (C-C motif) ligand (CCL)-3, CCL-4, and CXCL-2 and decreased the expression of the macrophage colony-stimulating factor [81]. In addition, probiotics *Lactobacillus acidophilus* significantly enhanced the number of $CD8^+$ and $CD4^+$ cells in mice with AOM-induced colon cancer [82]. Along with an increase in the number of $CD8^+$ cells, apoptosis in the tumor tissue also demonstrated an increase in mice with tumors treated with probiotics [83]. Furthermore, *Lactobacillus acidophilus* substantially reduced the expression levels of the tumor markers carbohydrate antigen 19-9 and carcinoembryonic antigen (CEA) [82]. This finding alludes that A probiotic can inhibit colon cancer via T cell-mediated immune response. Also, probiotics stimulate the body to produce immunoglobulin A (IgA), which reduces the exposure of colon cells to certain carcinogens [84]. In a clinical trial, administering oral bifid triple viable probiotics to patients undergoing radical colorectal resection was shown to bolster host immunity, elevate serum IgG and IgA levels, and diminish postoperative infective complications [85]. Patients undergoing colon resection who received preoperative oral *Saccharomyces boulardii* exhibited lowered mucosal levels of IL-1β, IL-10, and IL-23A [86]. Additionally, probiotics modulated the immune response via the induction of phagocytosis, which inhibited cancer cells at an early stage of development [87].

4.3. Regulation of Intestinal Microorganisms and Their Metabolism

Dysregulation of gut microbial ecology can lead to alterations associated with colon cancer in the host physiology. Studies have established that probiotics can ameliorate colon cancer by modulating the gut microbiota. Among other modes of action, probiotics mitigate colon cancer by augmenting the number of salutary microorganisms, inhibiting pernicious microorganisms, and altering the metabolic activity of intestinal metabolites and intestinal flora.

In mice models of colon cancer induced by DMH or AOM/dextran sulfate sodium (DSS), probiotic treatment resulted in changes to microbial diversity and abundance. It also induced alterations in microbiota composition, facilitating the restoration of a healthy gut microbiota state [69,76,88]. Supplementation with *Loigolactobacillus coryniformis* MXJ32 augmented *Lactobacillus*, *Bifidobacterium*, *Akkermansia*, and *Faecalibaculum*. These genera exhibit a positive correlation with colon length, goblet cells, and short-chain fatty acids, potentially contributing to the alleviation of colon cancer. Conversely, the quantities of *Desulfovibrio* and *Helicobacter* decreased, which can exacerbate intestinal inflammation [79]. Likewise, supplementation with *Clostridium butyricum* decreased the *Firmicutes/Bacteroidetes* ratio at the phylum level [77]. *Firmicutes* and *Bacteroidetes* represent the two predominant phyla within the gut microbiota and serve as indicators of microbial equilibrium. The application of *Ligilactobacillus salivarius* Ren decreased the abundance of *Ruminococcus* species, *Clostridiales*, and *Bacteroides dorei* [88]. *Bacteroides* and *Clostridiales* have been linked to the onset of colon cancer. Similar results were also observed in clinical trials. Patients who underwent the resection of anterior sigmoid colon cancer were treated with a probiotic powder containing *Bifidobacterium animalis* subsp. *lactis* HY8002 (1×10^8 CFU), Lacticaseibacillus casei HY2782 (5×10^7 CFU), and *Lactiplantibacillus plantarum* HY7712 (5×10^7 CFU) for 4 weeks. The findings showed that harmful bacteria related to colon cancer (*Alloprevotella* and *Porphyromonas*) decreased after the probiotic treatment [89]. These two detrimental bacteria are linked to an elevation in pro-inflammatory factors.

After the induction of colon cancer in mice with AOM/DSS, the gavage of *Bifidobacterium bifidum* CGMCC 15068 led to a reduction in tumor incidence by altering not only the microbiota composition but also intestinal metabolites [90]. In this study, some metabolites involved in the tricarboxylic acid (TCA) cycle, glycolysis, butyric acid metabolism, fatty acid biosynthesis, and other processes were considerably altered [90]. Of the various metabolic pathways, the TCA cycle and glycolysis play key roles in cancer. Butyrate is a vital metabolite for maintaining intestinal homeostasis and exerts anticancer and anti-inflammatory activities. The abundance of butyrate-producing bacteria-*Akkermansia* increased after the administration of CGMCC 15068 [90]. These findings imply that the gut microbiota may play a role in modulating changes in the metabolome, potentially influencing both colon health and the progression of colon cancer. Similar outcomes have been observed in clinical trials. For instance, in a 12-week trial administering the probiotic *Lactobacillus gasseri* OLL2716: LG21, researchers noted a reduction in fecal pH and fecal spoilage products, along with an increase in the short-chain fatty acid isobutyrate [91]. Additionally, oral probiotics have been shown to reduce serum endotoxin and D-lactic acid levels [85]. Gut microbiota produce enzymes such as β-glucosidase and β-glucuronidase, which catalyze the conversion reactions of polycyclic aromatic hydrocarbons, heterocyclic aromatic amines, primary bile acids, etc., to produce active carcinogens and generate several toxic metabolites that can lead to cancer [92]. Urease catalyzes the hydrolysis of urea to produce ammonia, whose derivatives are extremely poisonous and are capable of causing mutagenic DNA damage [93]. When healthy subjects were administered capsules containing live microorganisms (*Lacticaseibacillus rhamnosus* LC705 and *Propionibacterium freudenreichii* ssp. *shermanii* JS, 2×10^{10} CFU of each strain daily), the activities of fecal β-glucosidase and urease were reduced by 10% and 13%, respectively. Furthermore, the decrease in β-glucosidase activity was linked to an increase in *Propionibacterium* [94]. The reduced activity of β-glucuronidase was also observed after rats with colon cancer were gavaged with probiotic fermented milk [95]. This finding signifies that probiotics alter the growth and metabolism of microorganisms by inhibiting the activities of metabolic enzymes, thereby reducing the risk of colon cancer

4.4. Strengthening of the Intestinal Barrier

The intestinal barrier maintains intestinal homeostasis and prevents the displacement of the intestinal microbiota and leakage of the intestinal material. However, the intestinal barrier function is often impaired and the intestinal environment deteriorates in patients with colon cancer. Transmembrane proteins are present between intestinal epithelial cells and bind to the cytoskeleton to form tight junctions. Increased intestinal permeability during inflammation and cancer alters the expressions of these proteins [96]. The gavage of *Lactiplantibacillus plantarum* KX041 isolated from kimchi in AOM/DSS-induced colon cancer mice reduced the intestinal leakage and increased expression levels of occludin, claudin-1, and ZO-1 [97]. The possible reason is that probiotics reduce the absorption of carcinogenic and inflammatory substances and protect colon cells. Furthermore, treatment with the probiotic capsule containing *Lactiplantibacillus plantarum* CGMCC 1258, *Lactobacillus acidophilus* LA-11, and *Bifidobacterium longum* BL-88 augmented transepithelial resistance and mucosal tight junction protein expression levels [98]. The administration of *Loigolactobacillus coryniformis* MXJ32 significantly suppressed the total tumor number and mean tumor diameter in mice after the AOM/DSS induction of colon cancer. Moreover, MXJ32 restored the reduction in goblet cells [79]. *Lacticaseibacillus rhamnosus* LS8 significantly attenuated inflammatory infiltration and crypt damage in colon cancer mice, and the mortality rate was reduced from 35% to 25% and the colon length was increased by 9.21%. In another study, LS8 attenuated AOM/DSS-induced goblet cell loss and promoted the expression of tight junction proteins [99]. Mucins produced by goblet cells are key components of the intestinal barrier and reduce the possibility of carcinogens and pathogens coming into contact with colon cells. Therefore, decreased mucin production increases colon cancer risk [100]. Hence, probiotics may alleviate symptoms of colon cancer by enhancing the expression of tight junction proteins and restoring the loss of goblet cells.

4.5. Enhancement of Antioxidant Activity

ROS accumulation and oxidoreductase enzyme imbalance can lead to tumor development. Some ROS (e.g., hydrogen peroxide) influence colon cancer cell growth and invasion. However, tumor development can be altered by regulating the activity of antioxidant enzymes, thereby reducing the spread and growth of colon cancer. The probiotic fermented milk containing *Lactobacillus acidophilus* CL1285, *Lacticaseibacillus casei* LBC80R, and *Lacticaseibacillus rhamnosus* CLR2 exerted a therapeutic effect on colon cancer by reducing the number of abnormal crypt foci in colon cancer rats. The mechanism of action was related to an increase in the activity of glutathione-S-transferase with detoxifying effects [95]. The probiotic *Lactiplantibacillus plantarum* AS1, derived from the traditional fermented food Kallappam, was examined for its ability to inhibit peroxidation and scavenge free radicals in vitro. Results indicated that *Lactiplantibacillus plantarum* AS1 exhibited a 50.96% inhibition rate of linoleic acid peroxide and a 29.15% elimination rate of DPPH free radicals. Moreover, in a DMH-induced rat model of colon cancer, rats administered with *Lactiplantibacillus plantarum* AS1 demonstrated reduced tumor incidence and a decrease in the mean number of tumors. Comparatively, the control group exhibited significantly increased levels of lipid peroxidation in both colon and plasma following the DMH treatment. However, after probiotic treatment, lipid peroxidation levels returned to normal, suggesting that *Lactiplantibacillus plantarum* AS1 regulates colon cancer development through its antioxidant properties [101]. The mechanisms of probiotics in the treatment of colon cancer are summarized in Figure 2.

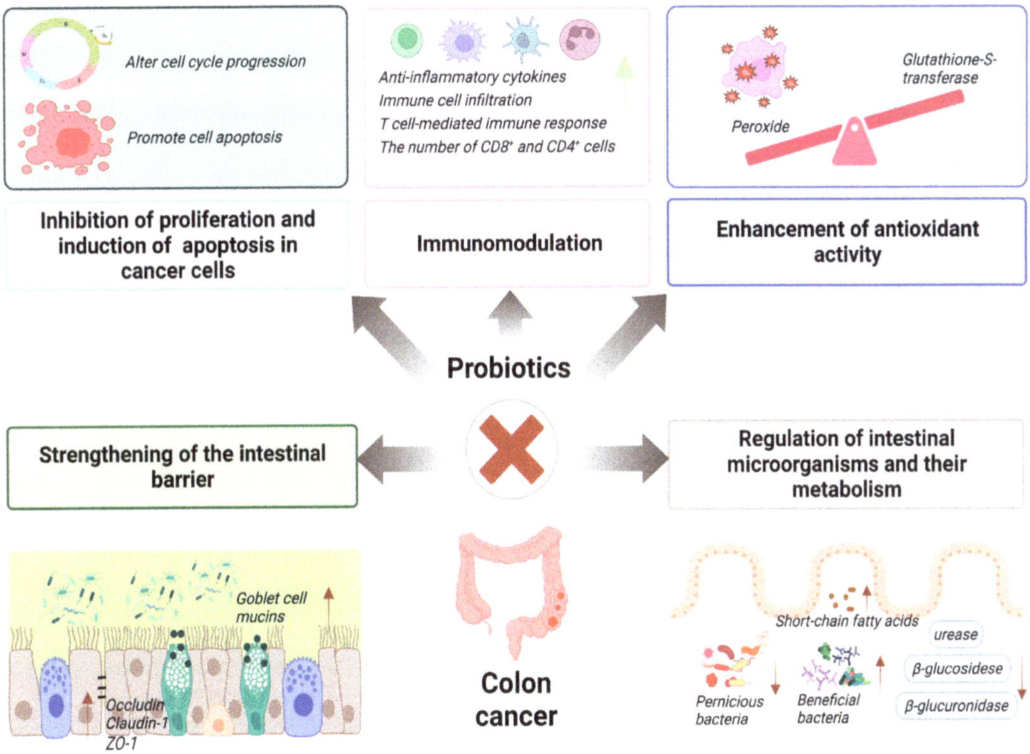

Figure 2. The action mechanisms of probiotics in the prevention of colon cancer.

5. Curative Effects of Probiotic Components and Metabolites on Colon Cancer

5.1. Probiotic Metabolites

The cell-free supernatant of *Bifidobacterium* caused cytopathic effects, such as fragmentation, cell destruction, and decreased cell density in SW742 cells, significantly inhibiting cell proliferation [102]. *Kluyveromyces marxianus* AS41 was isolated and characterized from cheese. Treatment with its cell-free supernatant resulted in 26.7% and 33.1% survival of HT-29 and Caco-2 cells, respectively, but no significant toxicity to KDR/293 cells. Additionally, the proportion of inchoate and advanced apoptotic cells increased [103]. The probiotic *Pichia kudriavzevii* AS-12 supernatant acted on HT-29 and Caco-2 cells and induced apoptosis. The reduction in cell volume, condensation, or fragmentation of nuclei, and the formation of membrane vesicles and apoptotic vesicles were observed. Apoptosis was induced by upregulating the expressions of Bad, caspase-9, caspase-3, caspase-8, and ligand Fas R and downregulating the expression of Bcl-2 [104]. In addition, probiotic supernatants inhibited the activity of matrix metalloproteinase 9 (MMP-9) linked to cancer risk. *Lacticaseibacillus rhamnosus*, *Lactobacillus crispatus*, *Lactobacillus delbrueckii*, and *Limosilactobacillus reuteri* supernatants inhibited cancer cell proliferation, promoted apoptosis, and downregulated the expression of MMP-9 to reduce cancer cell invasion [12,105,106]. Treating Caco-2 cells with different concentrations of *Saccharomyces cerevisiae* var. *boulardii* metabolites for 24 and 48 h showed that 2000 μg/mL exerted the best inhibitory effect after 24 h of treatment. Metabolites of *Saccharomyces boulardii* induced apoptosis and further inhibited the mRNA expressions of genes encoding IL-8 and NF-κB [107]. Thus, probiotic metabolites can arrest the progression of colon cancer. However, the exact compounds in these probiotic supernatants that exert apoptotic effects are yet to be elucidated, and further studies focusing on the supernatant components and evaluating their effects on cancer cells are required.

Nisin is a peptide substance produced by *Lactococcus lactis* and is an important metabolite. At 40–50 IU/mL, nisin significantly suppressed the proliferation of LS180 cells, and at 250–350 IU/mL, it inhibited the growth of SW48, HT-29, and Caco-2 cells. The expressions of the metastatic genes CEA, CEAM6, and MMP2F were significantly downregulated after the treatment with nisin [108]. A bacteriocin-producing probiotic strain, *Pediococcus acidilactici* K2a2-3, was isolated and identified from the intestine. HT-29 cells were treated with pediocin produced by the K2a2-3 strain. At concentrations of 800 AU/mL and 1600 AU/mL, the inhibitory rates against cancer cells were 55% and 53.7%, respectively [109]. Plantaricin P1050 was identified from the *Lactiplantibacillus plantarum* PBS067 strain, and used to treat healthy colon CCD 841 cells and E705 colon cancer cells at a concentration of 1 μg/mL for 48 h. The results revealed a 20% increase in the viability of healthy colon cells and a 30% decrease in colon cancer cell viability [110]. However, the molecular mechanism underlying the inhibition of colon cancer cell proliferation by plantaricin P1050 requires further investigation. Additionally, a novel peptide (Entap) with anti-proliferative effects on HT-29 cells was isolated from the bacteria of the *Enterococcus* genus. This peptide may inhibit colon cancer by disrupting the cell cycle and inducing autophagy [111]. The cell-free supernatant of the *Lacticaseibacillus casei* ATCC334 culture simultaneously suppressed the proliferation of Caco-2/bbe, SKCO-1, and SW620 cells. One of the tumor suppressor molecules was analyzed and identified as ferrochrome via mass spectrometry. Ferrochrome in the supernatant specifically inhibited the proliferation of cancer cells without affecting intestinal epithelial cells. The cancer inhibitory effects of 5-FU and cisplatin are well known, and the effects of ferrochrome were either superior to or comparable to these drugs. Ferrochrome treatment induced massive apoptosis, which was reduced when the c-Jun N-terminal kinase (JNK) signaling pathway was inhibited. This finding suggests that the probiotic metabolite ferrochrome acts by activating the JNK signaling pathway [112]. Butyrate is a key metabolite in microorganisms. Butyrate administration in the mice with AOM/DSS caused a decrease in the number of tumors, the disappearance of intestinal wall thickening, and a significant reduction in colon damage, bleeding, and inflammation. Furthermore, butyrate promoted the colonization of *Actinobacteriota* (antibiotic-producing bacteria), *Bifidobacteriales* (key symbiotic bacteria), and *Muribaculacea* (intestinal mucus-producing symbiotic bacteria), which are beneficial microorganisms [113]. The supernatant containing butyrate from *Lactiplantibacillus plantarum* S2T10D significantly inhibited the growth of HT-29 cells, arrested the cell cycle at the G_2/M phase, and reduced the expression levels of cyclin D_1 and cyclin B_1. Conversely, the non-butyrate-producing strain O2T60C exhibited a limited ability to inhibit the growth of HT-29 cells. Viable strains S2T10D and O2T60C inhibited HT-29 cell growth. But the inhibitory effect of S2T10D was greater than that of the other strain, probably due to its higher butyric acid-producing capacity [114].

5.2. Probiotic EPSs

EPSs have received much attention because of their antioxidant, immunomodulatory, and anticancer activities. The EPSs of probiotics have few side effects and therefore could be a potential new source of anticancer drugs. EPSs of probiotics exhibited anticancer potential in vitro, significantly restraining the viability of cancer cells and inducing apoptosis. The EPS-producing strain *Limosilactobacillus fermentum* YL-11 was isolated from dairy products, and the inhibition rate of EPSs at 600 μg/mL for 48 h on HT-29 cells was 46.5% [115]. EPSs from *Lactobacillus delbrueckii* subsp. *lactis* NCDC252 inhibited HCT-116 cells up to 67.1% after 48 h of action at 10 μg/mL [116]. The antiproliferative activity of EPSs from *Lacticaseibacillus paracasei* increased from 36% to 80% after treating HT-29 cells at 40 mg/mL for 24 and 72 h, and the maximum apoptosis-inducing effect was observed at 48 h [117]. This observation implies that the effects of probiotic EPSs on antiproliferative activity and apoptosis-inducing activity are dose-, time- and strain-dependent. EPSs from

Lacticaseibacillus paracasei caused remarkable early/late apoptotic cell death after incubation at 15 μg/mL for 24 h. The findings showed that 41.16%, 41.86%, and 44.74% of SW-480, HT-29, and HCT-116 cells, respectively, were in the early/late apoptotic or necrotic phase. EPSs from *Lacticaseibacillus paracasei* blocked the mRNA expression of Akt-1, mTOR, and Janus kinase (Jak)-1, reduced the level of Bcl-2, and enhanced the levels of Bax, caspase-3, and caspase-8 [118]. EPSs of *Streptococcus thermophilus* and *Lactobacillus delbrueckii* subsp. *bulgaricus* from yogurt displayed good anticancer activity. Moreover, the activities of EPSs were significantly enhanced when the concentration was increased within a certain range. The inhibitory effects of EPSs from *Streptococcus thermophilus* and *Lactobacillus delbrueckii* subsp. *bulgaricus* on HT-29 cells after 24 h of treatment at 5 mg/mL were 32.28% and 53.06%, respectively, and were 89.52% and 60.70% on HCT-116 cells, respectively. In addition, EPSs exerted good antioxidant activity, as evidenced by their ability to scavenge the free radicals DPPH and ABTS [13]. LHEPS-1, the EPS fraction purified from *Lactobacillus helveticus* MB2-1, acted at 600 μg/mL for 72 h to inhibit Caco-2 cells by 56.34%. Further analysis of EPSs from MB2-1 revealed that their structural characteristics, such as monosaccharide composition, molecular weight, conformation, and branching degree, affected their anti-colon cancer activity [119]. EPSs have also been observed to alleviate colon cancer in in vivo experiments. EPSs from *Limosilactobacillus fermentum* YL-11 inhibited tumor growth and reduced the tumor weight in colon cancer mice. The anti-colon cancer mechanism may be related to the activation of endogenous mitochondrial caspases and the inhibition of the PI3K/AKT signaling pathway as well as the regulation of cell cycle-related proteins [120]. The supplementation of EPSs from *Lactiplantibacillus plantarum*-12 alleviated colon cancer symptoms by restoring colon length and significantly reducing tumor numbers in mice with AOM/DSS-induced colon cancer [121].

5.3. Probiotic Slp

The Slp is situated in a paracrystalline layer outside the bacterial cell wall. Slp plays a crucial part in preserving the cellular morphology and structure, promoting intercellular adhesion, recognizing external receptors, preventing the attachment of pathogenic bacteria, and regulating the immune response. Furthermore, probiotic Slp has been reported to exert anticancer effects. The Slp from *Lactobacillus acidophilus* CICC 6074 inhibited the proliferation of HT-29 cells, and the survival rate of the cells was 54.68% when the concentration of Slp reached 100 mg/L. Upon Slp action, the levels of p53, p21, and p16 were upregulated, whereas those of cyclin-dependent kinase and cyclin B were downregulated, which arrested the cell cycle in the G_0/G_1 phase. After the Slp treatment, the cells showed obvious changes characteristic of apoptosis (e.g., chromatin condensation, nuclear fragmentation, and vacuole formation). In addition, Slp remarkably elevated the mRNA levels of Fas, FADD, caspase-3, caspase-8, Bax, and caspase-9 and decreased the level of Bcl-2. A reduction in cellular mitochondrial membrane potential was also noted. The results suggested that Slp may induce apoptosis by activating the death receptor pathway and interfering with the mitochondrial pathway [14]. The Slp of *Lactobacillus acidophilus* NCFM significantly suppressed the growth of HCT116 cells and further caused cellular autophagy. The expressions of autophagy proteins Atg7, LC3-II, and Beclin-1 were increased in HCT-116 cells, and autophagic vacuoles were formed. The results demonstrated that Slp promoted ROS accumulation, suppressed mTOR expression, and activated JNK activity, thereby inducing autophagy [122]. In view of its anticancer activity, the Slp of *Lactobacillus acidophilus* may serve as a potential anticancer drug. In the future, in vivo confirmation of its inhibitory effect on colon cancer is warranted.

Both live and inactivated forms of probiotics exhibit anti-colon cancer activity. Furthermore, other components with probiotic properties, such as secreted metabolites and EPSs, as well as Slp exert inhibitory effects on colon cancer. The anti-colon cancer effects of probiotic components and metabolites are summarized in Figure 3.

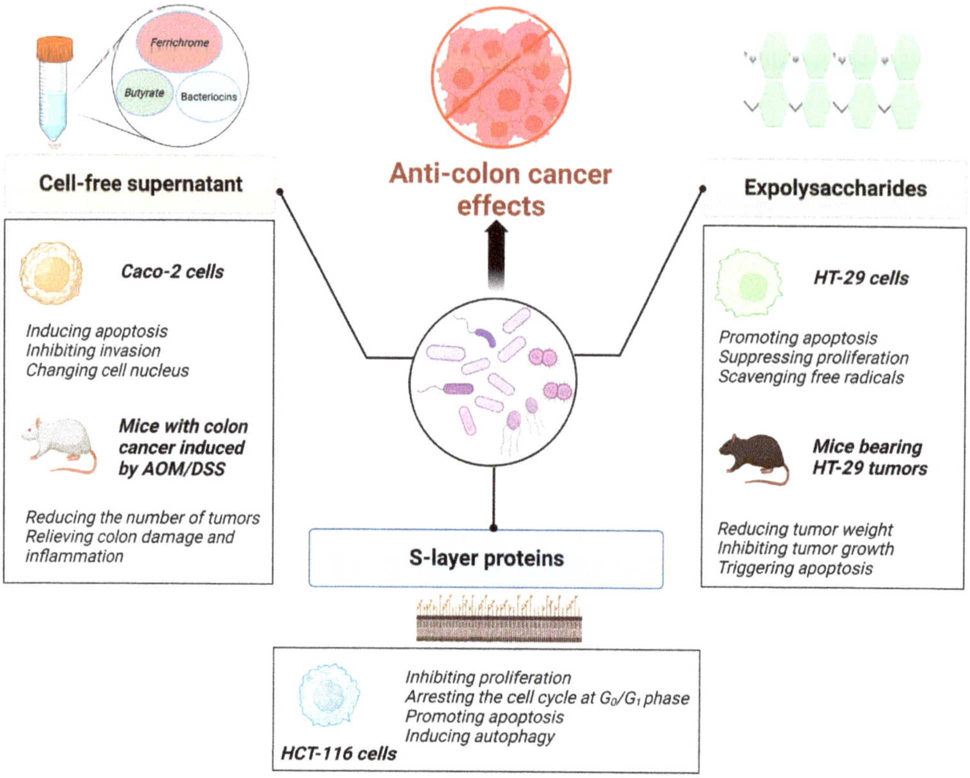

Figure 3. Anti-colon cancer effects of probiotic components and metabolites.

This review provides a summary of research on the inhibition of colon cancer by probiotics, probiotic components, and metabolites. Cellular experiments are detailed in Table 1, animal experiments are detailed in Table 2, and clinical studies are detailed in Table 3.

Table 1. In vitro studies of probiotics, their components, and metabolites inhibiting colon cancer.

Cells	Probiotics, Their Components, and Metabolites	Food Sources of Probiotics	Treatments	Results	Ref.
Caco-2 cells	Conditioned media for *Lactiplantibacillus pentosus* B281 and *Lactiplantibacillus plantarum* B282	Naturally fermented olives	Cells were treated with different concentrations of probiotic-conditioned media for 24, 48, and 72 h at 37 °C.	Inhibition of cell proliferation; decreased colony formation; cell cycle arrest in the G_1 phase; decreased expression of cyclin A, cyclin B_1, cyclin B_2, and cyclin E, and increased expression of cyclin D.	[75]
CT26 and HT-29 cells	*Lacticaseibacillus casei* ATCC393	NA	Cells were treated at a concentration of 10^8 CFU/mL for 24, 48, and 72 h at 37 °C.	Inhibition of cell proliferation; promotion of apoptosis; increased expression of the apoptosis-inducing ligand TRAIL and decreased expression of cyclin D_1.	[68]
HT-29 cells	Heat-inactivated *Levilactobacillus brevis* and *Lacticaseibacillus paracasei*	Iranian food "Terxine"	Cells were treated with different concentrations of probiotics for 24, 48, and 72 h at 37 °C.	Inhibition of cell proliferation; activation of mitochondrial pathway to promote apoptosis.	[72]

Table 1. *Cont.*

Cells	Probiotics, Their Components, and Metabolites	Food Sources of Probiotics	Treatments	Results	Ref.
SW480 cells	Heat-inactivated *Saccharomyces cerevisiae*	NA	Cells were treated with different concentrations of probiotics for 24 and 48 h at 37 °C.	Inhibition of cell proliferation; activation of Akt/NF-κB signaling pathway to promote apoptosis.	[73]
HT-29 and HCT-116 cells	*Lactobacillus sporogenes* and *Clostridium butyricum* TO-A	NA	Cells were treated at 10^7 CFU/mL for 24, 48, and 72 h at 37 °C.	Inhibition of cell proliferation; promotion of apoptosis; increased expression of Bax, Bid, Bad, and Bak and decreased expression of Bcl-2 and Bcl-XL.	[70]
Caco-2 cells	*Saccharomyces cerevisiae* var. *boulardii* CNCM I-745 metabolites	NA	Cells were treated with different concentrations of metabolites for 24 and 48 h at 37 °C.	Inhibition of cell proliferation; promotion of apoptosis; inhibition of IL-8 and NF-κB gene expression.	[107]
HT-29 and Caco-2 cells	*Lactobacillus acidophilus* KLDS1.0901	NA	Cells were treated with different concentrations of probiotics for 12, 24, and 48 h at 37 °C.	Inhibition of cell proliferation; promotion of apoptosis; decreased cellular mitochondrial membrane potential; increased ROS levels.	[71]
CT26 cells	*Bifidobacterium longum*, *Bifidobacterium bifidum*, *Lactobacillus acidophilus*, and *Lactiplantibacillus plantarum*	NA	Cells were treated at 10^8 CFU/mL for 24, 48, 72, and 96 h at 37 °C.	Inhibition of cell proliferation, migration, and invasion.	[83]
Caco-2, HT-29 and HCT-116 cells	*Lacticaseibacillus casei* JY300-8	Fresh corn stalks	Cells were treated with different concentrations of probiotics for 24, 48, 72, and 96 h at 37 °C.	Inhibition of cell proliferation; promotion of apoptosis.	[69]
SW48, HT-29, LS180 and Caco-2 cells	Nisin	NA	Cells were treated with different concentrations of nisin for 24 and 48 h at 37 °C.	Inhibition of cell proliferation; inhibition of cancer cell metastasis genes.	[108]
HT-29 and Caco-2 cells	*Pichia kudriavzevii* AS-12 cell-free supernatant	NA	Cells were treated with different concentrations of supernatants for 24 and 48 h at 37 °C.	Inhibition of cell proliferation; promotion of apoptosis; increased expression of Bad, caspase-3, caspase-8, caspase-9, and Fas-R and decreased expression of Bcl-2.	[104]
HT29-ShE cells	Heat inactivation of sonicated *Limosilactobacillus reuteri* and cell-free supernatants	NA	The 10% (v/v) heat-inactivated sonicated fraction and 5% supernatant treated cells for 24 h.	Inhibition of cell proliferation; promotion of apoptosis; inhibition of cancer cell metastasis; downregulation of MMP-9 and COX-2 expression.	[106]
Caco-2 and SW620 cells	Ferrochrome in the cell-free supernatant of *Lacticaseibacillus casei* ATCC334	NA	Treatment of cells with different concentrations of ferrochrome.	Inhibition of cell proliferation; induction of apoptosis through the activation of the JNK-DDTI3 pathway.	[112]
HT-29 and HCT-116 cells	EPSs of *Streptococcus pyogenes* and *Lactobacillus delbrueckii* subsp. *bulgaricus*	Labaneh (yogurt-like product)	Cells were treated with different concentrations of EPSs for 24, 48, and 72 h at 37 °C.	Inhibition of cell proliferation; facilitation of scavenging of the free-radicals DPPH and ABTS.	[13]
SW-480, HT-29 and HCT-116 cells	EPSs of *Lacticaseibacillus paracasei*	Dairy products	Cells were treated with different concentrations of EPSs for 24 and 48 h at 37 °C.	Inhibition of cell proliferation; promotion of apoptosis; increased expression of Bax, caspase-3, and caspase-8 and inhibition of AKT-1, JAK-1, and mTOR gene expression.	[118]

Table 1. *Cont.*

Cells	Probiotics, Their Components, and Metabolites	Food Sources of Probiotics	Treatments	Results	Ref.
HT-29 cells	Slp of *Lactobacillus acidophilus* CICC 6074	NA	Cells were treated with different concentrations of Slp for 24 h at 37 °C.	Inhibition of cell proliferation; cell cycle block; promotion of apoptosis via death receptor pathway and mitochondrial pathway; inhibition of cell invasion.	[14]
HCT-116 cells	Slp of *Lactobacillus acidophilus* NCFM	NA	Cells were treated with different concentrations of Slp for 24 h at 37 °C	Inhibition of cell proliferation; cell cycle block; induction of autophagic death through inhibition of mTOR activity; activation of the JNK signaling pathway.	[122]

Table 2. In vivo animal model studies of probiotics, their components, and metabolites inhibiting colon cancer.

Animal Model	Probiotics, Their Components, and Metabolites	Food Sources of Probiotics	Treatments	Results	Ref.
A model of colon cancer induced by subcutaneous injection of CT26 cells; female BALB/c mice	*Lacticaseibacillus casei* ATCC393	NA	Probiotics (150 μL) at a dose of 10^9 CFU/L were administered daily 10 days before subcutaneous injection of CT26 cells, for 13 days.	Tumor volume reduction; upregulation of TRAIL protein expression and downregulation of survivor protein expression in the tumor tissues.	[68]
DMH-induced colon cancer model in rats	*Lactiplantibacillus plantarum and Lacticaseibacillus rhamnosus* GG	Traditional fermented food of Himachal Pradesh	Probiotics were administered by gavage at a dose of 2×10^{10} CFU/day, once a week for 16 weeks, for 6 consecutive days.	Reduced tumor incidence, tumor diversity, and the mean tumor size; reduced total sialic acid and COX-2 expression.	[80]
DMH-induced colon cancer model; male Sprague–Dawley rats	*Lacticaseibacillus rhamnosus* GG and *Lactobacillus acidophilus*	NA	Probiotics were administered by gavage at a dose of 1×10^{10} CFU/day, one week prior to DMH administration for 18 weeks.	Decreased tumor load, tumor diversity; downregulation of Bcl-2 and K-ras gene expression, and upregulation of Bax and p53 gene expression.	[10]
Subcutaneous injection of CT26 cells induced colon cancer model; female BALB/c mice	*Lacticaseibacillus casei*	NA	Probiotics 150 μL at a dose of 10^9 CFU/L were administered daily 10 days before subcutaneous injection of CT26 cells, for 13 days.	Decreased tumor volume; increased levels of IFN-γ, IL-12, and IL-10; increased tumor-infiltrating lymphocytes; increased expression of cleaved caspase-3 and cleaved PARP1.	[78]
AOM/DSS-induced colon cancer model; male C57BL/6 mice	*Loigolactobacillus coryniformis* MXJ32	Traditional fermented vegetable (Jiangshui Cai)	Probiotics were administered by gavage at a dose of 1×10^9 CFU/day, for 15 weeks.	Suppression of total tumor number and average tumor diameter; restoration of intestinal barrier function; downregulation of inflammatory cytokine expression; increased amount of salutary bacteria and decreased amount of pernicious bacteria.	[79]
Colon cancer model induced by injection of CT26 cells; female BALB/c mice	*Bifidobacterium longum*, *Bifidobacterium bifidum*, *Lactobacillus acidophilus* and *Lactiplantibacillus plantarum*	NA	Probiotic mixture (1×10^9 CFU/200 μL) administered daily via tube feeding, for 3 weeks.	Inhibition of tumor growth; increased apoptotic cells in tumor tissues; increased number of $CD8^+$ cells.	[83]

Table 2. *Cont.*

Animal Model	Probiotics, Their Components, and Metabolites	Food Sources of Probiotics	Treatments	Results	Ref.
AOM-induced colon cancer model; male BALB/c mice	*Lactobacillus acidophilus* and *Bifidobacterium bifidum*	Traditional yogurt	Probiotics (1×10^9 CFU/g body weight per strain) were mixed in drinking water and gavaged daily for 10 days before AOM, for 5 months.	Decreased incidence of colonic lesions and smaller tumor size; decreased levels of tumor markers; increased levels of IFN-γ and IL-10; increased numbers of $CD8^+$ and $CD4^+$ cells.	[82]
AOM-induced colon cancer model; male C57BL/6 mice	*Bifidobacterium bifidum* CGMCC 15068	NA	Probiotics were administered by gavage at a dose of 3×10^8 CFU/day during the recovery period.	Reduced tumor incidence; increased abundance of beneficial probiotics in the gut; altered gut metabolites.	[90]
Subcutaneous injection of CT26 cells induced colon cancer model; BABL/C male mice	*Lacticaseibacillus casei* JY300-8	Fresh corn stalks	Probiotics were administered by half-day gavage at a dose of 1×10^9 CFU/mL until the end of the experiment.	Reduced tumor size and tumor incidence; decreased abundance of *Enterobacteriaceae* and *Clostridium perfringens*; increased abundance of *Lactobacillaceae* and *Bifidobacteriaceae*.	[69]
AOM/DMH-induced colon cancer model; male C57BL/6 mice	*Lactiplantibacillus plantarum* KX041	Traditional Chinese pickle juice	Oral administration of 200 µL probiotic suspension (1×10^9 CFU/day), for 13 weeks.	Decrease in the total number and mean diameter of colon tumors; attenuation of inflammatory infiltration and crypt damage; increase in short-chain fatty acid content of feces; increased the number of salutary bacteria and decreased the number of pernicious bacteria.	[97]
DMH-induced colon cancer model; F344 male rats	*Lactobacillus acidophilus* CL1285, *Lacticaseibacillus casei* LBC80R and *Lacticaseibacillus rhamnosus* CLR2	NA	Gavage of 2, 1.5, 1, 0.5, and 0.25 mL of probiotic fermented milk daily, for 12 weeks.	Decreased counts of abnormal crypts in the colon; increased activity of glutathione-S-transferase and decreased activity of quinone reductase and β-glucuronide glycosidase.	[95]
Subcutaneous injection of CT26 cells induced colon cancer model; female BALB/c mice	EPSs of *Limosilactobacillus fermentum* YL-11	Fermented milk	Intravenous EPSs at 90 mg/kg body weight every alternate day, for 3 weeks.	Decreased tumor weight; nuclear consolidation; increased intracellular space; and fatty degeneration of tumor cells.	[120]
AOM/DSS-induced colon cancer model; male C57BL/6 mice	EPSs of *Lactiplantibacillus plantarum*	NA	EPSs at 200 mg/kg body weight/day via gavage, for 12 weeks.	Restoration of colon length and reduction in the tumor number; enhancement of intestinal barrier function; inhibition of the NF-κB pathway and activation of the caspase cascade reaction to promote apoptosis; alteration of the intestinal microbiota and metabolites.	[121]

Table 3. The clinical studies of probiotics inhibiting colon cancer.

Subjects	Probiotics	Food Sources of Probiotics	Treatments	Results	Ref.
22 patients undergoing radical colectomy	*Bifidobacterium longum*, *Lactobacillus acidophilus*, and *Enterococcus faecalis*	NA	Subjects received capsules containing live microorganisms 3 times/day (total daily dose of 6×10^7 CFU).	Increased abundance and diversity of colonic mucosal microorganisms; decreased abundance of *Fusobacterium* and *Peptostreptococcus*.	[123]
60 patients who underwent resection of anterior sigmoid colon cancer	*Bifidobacterium animalis* subsp. *lactis* HY8002, *Lacticaseibacillus casei* HY2782, and *Lactiplantibacillus plantarum* HY7712	NA	Subjects received probiotic powder treatment twice daily, for 4 weeks.	Improved control of flatulence; increased abundance of beneficial bacteria; and decreased abundance of harmful bacteria in the gut.	[89]
10 colon cancer patients and 20 healthy subjects	*Lactobacillus gasseri* OLL2716: LG21	NA	Subjects administered probiotics daily, for 12 weeks.	Increased levels of *Lactobacillus* and decreased total *Clostridium perfringens* in the intestines; increased short-chain fatty acid isobutyrate in the feces; decreased pH; and decreased synthesis of spoilage products; increased NK cell activity.	[91]
38 men aged 24-55	*Lacticaseibacillus rhamnosus* LC705 and *Propionibacterium freudenreichii* ssp. *shermanii* JS	NA	Subjects consumed 2 capsules containing live microorganisms (2×10^{10} CFU per strain) per day, for 4 weeks.	Increased levels of *Lactobacillus* and *Propionibacterium* in the intestinal tract; decreased β-glucosidase and urease activity.	[94]
68 patients with treated colon cancer	*Lacticaseibacillus rhamnosus* R0011 and *Lactobacillus acidophilus* R0052	NA	Subjects were treated with probiotics (2×10^9 CFU) daily, for 12 weeks.	Improved bowel symptoms and quality of life.	[11]
33 patients who underwent colectomy	*Saccharomyces boulardii*	NA	Subjects received preoperative probiotic oral therapy for at least 7 days.	Downregulation of mucosal IL-1β, IL-10, and IL-23A levels	[86]
60 patients who underwent radical colorectal resection	Bifid triple viable probiotics	NA	Subjects received oral probiotic treatment 3 days before surgery.	Reduced serum endotoxin, D-lactic acid, and IL-6 levels; elevated IgG and IgA levels; reduced incidence of postoperative infectious complications.	[85]

6. Conclusions

The factors influencing the occurrence of colon cancer are complex and multifaceted. Key contributors include diet, obesity, the composition of the intestinal microbiome, and genetic instability, all of which promote colon cancer. Current mainstream treatment options for colon cancer patients encompass surgery, chemotherapy, immunotherapy, and radiation therapy. However, it is important to acknowledge that these treatments come with limitations and side effects. Substantial research evidence highlights the pivotal role of probiotics in preventing and alleviating colon cancer. This review illuminates the diverse mechanisms through which probiotics contribute to mitigating colon cancer, including inhibiting cell proliferation and promoting apoptosis, enhancing the immune response, regulating the intestinal microbiota and their metabolic processes, strengthening the intestinal barrier, and boosting antioxidant effects. Notably, probiotics themselves exhibit anticancer activity, and the metabolites they produce, such as bacteriocins and short-chain fatty acids, possess the capacity to inhibit colon cancer. Additionally, certain bacterial components, such as EPSs and Slp, also play an inhibitory role in colon cancer.

7. Future Prospects

Probiotics exert their anticancer activity via physiological mechanisms that are usually strain specific. Therefore, further investigations should target these strains with anti-cancer activity to determine the optimal dose, duration, and frequency of treatment. Research into the development of probiotic products (e.g., probiotic capsules and yogurt) for individuals at a high risk of developing colon cancer is one of the avenues for future work in the field. In the past, there have been few reports on whether oral probiotics have side effects, which is a promising direction for further research. In addition, more randomized, double-blind, and placebo-controlled clinical trials should be conducted to validate the efficacy of probiotics in treating colon cancer.

Author Contributions: X.D.: Writing—original draft; J.Y.: Writing—original draft, Writing—review and editing; Y.Z.: Writing—review and editing; X.C.: Writing—review and editing; C.W.: Writing—review and editing; H.S.: Funding acquisition; J.S.: Funding acquisition, Supervision, Writing—original draft, Writing—review and editing. All authors have read and agreed to the published version of the manuscript.

Funding: This work was supported by the Natural Science Foundation of Chongqing (cstc2021jcyj-msxmX0772), Science and Technology Research Program of Chongqing Municipal Education Commission (KJQN202300204), Major Science and Technology Special Projects in Tibet Autonomous Region (XZ202201ZD0001N), Key Construction Disciplines of Traditional Chinese Medicine in Chongqing (2021-4322190044), and University Innovation Research Group in Chongqing (CXQT21007).

Data Availability Statement: No data were used for the research described in this article.

Acknowledgments: We would like to acknowledge Qinchun Rao's (Department of Nutrition & Integrative Physiology, Florida State University, Tallahassee, Florida 32306, United States) significant contribution to the creation of figures.

Conflicts of Interest: The authors declare no conflict of interest.

Abbreviations

High microsatellite instability: MSI-H; extracellular polysaccharides: EPSs; S-layer protein: Slp; tumor necrosis factor-α: TNF-α; interleukin: IL; insulin-like growth factor 1: IGF-1; phosphatidylinositol 3-kinase: PI3K; mammalian target of rapamycin: mTOR; reactive oxygen species: ROS; enterotoxigenic *Bacteroides fragilis*: ETBF; adenomatous polyposis coli: APC; fluorouracil combined with leucovorin: FL; 5-fluorouracil: 5-FU; active specific immunotherapy: ASI; mismatch repair-deficient: dMMR; methylthiazolyldiphenyl-tetrazolium: MTT; multiplicities of infection: MOI; 1,2-dimethylhydrazine: DMH; azoxymethane: AOM; nuclear factor κB: NF-κB; interferon: IFN; cyclooxygenase-2: COX-2; CC chemokine receptor: CCR; CXC chemokine receptor: CXCR; chemokine (C-X-C motif) ligand: CXCL; chemokine (C-C motif) ligand: CCL; carcinoembryonic antigen: CEA; immunoglobulin A: IgA; dextran sulfate sodium: DSS; tricarboxylic acid: TCA; lipid peroxidation: LPO; matrix metalloproteinase: MMP; c-Jun N-terminal kinase: JNK; Janus kinase: Jak.

References

1. Keum, N.; Giovannucci, E. Global burden of colorectal cancer: Emerging trends, risk factors and prevention strategies. *Nat. Rev. Gastroenterol.* **2019**, *16*, 713–732. [CrossRef] [PubMed]
2. Sung, H.; Ferlay, J.; Siegel, R.L.; Laversanne, M.; Soerjomataram, I.; Jemal, A.; Bray, F. Global cancer statistics 2020: GLOBOCAN estimates of incidence and mortality worldwide for 36 cancers in 185 countries. *CA Cancer J. Clin.* **2021**, *71*, 209–249. [CrossRef] [PubMed]
3. Kulshrestha, R.; Tiwari, S. Diet and colon cancer: A comprehensive review. In *Colon Cancer Diagnosis and Therapy*; Springer: Cham, Switzerland, 2021; pp. 53–71.
4. Wu, N.; Yang, X.; Zhang, R.; Li, J.; Xiao, X.; Hu, Y.; Chen, Y.; Yang, F.; Lu, N.; Wang, Z.; et al. Dysbiosis signature of fecal microbiota in colorectal cancer patients. *Microb. Ecol.* **2013**, *66*, 462–470. [CrossRef] [PubMed]
5. Mishra, J.; Drummond, J.; Quazi, S.H.; Karanki, S.S.; Shaw, J.; Chen, B.; Kumar, N. Prospective of colon cancer treatments and scope for combinatorial approach to enhanced cancer cell apoptosis. *Crit. Rev. Oncol. Hematol.* **2013**, *86*, 232–250. [CrossRef]
6. Baskar, R.; Lee, K.A.; Yeo, R.; Yeoh, K.-W. Cancer and radiation therapy: Current advances and future directions. *Int. J. Med. Sci.* **2012**, *9*, 193. [CrossRef]
7. André, T.; Shiu, K.-K.; Kim, T.W.; Jensen, B.V.; Jensen, L.H.; Punt, C.; Smith, D.; Garcia-Carbonero, R.; Benavides, M.; Gibbs, P. Pembrolizumab in microsatellite-instability–high advanced colorectal cancer. *N. Engl. J. Med.* **2020**, *383*, 2207–2218. [CrossRef]

8. Mantzourani, I.; Chondrou, P.; Bontsidis, C.; Karolidou, K.; Terpou, A.; Alexopoulos, A.; Bezirtzoglou, E.; Galanis, A.; Plessas, S. Assessment of the probiotic potential of lactic acid bacteria isolated from kefir grains: Evaluation of adhesion and antiproliferative properties in in vitro experimental systems. *Ann. Microbiol.* **2019**, *69*, 751–763. [CrossRef]
9. Dubey, V.; Ghosh, A.R.; Bishayee, K.; Khuda-Bukhsh, A.R. Appraisal of the anti-cancer potential of probiotic Pediococcus pentosaceus GS4 against colon cancer: In vitro and in vivo approaches. *J. Funct. Foods* **2016**, *23*, 66–79. [CrossRef]
10. Sharaf, L.K.; Sharma, M.; Chandel, D.; Shukla, G. Prophylactic intervention of probiotics (*L. acidophilus*, *L. rhamnosus* GG) and celecoxib modulate Bax-mediated apoptosis in 1,2-dimethylhydrazine-induced experimental colon carcinogenesis. *BMC Cancer* **2018**, *18*, 1111. [CrossRef]
11. Lee, J.-Y.; Chu, S.-H.; Jeon, J.Y.; Lee, M.-K.; Park, J.-H.; Lee, D.-C.; Lee, J.-W.; Kim, N.-K. Effects of 12 weeks of probiotic supplementation on quality of life in colorectal cancer survivors: A double-blind, randomized, placebo-controlled trial. *Dig. Liver Dis.* **2014**, *46*, 1126–1132. [CrossRef]
12. Nouri, Z.; Karami, F.; Neyazi, N.; Modarressi, M.H.; Karimi, R.; Khorramizadeh, M.R.; Taheri, B.; Motevaseli, E. Dual anti-metastatic and anti-proliferative activity assessment of two probiotics on HeLa and HT-29 cell lines. *Cell J.* **2016**, *18*, 127. [PubMed]
13. Al-Nabulsi, A.A.; Jaradat, Z.W.; Al Qudsi, F.R.; Elsalem, L.; Osaili, T.M.; Olaimat, A.N.; Esposito, G.; Liu, S.-Q.; Ayyash, M.M. Characterization and bioactive properties of exopolysaccharides produced by *Streptococcus thermophilus* and *Lactobacillus bulgaricus* isolated from labaneh. *LWT* **2022**, *167*, 113817. [CrossRef]
14. Zhang, T.; Pan, D.; Yang, Y.; Jiang, X.; Zhang, J.; Zeng, X.; Wu, Z.; Sun, Y.; Guo, Y. Effect of *Lactobacillus acidophilus* CICC 6074 S-layer protein on colon cancer HT-29 cell proliferation and apoptosis. *J. Agric. Food Chem.* **2020**, *68*, 2639–2647. [CrossRef] [PubMed]
15. Um, C.Y.; Prizment, A.; Hong, C.-P.; Lazovich, D.; Bostick, R.M. Associations of calcium, vitamin D, and dairy product intakes with colorectal cancer risk among older women: The Iowa women's health study. *Nutr. Cancer* **2019**, *71*, 739–748. [CrossRef] [PubMed]
16. Giovannucci, E. The epidemiology of vitamin D and cancer incidence and mortality: A review (United States). *Cancer Cause Control* **2005**, *16*, 83–95. [CrossRef]
17. Xu, Y.; Qian, M., Hong, J.; Ng, D.M.; Yang, T.; Xu, L.; Ye, X. The effect of vitamin D on the occurrence and development of colorectal cancer: A systematic review and meta-analysis. *Int. J. Color. Dis.* **2021**, *36*, 1329–1344. [CrossRef]
18. Hullings, A.G.; Sinha, R.; Liao, L.M.; Freedman, N.D.; Graubard, B.I.; Loftfield, E. Whole grain and dietary fiber intake and risk of colorectal cancer in the NIH-AARP Diet and Health Study cohort. *Am. J. Clin. Nutr.* **2020**, *112*, 603–612. [CrossRef]
19. Di, Y.; Ding, L.; Gao, L.; Huang, H. Association of meat consumption with the risk of gastrointestinal cancers: A systematic review and meta-analysis. *BMC Cancer* **2023**, *23*, 782. [CrossRef]
20. Santarelli, R.L.; Pierre, F.; Corpet, D.E. Processed meat and colorectal cancer: A review of epidemiologic and experimental evidence. *Nutr. Cancer* **2008**, *60*, 131–144. [CrossRef]
21. Corpet, D.E. Red meat and colon cancer: Should we become vegetarians, or can we make meat safer? *Meat Sci.* **2011**, *89*, 310–316. [CrossRef]
22. Tayyem, R.F.; Bawadi, H.A.; Shehadah, I.; Bani-Hani, K.E.; Takruri, H.; Al-Jaberi, T.; Heath, D.D. Fast foods, sweets and beverage consumption and risk of colorectal cancer: A case-control study in Jordan. *Asian Pac. J. Cancer Prev.* **2018**, *19*, 261. [PubMed]
23. Castelló, A.; Rodríguez-Barranco, M.; Fernández de Larrea, N.; Jakszyn, P.; Dorronsoro, A.; Amiano, P.; Chirlaque, M.-D.; Colorado-Yohar, S.; Guevara, M.; Moreno-Iribas, C. Adherence to the Western, Prudent and Mediterranean Dietary Patterns and Colorectal Cancer Risk: Findings from the Spanish Cohort of the European Prospective Investigation into Cancer and Nutrition (EPIC-Spain). *Nutrients* **2022**, *14*, 3085. [CrossRef] [PubMed]
24. Sharma, I.; Roebothan, B.; Zhu, Y.; Woodrow, J.; Parfrey, P.S.; Mclaughlin, J.R.; Wang, P.P. Hypothesis and data-driven dietary patterns and colorectal Cancer survival: Findings from Newfoundland and Labrador colorectal Cancer cohort. *Nutr. J.* **2018**, *17*, 55. [CrossRef] [PubMed]
25. Schwingshackl, L.; Schwedhelm, C.; Hoffmann, G.; Knüppel, S.; Laure Preterre, A.; Iqbal, K.; Bechthold, A.; De Henauw, S.; Michels, N.; Devleesschauwer, B. Food groups and risk of colorectal cancer. *Int. J. Cancer* **2018**, *142*, 1748–1758. [CrossRef] [PubMed]
26. Veettil, S.K.; Wong, T.Y.; Loo, Y.S.; Playdon, M.C.; Lai, N.M.; Giovannucci, E.L.; Chaiyakunapruk, N. Role of diet in colorectal cancer incidence: Umbrella review of meta-analyses of prospective observational studies. *JAMA Netw. Open* **2021**, *4*, e2037341. [CrossRef] [PubMed]
27. Safari, A.; Shariff, Z.M.; Kandiah, M.; Rashidkhani, B.; Fereidooni, F. Dietary patterns and risk of colorectal cancer in Tehran Province: A case–control study. *BMC Public Health* **2013**, *13*, 222. [CrossRef]
28. Hussan, H.; Gray, D.M.; Hinton, A.; Krishna, S.G.; Conwell, D.L.; Stanich, P.P. Morbid obesity is associated with increased mortality, surgical complications, and incremental health care utilization in the peri-operative period of colorectal cancer surgery. *World J. Surg.* **2016**, *40*, 987–994. [CrossRef]
29. Tarasiuk, A.; Mosińska, P.; Fichna, J. The mechanisms linking obesity to colon cancer: An overview. *Obes. Res. Clin. Pract.* **2018**, *12*, 251–259. [CrossRef]
30. Divella, R.; De Luca, R.; Abbate, I.; Naglieri, E.; Daniele, A. Obesity and cancer: The role of adipose tissue and adipo-cytokines-induced chronic inflammation. *J. Cancer* **2016**, *7*, 2346. [CrossRef]

31. Huang, X.F.; Chen, J.Z. Obesity, the PI3K/Akt signal pathway and colon cancer. *Obes. Rev.* **2009**, *10*, 610–616. [CrossRef]
32. Hvid, H.; Blouin, M.-J.; Birman, E.; Damgaard, J.; Poulsen, F.; Fels, J.J.; Fledelius, C.; Hansen, B.F.; Pollak, M. Treatment with insulin analog X10 and IGF-1 increases growth of colon cancer allografts. *PLoS ONE* **2013**, *8*, e79710. [CrossRef]
33. Shi, B.; Sepp-Lorenzino, L.; Prisco, M.; Linsley, P.; DeAngelis, T.; Baserga, R. Micro RNA 145 targets the insulin receptor substrate-1 and inhibits the growth of colon cancer cells. *J. Biol. Chem.* **2007**, *282*, 32582–32590. [CrossRef] [PubMed]
34. Mullen, M.; Gonzalez-Perez, R.R. Leptin-induced JAK/STAT signaling and cancer growth. *Vaccines* **2016**, *4*, 26. [CrossRef] [PubMed]
35. Jaffe, T.; Schwartz, B. Leptin promotes motility and invasiveness in human colon cancer cells by activating multiple signal-transduction pathways. *Int. J. Cancer* **2008**, *123*, 2543–2556. [CrossRef]
36. Fujisawa, T.; Endo, H.; Tomimoto, A.; Sugiyama, M.; Takahashi, H.; Saito, S.; Inamori, M.; Nakajima, N.; Watanabe, M.; Kubota, N. Adiponectin suppresses colorectal carcinogenesis under the high-fat diet condition. *Gut* **2008**, *57*, 1531–1538. [CrossRef] [PubMed]
37. Sears, C.L.; Pardoll, D.M. Perspective: Alpha-bugs, their microbial partners, and the link to colon cancer. *J. Infect. Dis.* **2011**, *203*, 306–311. [CrossRef]
38. Chattopadhyay, I.; Dhar, R.; Pethusamy, K.; Seethy, A.; Srivastava, T.; Sah, R.; Sharma, J.; Karmakar, S. Exploring the role of gut microbiome in colon cancer. *Appl. Biochem. Biotechnol.* **2021**, *193*, 1780–1799. [CrossRef]
39. Sears, C.L.; Geis, A.L.; Housseau, F. Bacteroides fragilis subverts mucosal biology: From symbiont to colon carcinogenesis. *J. Clin. Investig.* **2014**, *124*, 4166–4172. [CrossRef]
40. Tjalsma, H.; Boleij, A.; Marchesi, J.R.; Dutilh, B.E. A bacterial driver–passenger model for colorectal cancer: Beyond the usual suspects. *Nat. Rev. Microbiol.* **2012**, *10*, 575–582. [CrossRef]
41. Brennan, C.A.; Garrett, W.S. Fusobacterium nucleatum—Symbiont, opportunist and oncobacterium. *Nat. Rev. Microbiol.* **2019**, *17*, 156–166. [CrossRef]
42. Dziubańska-Kusibab, P.J.; Berger, H.; Battistini, F.; Bouwman, B.A.; Iftekhar, A.; Katainen, R.; Cajuso, T.; Crosetto, N.; Orozco, M.; Aaltonen, L.A. Colibactin DNA-damage signature indicates mutational impact in colorectal cancer. *Nat. Med.* **2020**, *26*, 1063–1069. [CrossRef] [PubMed]
43. Ren, L.; Ye, J.; Zhao, B.; Sun, J.; Cao, P.; Yang, Y. The role of intestinal microbiota in colorectal cancer. *Front. Pharmacol.* **2021**, *12*, 674807. [CrossRef] [PubMed]
44. Gao, Z.; Guo, B.; Gao, R.; Zhu, Q.; Qin, H. Microbiota disbiosis is associated with colorectal cancer. *Front. Microbiol.* **2015**, *6*, 20. [CrossRef] [PubMed]
45. Gubatan, J.; Boye, T.L.; Temby, M.; Sojwal, R.S.; Holman, D.R.; Sinha, S.R.; Rogalla, S.R.; Nielsen, O.H. Gut microbiome in inflammatory bowel disease: Role in pathogenesis, dietary modulation, and colitis-associated colon cancer. *Microorganisms* **2022**, *10*, 1371. [CrossRef]
46. Attene-Ramos, M.S.; Wagner, E.D.; Plewa, M.J.; Gaskins, H.R. Evidence that hydrogen sulfide is a genotoxic agent. *Mol. Cancer Res.* **2006**, *4*, 9–14. [CrossRef] [PubMed]
47. Bultman, S.J. Interplay between diet, gut microbiota, epigenetic events, and colorectal cancer. *Mol. Nutr. Food Res.* **2017**, *61*, 1500902. [CrossRef]
48. Ocvirk, S.; O'Keefe, S.J. Influence of bile acids on colorectal cancer risk: Potential mechanisms mediated by diet-gut microbiota interactions. *Curr. Nutr. Rep.* **2017**, *6*, 315–322. [CrossRef]
49. Zou, S.; Fang, L.; Lee, M.-H. Dysbiosis of gut microbiota in promoting the development of colorectal cancer. *Gastroenterol. Rep.* **2018**, *6*, 1–12. [CrossRef]
50. Pino, M.S.; Chung, D.C. The chromosomal instability pathway in colon cancer. *Gastroenterology* **2010**, *138*, 2059–2072. [CrossRef]
51. Hope, M.E.; Hold, G.L.; Kain, R.; El-Omar, E.M. Sporadic colorectal cancer–role of the commensal microbiota. *FEMS Microbiol. Lett.* **2005**, *244*, 1–7. [CrossRef]
52. Fearon, E.R. Molecular genetics of colorectal cancer. *Annu. Rev. Pathol. Mech. Dis.* **2011**, *6*, 479–507. [CrossRef] [PubMed]
53. Sansom, O.J.; Reed, K.R.; Hayes, A.J.; Ireland, H.; Brinkmann, H.; Newton, I.P.; Batlle, E.; Simon-Assmann, P.; Clevers, H.; Nathke, I.S. Loss of Apc in vivo immediately perturbs Wnt signaling, differentiation, and migration. *Genes Dev.* **2004**, *18*, 1385–1390. [CrossRef] [PubMed]
54. Tan, B.L.; Norhaizan, M.E.; Huynh, K.; Heshu, S.R.; Yeap, S.K.; Hazilawati, H.; Roselina, K.J.B.C.; Medicine, A. Water extract of brewers' rice induces apoptosis in human colorectal cancer cells via activation of caspase-3 and caspase-8 and downregulates the Wnt/β-catenin downstream signaling pathway in brewers' rice-treated rats with azoxymethane-induced colon carcinogenesis. *BMC Complement. Altern. Med.* **2015**, *15*, 205.
55. Zhang, T.; Otevrel, T.; Gao, Z.; Gao, Z.; Ehrlich, S.M.; Fields, J.Z.; Boman, B.M. Evidence that APC regulates survivin expression: A possible mechanism contributing to the stem cell origin of colon cancer. *Cancer Res.* **2001**, *61*, 8664–8667.
56. Mastalier, B.; Tihon, C.; Ghita, B.; Botezatu, C.; Deaconescu, V.; Mandisodza, P.; Draghici, C.; Simion, S. Surgical treatment of colon cancer: Colentina surgical clinic experience. *J. Med. Life* **2012**, *5*, 348.
57. Schippinger, W.; Samonigg, H.; Schaberl-Moser, R.; Greil, R.; Thödtmann, R.; Tschmelitsch, J.; Jagoditsch, M.; Steger, G.; Jakesz, R.; Herbst, F. A prospective randomised phase III trial of adjuvant chemotherapy with 5-fluorouracil and leucovorin in patients with stage II colon cancer. *Br. J. Cancer* **2007**, *97*, 1021–1027. [CrossRef]

58. André, T.; Boni, C.; Mounedji-Boudiaf, L.; Navarro, M.; Tabernero, J.; Hickish, T.; Topham, C.; Zaninelli, M.; Clingan, P.; Bridgewater, J. Oxaliplatin, fluorouracil, and leucovorin as adjuvant treatment for colon cancer. *N. Engl. J. Med.* **2004**, *350*, 2343–2351. [CrossRef]
59. Cohen, R.; Taieb, J.; Fiskum, J.; Yothers, G.; Goldberg, R.; Yoshino, T.; Alberts, S.; Allegra, C.; De Gramont, A.; Seitz, J.-F. Microsatellite instability in patients with stage III colon cancer receiving fluoropyrimidine with or without oxaliplatin: An ACCENT pooled analysis of 12 adjuvant trials. *J. Clin. Oncol.* **2021**, *39*, 642. [CrossRef]
60. Chen, W.; Dong, H.; Wang, G.; Chen, J.; Wang, W. Effect of the duration of the capecitabine regimen following colon cancer surgery in an elderly population: A retrospective cohort study. *World J. Surg. Oncol.* **2021**, *19*, 238. [CrossRef]
61. Kalidindi, A.V.; Dubashi, B.; Jayanthi, M.; Shewade, D. Efficacy and safety of capecitabine and oxaliplatin (CAPOX) treatment in colorectal cancer: An observational study from a tertiary cancer center in South India. *Indian J. Cancer* **2022**, *59*, 73–79.
62. Jiang, W.; Liu, Q.; Yang, D.; Yang, S.-b. The efficacy of irinotecan supplementation for colorectal cancer: A meta-analysis of randomized controlled studies. *Medicine* **2022**, *101*, e28090. [CrossRef] [PubMed]
63. Salvatore, L.; Bria, E.; Sperduti, I.; Hinke, A.; Hegewisch-Becker, S.; Aparicio, T.; Le Malicot, K.; Boige, V.; Koeberle, D.; Baertschi, D. Bevacizumab as maintenance therapy in patients with metastatic colorectal cancer: A meta-analysis of individual patients' data from 3 phase III studies. *Cancer Treat. Rev.* **2021**, *97*, 102202. [CrossRef] [PubMed]
64. Uyl-de Groot, C.; Vermorken, J.; Hanna, M., Jr.; Verboom, P.; Groot, M.; Bonsel, G.; Meijer, C.; Pinedo, H. Immunotherapy with autologous tumor cell-BCG vaccine in patients with colon cancer: A prospective study of medical and economic benefits. *Vaccine* **2005**, *23*, 2379–2387. [CrossRef] [PubMed]
65. Riethmüller, G.; Holz, E.; Schlimok, G.; Schmiegel, W.; Raab, R.; Höffken, K.; Gruber, R.; Funke, I.; Pichlmaier, H.; Hirche, H. Monoclonal antibody therapy for resected Dukes' C colorectal cancer: Seven-year outcome of a multicenter randomized trial. *J. Clin. Oncol.* **1998**, *16*, 1788–1794. [CrossRef]
66. Wegner, R.E.; Abel, S.; Monga, D.; Raj, M.; Finley, G.; Nosik, S.; McCormick, J.; Kirichenko, A.V. Utilization of adjuvant radiotherapy for resected colon cancer and its effect on outcome. *Ann. Surg. Oncol.* **2020**, *27*, 825–832. [CrossRef] [PubMed]
67. Martenson, J.A., Jr.; Willett, C.G.; Sargent, D.J.; Mailliard, J.A.; Donohue, J.H.; Gunderson, L.L.; Thomas, C.R., Jr.; Fisher, B.; Benson, A.B., III; Myerson, R. Phase III study of adjuvant chemotherapy and radiation therapy compared with chemotherapy alone in the surgical adjuvant treatment of colon cancer: Results of intergroup protocol 0130. *J. Clin. Oncol.* **2004**, *22*, 3277–3283. [CrossRef]
68. Tiptiri-Kourpeti, A.; Spyridopoulou, K.; Santarmaki, V.; Aindelis, G.; Tompoulidou, E.; Lamprianidou, E.E.; Saxami, G.; Ypsilantis, P.; Lampri, E.S.; Simopoulos, C. *Lactobacillus casei* exerts anti-proliferative effects accompanied by apoptotic cell death and up-regulation of TRAIL in colon carcinoma cells. *PLoS ONE* **2016**, *11*, e0147960. [CrossRef]
69. Xu, F.; Li, Q.; Wang, S.; Bai, J.; Dong, M.; Xiao, G.; Wang, J. *Lactobacillus casei* JY300-8 generated by 12C6+ beams mutagenesis inhibits tumor progression by modulating the gut microbiota in mice. *J. Funct. Foods* **2021**, *87*, 104779. [CrossRef]
70. Budu, O.; Banciu, C.; Pinzaru, I.; Sarău, C.; Lighezan, D.; Șoica, C.; Dehelean, C.; Drăghici, G.; Dolghi, A.; Prodea, A. A combination of two probiotics, *Lactobacillus sporogenes* and *Clostridium butyricum*, inhibits Colon Cancer Development: An In Vitro Study. *Microorganisms* **2022**, *10*, 1692. [CrossRef]
71. Yue, Y.; Wang, S.; Shi, J.; Xie, Q.; Li, N.; Guan, J.; Evivie, S.E.; Liu, F.; Li, B.; Huo, G. Effects of *Lactobacillus acidophilus* KLDS1. 0901 on proliferation and apoptosis of colon cancer cells. *Front. Microbiol.* **2022**, *12*, 788040. [CrossRef]
72. Karimi Ardestani, S.; Tafvizi, F.; Tajabadi Ebrahimi, M. Heat-killed probiotic bacteria induce apoptosis of HT-29 human colon adenocarcinoma cell line via the regulation of Bax/Bcl2 and caspases pathway. *Hum. Exp. Toxicol.* **2019**, *38*, 1069–1081. [CrossRef] [PubMed]
73. Shamekhi, S.; Abdolalizadeh, J.; Ostadrahimi, A.; Mohammadi, S.A.; Barzegari, A.; Lotfi, H.; Bonabi, E.; Zarghami, N. Apoptotic Effect of Saccharomyces cerevisiae on human colon cancer SW480 cells by regulation of Akt/NF-κB signaling pathway. *Probiotics Antimicro.* **2020**, *12*, 311–319. [CrossRef] [PubMed]
74. Huang, L.; Shan, Y.-J.; He, C.-X.; Ren, M.-H.; Tian, P.-J.; Song, W. Effects of L. paracasei subp. paracasei X12 on cell cycle of colon cancer HT-29 cells and regulation of mTOR signalling pathway. *J. Funct. Foods* **2016**, *21*, 431–439. [CrossRef]
75. Saxami, G.; Karapetsas, A.; Lamprianidou, E.; Kotsianidis, I.; Chlichlia, A.; Tassou, C.; Zoumpourlis, V.; Galanis, A. Two potential probiotic lactobacillus strains isolated from olive microbiota exhibit adhesion and anti-proliferative effects in cancer cell lines. *J. Funct. Foods* **2016**, *24*, 461–471. [CrossRef]
76. Gamallat, Y.; Ren, X.; Walana, W.; Meyiah, A.; Xinxiu, R.; Zhu, Y.; Li, M.; Song, S.; Xie, L.; Jamalat, Y. Probiotic *Lactobacillus rhamnosus* modulates the gut microbiome composition attenuates preneoplastic colorectal Aberrant crypt foci. *J. Funct. Foods* **2019**, *53*, 146–156. [CrossRef]
77. Liu, M.; Xie, W.; Wan, X.; Deng, T. *Clostridium butyricum* modulates gut microbiota and reduces colitis associated colon cancer in mice. *Int. Immunopharmacol.* **2020**, *88*, 106862. [CrossRef]
78. Aindelis, G.; Tiptiri-Kourpeti, A.; Lampri, E.; Spyridopoulou, K.; Lamprianidou, E.; Kotsianidis, I.; Ypsilantis, P.; Pappa, A.; Chlichlia, K. Immune responses raised in an experimental colon carcinoma model following oral administration of *Lactobacillus casei*. *Cancers* **2020**, *12*, 368. [CrossRef]
79. Wang, T.; Zhang, L.; Wang, P.; Liu, Y.; Wang, G.; Shan, Y.; Yi, Y.; Zhou, Y.; Liu, B.; Wang, X. Lactobacillus coryniformis MXJ32 administration ameliorates azoxymethane/dextran sulfate sodium-induced colitis-associated colorectal cancer via reshaping intestinal microenvironment and alleviating inflammatory response. *Eur. J. Nutr.* **2022**, *61*, 85–99. [CrossRef]

80. Walia, S.; Kamal, R.; Kanwar, S.S.; Dhawan, D.K. Cyclooxygenase as a target in chemoprevention by probiotics during 1,2-dimethylhydrazine induced colon carcinogenesis in rats. *Nutr. Cancer* **2015**, *67*, 603–611. [CrossRef]
81. Saxami, G.; Karapetsas, A.; Chondrou, P.; Vasiliadis, S.; Lamprianidou, E.; Kotsianidis, I.; Ypsilantis, P.; Botaitis, S.; Simopoulos, C.; Galanis, A. Potentially probiotic Lactobacillus strains with anti-proliferative activity induce cytokine/chemokine production and neutrophil recruitment in mice. *Benef. Microbes* **2017**, *8*, 615–623. [CrossRef]
82. Agah, S.; Alizadeh, A.M.; Mosavi, M.; Ranji, P.; Khavari-Daneshvar, H.; Ghasemian, F.; Bahmani, S.; Tavassoli, A. More protection of *Lactobacillus acidophilus* than *Bifidobacterium bifidum* probiotics on azoxymethane-induced mouse colon cancer. *Probiotics Antimicro.* **2019**, *11*, 857–864. [CrossRef] [PubMed]
83. Shang, F.; Jiang, X.; Wang, H.; Chen, S.; Wang, X.; Liu, Y.; Guo, S.; Li, D.; Yu, W.; Zhao, Z. The inhibitory effects of probiotics on colon cancer cells: In vitro and in vivo studies. *J. Gastrointest. Oncol.* **2020**, *11*, 1224. [CrossRef]
84. Corthésy, B.; Gaskins, H.R.; Mercenier, A. Cross-talk between probiotic bacteria and the host immune system. *J. Nutr.* **2007**, *137*, 781S–790S. [CrossRef]
85. Zhang, J.-W.; Du, P.; Yang, B.-R.; Gao, J.; Fang, W.-J.; Ying, C.-M. Preoperative probiotics decrease postoperative infectious complications of colorectal cancer. *Am. J. Med. Sci.* **2012**, *343*, 199–205. [CrossRef] [PubMed]
86. Consoli, M.L.D.; da Silva, R.S.; Nicoli, J.R.; Bruña-Romero, O.; da Silva, R.G.; de Vasconcelos Generoso, S.; Correia, M.I.T. Randomized clinical trial: Impact of oral administration of *Saccharomyces boulardii* on gene expression of intestinal cytokines in patients undergoing colon resection. *J. Parenter. Enteral Nutr.* **2016**, *40*, 1114–1121. [CrossRef] [PubMed]
87. Delcenserie, V.; Martel, D.; Lamoureux, M.; Amiot, J.; Boutin, Y.; Roy, D. Immunomodulatory effects of probiotics in the intestinal tract. *Curr. Issues Mol. Biol.* **2008**, *10*, 37–54.
88. Zhang, M.; Fan, X.; Fang, B.; Zhu, C.; Zhu, J.; Ren, F. Effects of Lactobacillus salivarius Ren on cancer prevention and intestinal microbiota in 1, 2-dimethylhydrazine-induced rat model. *J. Microbiol. Immunol. Infect.* **2015**, *53*, 398–405. [CrossRef]
89. Park, I.J.; Lee, J.-H.; Kye, B.-H.; Oh, H.-K.; Cho, Y.B.; Kim, Y.-T.; Kim, J.Y.; Sung, N.Y.; Kang, S.-B.; Seo, J.-M. Effects of probiotics on the symptoms and surgical outcomes after anterior resection of colon cancer (POSTCARE): A randomized, double-blind, placebo-controlled trial. *J. Clin. Med.* **2020**, *9*, 2181. [CrossRef]
90. Wang, Q.; Wang, K.; Wu, W.; Lv, L.; Bian, X.; Yang, L.; Wang, Q.; Li, Y.; Ye, J.; Fang, D. Administration of *Bifidobacterium bifidum* CGMCC 15068 modulates gut microbiota and metabolome in azoxymethane (AOM)/dextran sulphate sodium (DSS)-induced colitis-associated colon cancer (CAC) in mice. *Appl. Microbiol. Biotechnol* **2020**, *104*, 5915–5928. [CrossRef]
91. Ohara, T.; Yoshino, K.; Kitajima, M. Possibility of preventing colorectal carcinogenesis with probiotics. *Hepatogastroenterology* **2010**, *57*, 1411–1415.
92. Zhu, Q.; Gao, R.; Wu, W.; Qin, H. The role of gut microbiota in the pathogenesis of colorectal cancer. *Tumor Biol.* **2013**, *34*, 1285–1300. [CrossRef]
93. Burne, R.A.; Chen, Y.-Y.M. Bacterial ureases in infectious diseases. *Microb. Infect.* **2000**, *2*, 533–542. [CrossRef] [PubMed]
94. Hatakka, K.; Holma, R.; El-Nezami, H.; Suomalainen, T.; Kuisma, M.; Saxelin, M.; Poussa, T.; Mykkänen, H.; Korpela, R. The influence of *Lactobacillus rhamnosus* LC705 together with *Propionibacterium freudenreichii* ssp. *shermanii* JS on potentially carcinogenic bacterial activity in human colon. *Int. J. Food Microbiol.* **2008**, *128*, 406–410. [CrossRef] [PubMed]
95. Desrouillères, K.; Millette, M.; Vu, K.D.; Touja, R.; Lacroix, M. Cancer preventive effects of a specific probiotic fermented milk containing *Lactobacillus acidophilus* CL1285, L. casei LBC80R and L. rhamnosus CLR2 on male F344 rats treated with 1,2-dimethylhydrazine. *J. Funct. Foods* **2015**, *17*, 816–827. [CrossRef]
96. Ohland, C.L.; MacNaughton, W.K. Probiotic bacteria and intestinal epithelial barrier function. *Am. J. Physiol. Gastrointest. Liver Physiol.* **2010**, *298*, G807–G819. [CrossRef]
97. Wang, T.; Wang, P.; Yin, L.; Wang, X.; Shan, Y.; Yi, Y.; Zhou, Y.; Liu, B.; Wang, X.; Lü, X. Dietary *Lactiplantibacillus plantarum* KX041 attenuates colitis-associated tumorigenesis and modulates gut microbiota. *Food Sci. Hum. Wellness* **2023**, *12*, 1626–1636. [CrossRef]
98. Liu, Z.; Qin, H.; Yang, Z.; Xia, Y.; Liu, W.; Yang, J.; Jiang, Y.; Zhang, H.; Yang, Z.; Wang, Y. Randomised clinical trial: The effects of perioperative probiotic treatment on barrier function and post-operative infectious complications in colorectal cancer surgery–a double-blind study. *Aliment. Pharmacol. Ther.* **2011**, *33*, 50–63. [CrossRef]
99. Wang, T.; Zheng, J.; Dong, S.; Ismael, M.; Shan, Y.; Wang, X.; Lü, X. *Lacticaseibacillus rhamnosus* LS8 Ameliorates Azoxymethane/Dextran Sulfate Sodium-Induced Colitis-Associated Tumorigenesis in Mice via Regulating Gut Microbiota and Inhibiting Inflammation. *Probiotics Antimicro.* **2022**, *14*, 947–959. [CrossRef]
100. Boleij, A.; Tjalsma, H. Gut bacteria in health and disease: A survey on the interface between intestinal microbiology and colorectal cancer. *Biol. Rev.* **2012**, *87*, 701–730. [CrossRef]
101. Kumar, R.S.; Kanmani, P.; Yuvaraj, N.; Paari, K.; Pattukumar, V.; Thirunavukkarasu, C.; Arul, V. *Lactobacillus plantarum* AS1 isolated from south Indian fermented food Kallappam suppress 1,2-dimethyl hydrazine (DMH)-induced colorectal cancer in male Wistar rats. *Appl. Biochem. Biotechnol.* **2012**, *166*, 620–631. [CrossRef]
102. Bahmani, S.; Azarpira, N.; Moazamian, E. Anti-colon cancer activity of *Bifidobacterium* metabolites on colon cancer cell line SW742. *Turk. J. Gastroenterol.* **2019**, *30*, 835–842. [CrossRef] [PubMed]
103. Saber, A.; Alipour, B.; Faghfoori, Z.; Khosroushahi, A.Y. Secretion metabolites of dairy *Kluyveromyces marxianus* AS41 isolated as probiotic, induces apoptosis in different human cancer cell lines and exhibit anti-pathogenic effects. *J. Funct. Foods* **2017**, *34*, 408–421. [CrossRef]

104. Saber, A.; Alipour, B.; Faghfoori, Z.; Khosroushahi, A.Y. Secretion metabolites of probiotic yeast, *Pichia kudriavzevii* AS-12, induces apoptosis pathways in human colorectal cancer cell lines. *Nutr. Res.* **2017**, *41*, 36–46. [CrossRef] [PubMed]
105. Wan, Y.; Xin, Y.; Zhang, C.; Wu, D.; Ding, D.; Tang, L.; Owusu, L.; Bai, J.; Li, W. Fermentation supernatants of *Lactobacillus delbrueckii* inhibit growth of human colon cancer cells and induce apoptosis through a caspase 3-dependent pathway. *Oncol. Lett.* **2014**, *7*, 1738–1742. [CrossRef] [PubMed]
106. Maghsood, F.; Johari, B.; Rohani, M.; Madanchi, H.; Saltanatpour, Z.; Kadivar, M. Anti-proliferative and anti-metastatic potential of high molecular weight secretory molecules from probiotic *Lactobacillus reuteri* cell-free supernatant against human colon cancer stem-like cells (HT29-ShE). *Int. J. Pept. Res. Ther.* **2020**, *26*, 2619–2631. [CrossRef]
107. Pakbin, B.; Allahyari, S.; Dibazar, S.P.; Peymani, A.; Haghverdi, M.K.; Taherkhani, K.; Javadi, M.; Mahmoudi, R. Anticancer Properties of *Saccharomyces boulardii* Metabolite Against Colon Cancer Cells. *Probiotics Antimicro. Proteins* **2022**. *online ahead of print*. [CrossRef]
108. Norouzi, Z.; Salimi, A.; Halabian, R.; Fahimi, H. Nisin, a potent bacteriocin and anti-bacterial peptide, attenuates expression of metastatic genes in colorectal cancer cell lines. *Microb. Pathog.* **2018**, *123*, 183–189. [CrossRef]
109. Villarante, K.I.; Elegado, F.B.; Iwatani, S.; Zendo, T.; Sonomoto, K.; de Guzman, E.E. Purification, characterization and in vitro cytotoxicity of the bacteriocin from *Pediococcus acidilactici* K2a2-3 against human colon adenocarcinoma (HT29) and human cervical carcinoma (HeLa) cells. *World J. Microbiol. Biotechnol.* **2011**, *27*, 975–980. [CrossRef]
110. De Giani, A.; Bovio, F.; Forcella, M.; Fusi, P.; Sello, G.; Di Gennaro, P. Identification of a bacteriocin-like compound from *Lactobacillus plantarum* with antimicrobial activity and effects on normal and cancerogenic human intestinal cells. *AMB Express* **2019**, *9*, 88. [CrossRef]
111. Karpiński, T. *New Peptide (Entap) with Anti-Proliferative Activity Produced by Bacteria of Enterococcus Genus*; Poznań University of Medical Sciences Poznań: Poznań, Poland, 2012.
112. Konishi, H.; Fujiya, M.; Tanaka, H.; Ueno, N.; Moriichi, K.; Sasajima, J.; Ikuta, K.; Akutsu, H.; Tanabe, H.; Kohgo, Y. Probiotic-derived ferrichrome inhibits colon cancer progression via JNK-mediated apoptosis. *Nat. Commun.* **2016**, *7*, 12365. [CrossRef]
113. Kang, J.; Sun, M.; Chang, Y.; Chen, H.; Zhang, J.; Liang, X.; Xiao, T. Butyrate ameliorates colorectal cancer through regulating intestinal microecological disorders. *Anti-Cancer Drugs* **2023**, *34*, 227. [CrossRef] [PubMed]
114. Botta, C.; Spyridopoulou, K.; Bertolino, M.; Rantsiou, K.; Chlichlia, K.; Cocolin, L. *Lactiplantibacillus plantarum* inhibits colon cancer cell proliferation as function of its butyrogenic capability. *Biomed. Pharmacother.* **2022**, *149*, 112755. [CrossRef] [PubMed]
115. Wei, Y.; Li, F.; Li, L.; Huang, L.; Li, Q. Genetic and biochemical characterization of an exopolysaccharide with in vitro antitumoral activity produced by *Lactobacillus fermentum* YL-11. *Front. Microbiol.* **2019**, *10*, 2898. [CrossRef]
116. Kumar, R.; Bansal, P.; Singh, J.; Dhanda, S. Purification, partial structural characterization and health benefits of exopolysaccharides from potential probiotic *Pediococcus acidilactici* NCDC 252. *Process Biochem.* **2020**, *99*, 79–86. [CrossRef]
117. Mojibi, P.; Tafvizi, F.; Torbati, M.B. Cell-bound exopolysaccharide extract from indigenous probiotic bacteria induce apoptosis in HT–29 cell-line. *Iran. J. Pathol.* **2019**, *14*, 41. [CrossRef]
118. Saadat, Y.R.; Gargari, B.P.; Shahabi, A.; Nami, Y.; Khosroushahi, A.Y. Prophylactic role of Lactobacillus paracasei exopolysaccharides on colon cancer cells through apoptosis not ferroptosis. Unpublished work. 2020. [CrossRef]
119. Li, W.; Tang, W.; Ji, J.; Xia, X.; Rui, X.; Chen, X.; Jiang, M.; Zhou, J.; Dong, M. Characterization of a novel polysaccharide with anti-colon cancer activity from *Lactobacillus helveticus* MB2-1. *Carbohydr. Res.* **2015**, *411*, 6–14. [CrossRef]
120. Li, F.; Jiao, X.; Zhao, J.; Liao, X.; Wei, Y.; Li, Q. Antitumor mechanisms of an exopolysaccharide from *Lactobacillus fermentum* on HT-29 cells and HT-29 tumor-bearing mice. *Int. J. Biol. Macromol.* **2022**, *209*, 552–562. [CrossRef]
121. Ma, F.; Song, Y.; Sun, M.; Wang, A.; Jiang, S.; Mu, G.; Tuo, Y. Exopolysaccharide produced by *Lactiplantibacillus plantarum*-12 alleviates intestinal inflammation and colon cancer symptoms by modulating the gut microbiome and metabolites of C57BL/6 mice treated by azoxymethane/dextran sulfate sodium salt. *Foods* **2021**, *10*, 3060. [CrossRef] [PubMed]
122. Wang, H.; Cheng, X.; Zhang, L.; Xu, S.; Zhang, Q.; Lu, R. A surface-layer protein from *Lactobacillus acidophilus* NCFM induces autophagic death in HCT116 cells requiring ROS-mediated modulation of mTOR and JNK signaling pathways. *Food Funct.* **2019**, *10*, 4102–4112. [CrossRef]
123. Gao, Z.; Guo, B.; Gao, R.; Zhu, Q.; Wu, W.; Qin, H. Probiotics modify human intestinal mucosa-associated microbiota in patients with colorectal cancer. *Mol. Med. Rep.* **2015**, *12*, 6119–6127. [CrossRef]

foods

MDPI

Article

Enhancing Phytochemical Compounds, Functional Properties, and Volatile Flavor Profiles of Pomelo (*Citrus grandis* (L.) Osbeck) Juices from Different Cultivars through Fermentation with *Lacticaseibacillus paracasei*

Vernabelle Balmori [1,2], Marisa Marnpae [1,3], Charoonsri Chusak [1], Kritmongkhon Kamonsuwan [1], Kasinee Katelakha [3], Suvimol Charoensiddhi [4] and Sirichai Adisakwattana [1,*]

[1] Center of Excellence in Phytochemical and Functional Food for Clinical Nutrition, Department of Nutrition and Dietetics, Faculty of Allied Health Sciences, Chulalongkorn University, Bangkok 10330, Thailand; balmorivernabelle@gmail.com (V.B.); mmarisa.hsc@gmail.com (M.M.); charoonsri.c@gmail.com (C.C.); kritmongkhon.kam@gmail.com (K.K.)

[2] Department of Food Science and Technology, Southern Leyte State University, Sogod 6606, Southern Leyte, Philippines

[3] The Halal Science Center, Chulalongkorn University, Bangkok 10330, Thailand; kkasinee.hsc@gmail.com

[4] Department of Food Science and Technology, Faculty of Agro-Industry, Kasetsart University, Bangkok 10900, Thailand; suvimol.ch@ku.th

* Correspondence: sirichai.a@chula.ac.th; Tel.: +66-2-218-1099 (ext. 111)

Abstract: The current study aimed to explore the effects of fermenting five different pomelo cultivars using *Lacticaseibacillus paracasei* on various physicochemical, phytochemical, and organoleptic attributes. Fermentation led to an increase in viable lactic acid bacteria count (8.80–9.28 log cfu/mL), organic acids, total polyphenols, and flavonoids, resulting in improved antioxidant activity, bile acid binding, cholesterol micellization disruption, and inhibition of pancreatic lipase activity. Additionally, some cultivars displayed higher levels of naringin, naringenin, and hesperetin after fermentation. The levels of volatile compounds were elevated after fermentation. The bitterness and overall acceptability scores were improved in the fermented samples of the Kao Numpueng cultivar. The principal component analysis (PCA) revealed that the Tubtim Siam cultivar demonstrated the highest functionality and health-related benefits among all fermented pomelos. Overall, the study suggests that pomelo exhibits potential as a valuable resource for creating a dairy-free probiotic drink enriched with bioactive phytochemical compounds and beneficial functional attributes.

Keywords: pomelo juice; *Lacticaseibacillus paracasei*; lactic acid bacteria; flavonoids; volatile compounds

1. Introduction

Consuming foods that contain a variety of functional components has gained popularity due to their potential benefits for human health. Functional foods, such as those containing probiotics, antioxidants, fiber, prebiotics, and other secondary plant metabolites, are an emerging trend in the field of food science and nutrition [1]. Probiotics are one of the fastest-growing fields of functional foods because they have been found to exert potential health effects when consumed in sufficient quantities [2]. Remarkably, probiotics have exhibited a remarkable ability to modulate and influence the specific composition and balance of the gut microbiota. This modulation leads to various advantages such as decreased instances of gastrointestinal infections, enhanced lactose metabolism, immune system stimulation, potential protection against cancer, relief from diarrhea, and alleviation of symptoms associated with irritable bowel syndrome [3–6]. Probiotic foods have traditionally been dairy based, but there is increasing interest in exploring alternative non-dairy matrices as carriers of probiotic bacteria. This is due to concerns among certain populations, including those who are allergic to milk proteins, lactose intolerant, have issues with milk cholesterol, or are strict vegetarians [7]. Plant-based foods are excellent sources of vitamins, antioxidants, minerals, prebiotics, and bioactive compounds, which offer additional benefits to the product. Non-dairy probiotic foods include fruit juices fermented with probiotic strains, predominantly lactic acid bacteria. *Lactobacillus* and *Bifidobacterium* starter cultures have been primarily used to ferment different food matrices to develop functional products [8]. However, the use of these strains in fruit matrices is still being investigated due to technological challenges in maintaining the growth and survival of the probiotics, caused by intrinsic fruit characteristics, such as low pH and a lack of nitrogen sources [8]. *Lacticaseibacillus paracasei* has been identified as a suitable species for survival in juices like orange and mulberry with good shelf-stability at refrigeration temperatures [9,10]. Interestingly, fruit juices fermented by lactic acid bacteria can alter their composition due to the synthesis of new compounds or the release of bioactive components through microbial metabolism and biochemical reactions. This procedure enhances the physical, chemical, sensory, and functional attributes of the fruit structure [11]. Moreover, scientists are consistently investigating the impact of probiotic fermentation on enhancing the biological properties and health advantages of food items. This indicates that non-dairy probiotic foods have great potential for providing a wide range of health benefits to individuals with specific dietary needs and preferences.

Pomelo, scientifically known as *Citrus grandis* (L.) Osbeck, is a tropical fruit that originates from Southeast Asia. It is abundantly composed of vitamin C (ascorbic acid) and phytochemical compounds, such as polyphenols and flavonoids [12]. In Thailand, various pomelo varieties are locally grown, including Kao Yai, Tong Dee, Kao Tangkwa, Kao Numpueng, Ta Koi, and Tubtim Siam. Research has demonstrated that Thai pomelo contains varying amounts of bioactive compounds, such as flavonoids, which differ depending on the specific cultivar [13,14]. This variability makes it a promising raw material for producing non-dairy probiotic beverages. However, the type of pomelo cultivar used can influence the resulting phytochemical content and biological activity of the fermented beverage. Hence, this study aimed to produce a fermented drink by employing various Thai pomelo cultivars and *Lacticaseibacillus paracasei*. Moreover, the influence of fermentation on the physicochemical properties, probiotic viability, volatile compounds, biological and functional characteristics, as well as sensory acceptability of the beverage were also examined.

2. Materials and Methods

2.1. Raw Materials, Bacteria, and Chemicals

Five different cultivars of pomelo fruits, namely Kao Yai (KY), Kao Numpueng (KN), Tubtim Siam (TS), Tong Dee (TD), and Ta Koi (TK), were purchased from the Talaad Thai Fruit Market located in Pathum Thani Province, Thailand.

The freeze-dried powder of *L. paracasei* subsp. *paracasei* CASEI 431 (Chr. Hansen A/S, Horsholm, Denmark) was obtained from Brenntag Ingredients (Thailand) Public Company Limited, Bangkok.

De Man–Rogosa–Sharpe (MRS) agar was procured from Oxoid (Basingstoke, UK). Food-grade sodium bicarbonate was obtained from Thai Food and Chemical Co., Ltd. (Bangkok, Thailand). Folin–Ciocalteu's reagent, phenol solution, sulfuric acid, potassium sodium tartrate, potassium dihydrogen phosphate, phosphoric acid, glucose standard, gallic acid, ascorbic acid, 1,1-diphenyl 2-picrylhydrazyl (DPPH), porcine bile extract, 6-hydroxyl-2,5,7,8-tetramethylchromane-2-carboxylic acid (Trolox), 2,4,6-tripyridyl-s-triazine (TPTZ), ferrous sulfate, hydrochloric acid, naringin, hesperidin, neohesperidin, naringenin, hesperitin, acetic acid, acetonitrile, oleic acid, bile extract, phosphatidylcholine, glycodeoxycholic acid (GDA), taurocholic acid, 4-methylumbelliferone, porcine pancreatic lipase, and sodium citrate were obtained from Sigma-Aldrich Co. (St. Louis, MO, USA). The cholesterol test kits (Cholesterol Liquicolor®) were purchased from Human Diagnostics (Wiesbaden, Germany), and the total bile acid assay kit was purchased from BIOBASE (Jinan, Shandong,

China). An alkane mixed standard of C6–C16 used for volatile compounds analysis was obtained from Restek Corporation (Bellefonte, PA, USA).

2.2. Fermentation of Pomelo Juices

The pomelo juice was obtained by removing the peel, seeds, and skin from the fruits, followed by extracting the fruit pulp using a low-speed juice extractor (Buono, BUO-143536). This pulp was then mixed with drinking water at a ratio of 4:1 (v/v). To eliminate solid particles, the resulting juice was filtered using cheesecloth, and sugar (5%) and salt (0.05% w/v) were added. Pasteurization was conducted by subjecting the juice to a temperature of 80 °C for 30 s. The pasteurized juice samples were collected and stored at −20 °C as control (non-fermented) juices. For the fermented samples, the pH of the pomelo juice was adjusted to 6.0 using 1 M sodium bicarbonate. Subsequently, *L. paracasei* powder was directly introduced at an initial microbial concentration of 7 log cfu/mL, followed by incubation at 37 °C for 24 h [14]. Once the fermentation period concluded, the juices were rapidly cooled down and stored at −20 °C for subsequent testing. Each experiment was performed in triplicate.

2.3. Determination of pH and Color Properties

The pH of the juice samples was measured using a digital pH meter (Seven Compact S220, Mettler Toledo, Columbus, OH, USA). The color properties of the juice samples, including the values for L^*, a^*, and b^*, were assessed utilizing a colorimeter (Colorflex, Hunter Associates Laboratory, Inc., Reston, VA, USA). The total color change (delta E) was calculated based on the changes in color values after fermentation. Each experiment was performed in triplicate.

2.4. Total Sugar and Reducing Sugar Content Analysis

The quantification of total sugar content in the samples was performed using the phenol-sulfuric acid method [15]. The fermented and non-fermented pomelo juices were diluted by adding distilled water at a 400-fold ratio. Subsequently, the diluted sample (400 μL) was combined with the working reagent in the glass tube (10 μL of an 80% phenol solution and 1 mL of a 95.5% sulfuric acid solution). Then, it was placed in a water bath set at 37 °C for 10 min. Following that, a period of 10 min was allotted for them to cool down to the ambient room temperature. The absorbance was then recorded at a wavelength of 490 nm. Each experiment was performed in triplicate.

The 3,5-dinitrosalicylic acid (DNS) method, as described by Jain et al. [16], was utilized to analyze the reduced sugar content. In this method, the juice sample (200 μL) was reacted with 200 μL of DNS reagent and subsequently boiled at 100 °C for 10 min. Once the mixture had reached a lower temperature, 200 μL of a 40% potassium sodium tartrate solution was introduced into the mixture. Subsequently, the measurement of absorbance was conducted at a wavelength of 540 nm. The glucose solutions were utilized to prepare a standard curve. Each experiment was performed in triplicate.

2.5. Analysis of Organic Acid

The quantification of organic acids was conducted using a well-established high-performance liquid chromatography (HPLC) method outlined in a prior publication [14]. First, the juice samples were subjected to centrifugation at 6500× g for 20 min at 4 °C. Subsequently, the supernatants were filtered through a 0.2 μm polyvinylidene fluoride syringe filter to prepare them for HPLC analysis. The mobile phase consisted of 10 mM potassium dihydrogen phosphate, pH 2.4, with phosphoric acid (A) and acetonitrile (B). The flow rate was set at 0.5 mL/min, and the gradient program consisted of the following: 0.01 min, 1% B; 12 min, 12% B; 15–24 min, 40% B; 25 min, 1% B, and held to 28 min, kept at 35 °C. Samples were injected at 5 μL per injection, and a mixed standard solution containing ascorbic, tartaric, lactic, citric, acetic, propionic, and malic acids was prepared to determine the retention times of each organic acid. External standards were used to quantify the peak area of the samples. Each experiment was performed in triplicate.

2.6. Enumeration of Viable Lactobacilli

To determine the total viable lactobacilli count, the standard plate count method was employed. The samples were plated on MRS agar and subjected to incubation at 37 °C for a duration of 48 h. Following incubation, the number of colony-forming units per mL was enumerated using the methodology described by a previous report [17]. Both analyses were performed in triplicate.

2.7. Determination of Phytochemical Compounds and Antioxidant Activity

The total phenolic content (TPC) was quantified by employing the Folin–Ciocalteau assay [18] and expressed as µg gallic acid equivalent (GAE) per mL of juice. The ferric-reducing antioxidant power (FRAP), DPPH radical scavenging activity, and Trolox equivalent antioxidant capacity (TEAC) were measured following the techniques described by Chayaratanasin et al. [19]. The FRAP values were presented as millimoles of ferrous sulfate ($FeSO_4$) equivalent per mL of juice, while the DPPH radical scavenging activity was reported as milligrams of ascorbic acid equivalent (AAE) per mL. Additionally, the TEAC values were reported as milligrams of trolox equivalent per mL of juice. The quantification of the total flavonoid content (TFC) was conducted through a colorimetric assay, employing the methodology established by Adisakwattana et al. [20]. All analyses were carried out in triplicate and the results were expressed as micrograms of quercetin equivalent (QE) per mL of juice.

2.8. Quantification of Major Flavonoid Constituents

In this study, the impact of fermentation on flavonoid constituents was examined through the application of HPLC, following a previously established method with slight modifications [21]. To begin, freeze-dried juice samples were carefully weighed and combined with an aqueous methanol solution (50% *v*/*v*) prior to homogenization for 1 min. Subsequently, the mixtures were subjected to extraction by immersing them in a water bath set at a constant temperature of 30 °C for a duration of 20 min. Following the extraction process, centrifugation was performed at 4500 rpm at 20 °C for a duration of 20 min. The resulting supernatants were collected and subjected to solid-phase extraction (SPE) using a Sep-Pak C_{18} cartridge. The column was preconditioned with 6 mL of absolute methanol and 6 mL of 0.01 N HCl solutions, followed by the introduction of the supernatant into the cartridge and washing with 20 mL of 0.01 N HCl to remove sugars and other organic acids. Final washing was carried out with a methanol/HCl solution (1% *v*/*v*) to elute the retained compounds. Washings were collected and subsequently dried using a rotary evaporator at 50 °C and 150 rpm until the volume was reduced to approximately 0.5 mL. Further drying was performed under nitrogen, and completely dried extracts were reconstituted with a water/acetic acid (99:1, *v*/*v*) and acetonitrile/acetic acid (99:1, *v*/*v*) mixture (1:1) before being filtered prior to HPLC injection. The mobile phase was composed of (A) water/acetic acid (99:1, *v*/*v*) and (B) acetonitrile/acetic acid (99:1, *v*/*v*) at a column temperature of 50 °C. The gradient flow was set as follows: 0 min, 5% B; 5 min, 8% B; 7 min, 12% B; 12 min, 18% B; 17 min, 22% B; 22 min, 25% B; 27 min, 35% B; 37 min, 53% B; 38 min, 53% B; 40 min, 55% B; 42 min, 60% B; 57 min, 80% B; 60 min, 85% B; and 65 min, 85% B. The flow rate was set at 1.2 mL/min, and chromatographic detection was carried out at 280 nm. Identification and quantification were performed through the use of an external standard composed of naringin, naringenin, hesperidin, neohesperidin, and hesperetin. The analysis was performed in triplicate, and the results were expressed in µg/g of freeze-dried juice.

2.9. Bile Acid Binding, Cholesterol Micellization, and Pancreatic Lipase Inhibition

In the bile acid binding assay, the experimental procedure involved incubating the samples with bile extract (2 mM) and glycodeoxycholic acid (GDA) (2 mM) in 0.1 M phosphate-buffered saline (PBS) at a temperature of 37 °C for a duration of 90 min [13]. Subsequently, the mixtures underwent filtration and were subjected to analysis using a bile acid assay kit. Analysis was performed in triplicate and the results were then reported as the percentage of bile acid binding. For comparative purposes, a positive control was incorporated using cholestyramine (3 mg/mL).

In the cholesterol micellization assay, synthetic micelles were prepared by combining 2 mM cholesterol, 1 mM oleic acid, 2.4 mM phosphatidylcholine, and 6.6 mM taurocholic acid in 15 mM PBS [13]. The samples were introduced to the mixed micelles and subjected to incubation at 37 °C for a duration of 2 h. Finally, the mixtures were centrifuged at 10,000 rpm for 20 min. The cholesterol concentration in the supernatants was assessed using a cholesterol test kit. Gallic acid (0.5 mg/mL) was used for the positive control. The analysis was repeated 3 times and the results were expressed as the percentage of cholesterol micellization inhibition.

In the pancreatic lipase activity assay, the samples were mixed with a solution of 0.2 mM 4-methylumbelliferone (4-MUO) in a microplate well [13]. Subsequently, the pancreatic lipase enzyme solution (7.5 mg/mL) was added to the mixtures and incubated at a temperature of 37 °C for a duration of 20 min. Following that, 0.1 M sodium citrate solution at pH 4.2 was added to halt the enzymatic reaction. The measurement of lipase activity was conducted using a fluorescence microplate reader, with excitation occurring at 355 nm and emission detected at 460 nm. Analysis was conducted in triplicate and orlistat was employed as a positive control.

2.10. Analysis of Volatile Compounds by Gas Chromatography

The volatile flavor compounds in the juice samples were analyzed using an electronic nose operating on ultrafast gas chromatography with two columns (Hera-cles II, Alpha M.O.S., Toulouse, France), following a previously established method with slight modifications [14]. The instrument was equipped with two columns of different polarities, namely, nonpolar MXT-5 and slightly polar MXT-1701, and two flame ionization detectors (FID).

To begin the analysis, juice samples were carefully transferred into sealed vials and incubated at a temperature of 60 °C for 300 s. Throughout the incubation process, agitation was maintained at 500 rpm within the auto-sampler. The gas produced in the headspace was then analyzed by gas chromatography with an injection volume of 5 mL and collected in a trap at 25 °C before desorption in two columns at 250 °C. The oven temperature initiated at 40 °C for 10 s, then increased to 60 °C at 4 °C/s, and finally to 250 °C at 1 °C/s. The FID detectors were set at 270 °C. Alkane standards C6-C16 analyzed at the same condition served as reference standards to convert retention times into Kovats retention indices (KI) using AlphaSoft software v14.1 (Alpha Software Co., Burlington, MA, USA). Volatile compounds were tentatively identified by comparing their KI values with the AroChemBase v5.0 reference volatile compound database (Alpha MOS, Toulouse, France), considering the possibility of a match (relevant index) at more than 80% [14].

Through untargeted qualitative analysis, the peak areas of each identified compound were compared to evaluate the influence of fermentation on alterations in volatile compounds. The analysis was performed in triplicate, with three independent repetitions (injections) carried out for each individual sample.

2.11. Organoleptic Test

A group of 50 individuals without specific training evaluated the organoleptic acceptability of the samples using a 9-point hedonic evaluation test. The assessment encompassed different attributes, including color, aroma, sweetness, sourness, bitterness (aftertaste), and overall acceptability. The protocol for this test was approved by the Research Ethics Review Committee for Research Involving Human Research Participants, Chulalongkorn University on 18 April 2022 (Certificate of Approval No. 090/65, Study Title No. 650032). Before commencing the experiment, written informed consent was obtained from all panelists participating in the evaluation.

2.12. Statistical Analysis

The obtained data were presented as the mean ± standard error of the mean (SEM). To examine the effects of cultivar variation, fermentation, and their interaction, a two-way analysis of variance (ANOVA) was conducted. In cases where significant main effects were detected, further analysis was performed using Duncan's Multiple Range Test (DMRT) to identify significant differences between groups. The paired t-test was utilized to determine the statistical significance of differences observed before and after fermentation for the parameters under investigation. The data were analyzed using IBM SPSS version 22.0 (Chicago, IL, USA), and the significance level was set at $p < 0.05$. The principal component analysis was conducted using the R software (Version 2023.09.1+494) to visualize the differences between the samples based on the identified factors.

3. Results and Discussion

3.1. Physicochemical Properties of Fermented Pomelo Juices

During this study, we made significant observations regarding the effects of fermentation on the pH levels and chemical composition of pomelo juices. After the fermentation process, the pH levels of the samples decreased notably, reaching a final pH range between 4.26 and 4.49, as indicated in Table 1. This decrease in pH can be attributed to the increased production of organic acids, particularly lactic acid and acetic acid. The rise in organic acids is directly responsible for the reduction in pH levels of fermented pomelo juices. Furthermore, this fermentation process led to a substantial decline in both the total and reducing sugar content of the juices obtained from all pomelo cultivars. The findings of our study revealed that the strain *L. paracasei* CASEI 431 played a key role in converting the glucose and fructose present in pomelo juice into lactic and acetic acid through an enzymatic process.

Table 1. pH and organic acids of pomelo juices before and after 24 h fermentation.

Samples		pH	Organic Acids (mg/mL)					
			Tartaric Acid	Ascorbic Acid	Citric Acid	Lactic Acid	Acetic Acid	Total Organic Acids
KN	NF	4.56 ± 0.12 ^{abA}	0.03 ± 0.00 ^{deB}	0.10 ± 0.01 ^{bcA}	1.50 ± 0.03 ^{cdA}	0.0 ^{cB}	0.0 ^{eB}	1.40 ± 0.03 ^{cB}
	F	4.49 ± 0.18 ^{aA}	0.06 ± 0.02 ^{bcA}	0.10 ± 0.01 ^{bcA}	1.27 ± 0.21 ^{bcA}	2.77 ± 0.26 ^{bA}	0.43 ± 0.04 ^{aA}	4.90 ± 0.85 ^{aA}
KY	NF	4.53 ± 0.16 ^{abA}	0.03 ± 0.00 ^{deB}	0.08 ± 0.00 ^{cdA}	1.22 ± 0.19 ^{cdA}	0.0 ^{cB}	0.0 ^{eB}	1.69 ± 0.19 ^{bB}
	F	4.37 ± 0.01 ^{abA}	0.08 ± 0.01 ^{abA}	0.10 ± 0.01 ^{bcA}	1.15 ± 0.16 ^{cdA}	3.65 ± 0.33 ^{aA}	0.19 ± 0.04 ^{cA}	5.12 ± 0.29 ^{aA}
TK	NF	4.57 ± 0.03 ^{bA}	0.08 ± 0.01 ^{abA}	0.06 ± 0.01 ^{dB}	2.13 ± 0.16 ^{aA}	0.0 ^{cB}	0.0 ^{eB}	2.63 ± 0.24 ^{bB}
	F	4.44 ± 0.02 ^{aA}	0.02 ± 0.01 ^{dfB}	0.08 ± 0.01 ^{cdA}	1.28 ± 0.06 ^{cdB}	2.89 ± 0.21 ^{bA}	0.29 ± 0.06 ^{bA}	4.31 ± 0.69 ^{aA}
TD	NF	4.46 ± 0.11 ^{bA}	0.09 ± 0.01 ^{aA}	0.13 ± 0.02 ^{abA}	1.15 ± 0.15 ^{cdA}	0.0 ^{cB}	0.0 ^{eB}	1.50 ± 0.17 ^{bcB}
	F	4.27 ± 0.17 ^{aA}	0.09 ± 0.01 ^{aA}	0.10 ± 0.01 ^{bcA}	1.01 ± 0.04 ^{dA}	2.85 ± 0.21 ^{bA}	0.17 ± 0.05 ^{cA}	5.39 ± 0.31 ^{aA}
TS	NF	4.49 ± 0.05 ^{aA}	0.05 ± 0.01 ^{cdA}	0.17 ± 0.00 ^{aA}	1.81 ± 0.04 ^{abA}	0.0 ^{cB}	0.0 ^{eB}	2.12 ± 0.01 ^{bB}
	F	4.26 ± 0.15 ^{aA}	0.06 ± 0.01 ^{bcA}	0.14 ± 0.01 ^{aA}	0.98 ± 0.03 ^{dB}	3.13 ± 0.26 ^{abA}	0.09 ± 0.02 ^{deA}	4.40 ± 0.29 ^{aA}

Data are expressed as Mean ± SEM, $n = 3$. Means with different uppercase letters (fermentation effects) at the same cultivar are significantly different ($p < 0.05$). Means with different lowercase letters (treatment effects) are significantly different ($p < 0.05$). Abbreviations: KN: Kao Numpueng; KY: Kao Yai; TK: Ta Koi; TD: Tong Dee; TS: Tubtim Siam; NF: Non-fermented juice; F: Fermented juice.

This enzymatic conversion resulted in a significant decrease in sugar content and a simultaneous increase in the organic acid content of the juice. Previous studies have also reported similar findings, demonstrating the ability of lactic acid bacteria to convert citric acid to lactate and acetate, resulting in reduced citric acid levels and higher lactic acid content [10,22]. The observed levels of organic acids in the non-fermented juices were lower than the typical organic acid content of pure pomelo juice as observed by a previous study [23] since the pomelo pulps were added with water in the juice extraction. Additionally, a previous study has observed that orange juice fermented with *L. paracasei* has led to an increase in lactic and acetic, as these acids are the major metabolites of LAB fermentation [9]. Pomelo fruits contain soluble sugars including glucose, fructose, and sucrose, which are utilizable by *L. paracasei* species, thus reducing the sugar content [24]. Fermentation of fruits and vegetables by lactic acid bacteria has been associated with enhanced health benefits due to the reduction in sugar content [25]. Similar outcomes have been observed in studies investigating the fermentation of other fruit juices, such as orange juice [9] and gac juice [14]. Interestingly, we found that the Tubtim Siam and Kao Yai cultivars exhibited the highest initial total sugar and reducing sugar content among all the cultivars studied (Table 2). This may explain their superior sweetness compared to other cultivars. Overall, our study underlines the transformative effects of fermentation on pomelo juice, leading to significant changes in pH, organic acid content, and sugar levels, which can impact the taste and potential health benefits of fermented juice.

Table 2. Total sugar, reducing sugar, color values, and viable lactic acid bacteria count of pomelo juices before and after 24 h fermentation.

Samples		Total Sugar (mg/mL)	Reducing Sugar (mg/mL)	Color Values				Lactic Acid Bacteria Count (log cfu/mL)
				L^*	a^*	b^*	ΔE	
KN	NF	109.04 ± 1.56 ^{abA}	13.54 ± 0.02 ^{abA}	27.79 ± 0.47 ^{bB}	−3.15 ± 0.15 ^{cA}	−2.67 ± 0.29 ^{cdA}	4.53 ± 0.41	7.09 ± 0.50 ^{aB}
	F	98.58 ± 3.39 ^{bB}	10.83 ± 0.79 ^{bB}	32.29 ± 0.76 ^{bA}	−2.78 ± 0.06 ^{cA}	−3.09 ± 0.02 ^{dA}		8.86 ± 0.48 ^{abA}
KY	NF	124.5 ± 3.50 ^{aA}	14.33 ± 0.09 ^{aA}	29.33 ± 0.30 ^{abB}	1.67 ± 0.13 ^{bB}	−3.71 ± 0.24 ^{dA}	4.53 ± 0.01	7.08 ± 0.01 ^{aB}
	F	120.40 ± 9.9 ^{aB}	8.50 ± 0.43 ^{cB}	33.77 ± 0.66 ^{abA}	2.45 ± 0.06 ^{bA}	−4.15 ± 0.16 ^{eA}		9.12 ± 0.51 ^{aA}
TK	NF	96.88 ± 6.88 ^{bA}	12.46 ± 0.05 ^{bA}	27.40 ± 0.79 ^{bB}	−3.41 ± 0.13 ^{cA}	−1.67 ± 0.19 ^{bcA}	4.84 ± 0.23	7.06 ± 0.05 ^{aB}
	F	92.45 ± 4.16 ^{bB}	9.76 ± 0.09 ^{bcB}	32.21 ± 0.48 ^{bA}	−3.04 ± 0.07 ^{cA}	−2.02 ± 0.33 ^{cA}		8.80 ± 0.52 ^{bA}
TD	NF	101.87 ± 0.39 ^{abA}	12.27 ± 0.13 ^{bA}	28.52 ± 1.10 ^{abB}	1.52 ± 0.10 ^{bB}	−1.65 ± 0.52 ^{bA}	5.26 ± 0.16	7.08 ± 0.15 ^{aB}
	F	92.55 ± 3.33 ^{abB}	11.02 ± 0.07 ^{bcB}	33.73 ± 0.29 ^{abA}	2.25 ± 0.25 ^{bA}	−1.82 ± 0.57 ^{bA}		9.01 ± 0.17 ^{abA}
TS	NF	126.19 ± 6.29 ^{aA}	14.41 ± 0.09 ^{aA}	32.53 ± 1.35 ^{aB}	4.25 ± 0.10 ^{aB}	1.54 ± 0.22 ^{aA}	4.31 ± 0.13	7.09 ± 0.02 ^{aB}
	F	112.8 ± 4.71 ^{abB}	8.44 ± 0.34 ^{cB}	36.71 ± 0.84 ^{aA}	5.29 ± 0.07 ^{aA}	1.67 ± 0.04 ^{aA}		9.28 ± 0.30 ^{aA}

Data are expressed as Mean ± SEM, $n = 3$. Means with different uppercase letters (fermentation effects) at the same cultivar are significantly different ($p < 0.05$). Means with different lowercase letters (treatment effects) are significantly different ($p < 0.05$). Abbreviations: KN: Kao Numpueng; KY: Kao Yai; TK: Ta Koi; TD: Tong Dee; TS: Tubtim Siam; NF: Non-fermented juice; F: Fermented juice.

The color of fruit juice plays a significant role in determining consumer acceptance, as it greatly influences sensory appeal and overall product quality [25]. In this study, the color values of fermented pomelo juice were evaluated, and it was found that the degree of lightness (L^*) significantly increased for all cultivars after fermentation. Additionally, the a^* values for Tong Dee, Ta Koi, and Tubtim Siam significantly increased after fermentation, indicating an increase in redness. Nevertheless, no significant difference in b^* values was observed after fermentation for all cultivars. The alterations in juice color following fermentation may be attributed to the intricate relationship among pH, pigments, and color parameters. Pink varieties of pomelo, containing pigments like lycopene and anthocyanin, exhibit a more stable red color at pH < 4, consequently elevating the a^* values of the juices. The reduction in lightness is linked to the oxidation and degradation of juice components during fermentation [26]. The calculated ΔE values across all cultivars (falling within the range of 4.31 to 5.26) suggest that the color changes post-fermentation are noticeable. It has been established that visual distinctions in color become apparent when ΔE exceeds 3 [27]. Consumers associate lighter-colored fruit juices with higher freshness and quality levels, which suggests that increasing lightness may improve consumer perception and acceptance of the product [26]. Therefore, it is possible that increasing the lightness of fermented pomelo by lactic acid bacteria may help to further improve consumer perception and acceptance of the product, although further research is needed to confirm this speculation.

3.2. Viable Lactobacilli Counts of Fermented Pomelo Juices

As demonstrated in Table 2, the viable lactobacilli count significantly increased after fermentation in all cultivars, with final counts ranging from 8.80 to 9.28 log cfu/mL, compared to an initial count of around 7 log cfu/mL. Among the cultivars, the highest viable counts were found in Tubtim Siam and Kao Yai, which are associated with a higher decrease in reducing sugar content in these cultivars. These findings indicate that pomelo cultivars are suitable carriers for *L. paracasei* bacteria, as they can proliferate in the juice matrix with viable counts higher than the recommended amount for beneficial effect ($\geq 10^6$ cfu/mL) [2]. This study represents the initial exploration into the viability of lactobacilli bacteria in fermented pomelo juice. Our findings are in line with earlier studies that demonstrated the survival of lactobacilli bacteria in citrus juices, including grapefruit [28] and orange [9], highlighting the possibility of using citrus fruits as a nutrient-rich source for the proliferation of lactic acid bacteria.

3.3. Phytochemicals and Biological Activities of Fermented Pomelo Juices

The beneficial antioxidant activities of pomelo juice can be attributed to the presence of ascorbic acid and phytochemical compounds. The findings from our study demonstrated a significant increase in the TPC and TFC of all pomelo cultivars following fermentation with *L. paracasei* (Table 3).

Table 3. Phytochemical and antioxidant properties of pomelo juices before and after 24 h fermentation.

Samples		TPC (µg GAE/mL)	TFC (µg QE/mL)	FRAP (mmol $FeSO_4$/mL)	TEAC (mg Trolox Eq/mL)	DPPH (mg AAE/mL)
KN	NF	910.76 ± 2.06 ^{bcB}	25.21 ± 0.09 ^{gB}	8.71 ± 0.18 ^{cB}	1.70 ± 0.08 ^{bB}	0.29 ± 0.06 ^{cB}
	F	963.89 ± 3.16 ^{bcA}	54.00 ± 0.26 ^{dA}	10.56 ± 0.08 ^{bcA}	1.94 ± 0.03 ^{abA}	0.82 ± 0.02 ^{abA}
KY	NF	913.43 ± 1.07 ^{bcB}	31.87 ± 0.27 ^{fB}	9.81 ± 0.28 ^{bcB}	1.56 ± 0.04 ^{bcB}	0.26 ± 0.02 ^{bB}
	F	1053 ± 6.09 ^{aA}	56.92 ± 0.46 ^{cA}	10.96 ± 0.29 ^{abA}	1.76 ± 0.01 ^{bA}	0.64 ± 0.12 ^{bA}
TK	NF	893.16 ± 3.07 ^{cB}	31.43 ± 0.26 ^{fB}	8.46 ± 0.11 ^{cB}	1.42 ± 0.06 ^{cB}	0.23 ± 0.01 ^{cB}
	F	962.78 ± 2.09 ^{bcA}	65.22 ± 0.17 ^{bA}	9.45 ± 0.19 ^{bcA}	1.73 ± 0.02 ^{bA}	0.93 ± 0.05 ^{aA}
TD	NF	957.45 ± 2.01 ^{bcB}	34.50 ± 0.18 ^{eB}	8.82 ± 0.24 ^{cB}	1.75 ± 0.02 ^{bB}	0.27 ± 0.02 ^{cB}
	F	1037.2 ± 3.02 ^{abA}	54.45 ± 0.74 ^{dA}	11.57 ± 0.31 ^{abA}	1.85 ± 0.02 ^{bA}	0.92 ± 0.03 ^{aA}
TS	NF	1063.14 ± 1.06 ^{abB}	81.71 ± 0.16 ^{bB}	11.85 ± 0.42 ^{abB}	2.21 ± 0.03 ^{aB}	0.34 ± 0.03 ^{cB}
	F	1288.67 ± 3.05 ^{aA}	118.99 ± 0.16 ^{aA}	13.06 ± 0.28 ^{aA}	2.56 ± 0.08 ^{aA}	0.98 ± 0.06 ^{aA}

Data are expressed as Mean ± SEM, n = 3. Means with different uppercase letters (fermentation effects) at the same cultivar are significantly different ($p < 0.05$). Means with different lowercase letters (treatment effects) are significantly different ($p < 0.05$). Abbreviations: KN: Kao Numpueng; KY: Kao Yai; TK: Ta Koi; TD: Tong Dee; TS: Tubtim Siam; NF: Non-fermented juice; F: Fermented juice: TPC: Total Phenolic Content; TFC: Total Flavonoid Content; FRAP: ferric-reducing antioxidant power; DPPH: DPPH radical scavenging activity; TEAC: Trolox equivalent antioxidant capacity (TEAC).

Among the cultivars, Tubtim Siam exhibited the highest TPC and TFC levels both before and after fermentation, with values increasing from 1063 to 1288 µg GAE/mL and from 81.71 to 118.99 µg QE/mL, respectively. In addition, fermentation significantly increased the FRAP, TEAC, and DPPH scavenging activity of all the pomelo cultivars. The FRAP values of fermented samples were increased to 9.45–13.06 (mmol $FeSO_4$/mL), which were 10.21–31.17% higher than the non-fermented juices. This suggests that fermentation with *L. paracasei* can improve the reducing power of pomelo juice. A similar outcome of increasing antioxidant activity was reported in a study conducted by Zhao and Shah [29] on fermented soymilk. Moreover, the TEAC and DPPH scavenging capacities of the pomelo juices exhibited a substantial increase subsequent to the fermentation process. The highest TEAC and DPPH capacities were observed in the fermented Tubtim Siam cultivar at 2.56 mg Trolox Eq/mL and 0.98 mg AAE/mL, respectively. Similar findings were reported in a study on fermented apple juice, indicating that lactic acid bacteria fermentation could enhance the accessibility of polyphenolic compounds characterized by their hydrogen-donating and oxygen-quenching properties [30]. We suggest that the variability in antioxidant activity among the different pomelo cultivars can be attributed to the varying levels of TPC and TFC, as well as the viability of *L. paracasei* in each cultivar.

Pomelo juice is abundant in phytochemical compounds, particularly belonging to the flavonoid family, specifically flavanones, which are the primary constituents. As presented in Table 4, the major flavanones presented in pomelo juice include naringin, which is the most abundant, followed by hesperidin and neohesperidin. A similar flavanone profile was reported in the study conducted by Makkumrai et al. [24]. Fermentation with lactic acid bacteria can transform conjugated flavonoid compounds into their free forms, which have improved bioavailability and higher antioxidant activity [31]. This results in enhanced health-related functionality of the flavonoids. Our findings indicated a significant increase in the levels of naringenin (52.83–91.37% increase) and hesperetin (39.16–341.18% increase) following fermentation. Naringenin is an aglycone of naringin and is naturally present in citrus fruits [28].

Table 4. Flavonoid constituents of fermented and non-fermented pomelo juices.

Samples		Naringin (μg/g)	Hesperidin (μg/g)	Neohesperidin (μg/g)	Naringenin (μg/g)	Hesperetin (μg/g)
KN	NF	442.80 ± 7.59 ^{cB}	77.57 ± 2.34 ^{bA}	41.21 ± 11.66 ^{bA}	1.86 ± 0.23 ^{bcB}	0.51 ± 0.03 ^{deB}
	F	1271.59 ± 15.34 ^{abA}	51.81 ± 9.74 ^{bA}	67.51 ± 13.89 ^{bA}	3.19 ± 0.62 ^{bA}	2.25 ± 0.08 ^{abcA}
KY	NF	208.99 ± 9.48 ^{cB}	45.78 ± 5.81 ^{bB}	56.83 ± 6.54 ^{bA}	0.53 ± 0.0.06 ^{cA}	0.33 ± 0.03 ^{eB}
	F	472.19 ± 5.27 ^{cA}	71.33 ± 8.46 ^{bA}	52.92 ± 5.45 ^{bA}	0.81 ± 0.07 ^{cA}	0.86 ± 0.04 ^{bcdeA}
TK	NF	1114.91 ± 32.12 ^{bcA}	217.85 ± 15.13 ^{abA}	99.21 ± 12.41 ^{abB}	1.00 ± 0.03 ^{cA}	0.38 ± 0.01 ^{eA}
	F	1291.22 ± 19.42 ^{abcA}	237.79 ± 18.06 ^{abA}	144.91 ± 23.08 ^{aA}	1.76 ± 0.08 ^{bcA}	0.58 ± 0.05 ^{cdeA}
TD	NF	710.77 ± 7.38 ^{cA}	333.98 ± 7.74 ^{aB}	73.69 ± 4.68 ^{bB}	1.18 ± 0.05 ^{cA}	1.66 ± 0.05 ^{bcdeB}
	F	717.54 ± 5.02 ^{cA}	460.09 ± 20.30 ^{aA}	105.68 ± 12.44 ^{abA}	1.95 ± 0.04 ^{cA}	2.31 ± 0.08 ^{abA}
TS	NF	2271.38 ± 43.04 ^{abA}	403.75 ± 8.65 ^{aA}	113.30 ± 9.13 ^{abA}	3.48 ± 0.06 ^{bB}	2.12 ± 0.24 ^{abcdA}
	F	2517.00 ± 30.51 ^{aA}	263.97 ± 4.14 ^{abA}	61.67 ± 5.84 ^{bB}	6.66 ± 0.78 ^{aA}	3.33 ± 0.15 ^{aA}

Data are expressed as Mean ± SEM, $n = 3$. Means with different uppercase letters (fermentation effects) at the same cultivar are significantly different ($p < 0.05$). Means with different lowercase letters (treatment effects) are significantly different ($p < 0.05$). Abbreviations: KN: Kao Numpueng; KY: Kao Yai; TK: Ta Koi; TD: Tong Dee; TS: Tubtim Siam; NF: Non-fermented juice; F: Fermented juice.

On the other hand, hesperetin is an aglycone of hesperidin and neohesperidin [32]. The increase in the content of aglycone flavonoids, which is associated with the rise in TPC and antioxidant activity, occurs as a result of the hydrolysis of glycosidic linkages by β-glucosidase enzymes produced by lactic acid bacteria [24]. This enzymatic process leads to the conversion of glycosylated flavonoids into their aglycone forms, which are more readily absorbed. By breaking down these glycosidic linkages, the lactic acid bacteria enhance the availability and bioactivity of flavonoids, thereby increasing the content of TPC and improving the antioxidant activity of the juice.

The findings show that the fermentation process of citrus fruits may lead to changes in the levels of certain compounds. Specifically, some cultivars experienced a decrease in the levels of hesperidin and neohesperidin after fermentation, but this decrease may not necessarily be negative, as the fermentation process could have led to the formation of other beneficial compounds. Interestingly, the levels of naringin, a flavonoid responsible for imparting bitterness to citrus fruits, increased in all cultivars after fermentation. The Tubtim Siam cultivar exhibited the highest naringin content among the tested cultivars (2271.38 and 2517.00 μg/g freeze-dried juice for non-fermented and fermented juice, respectively). The increase in naringin content observed in the fermented juices can be attributed to the degradation of citrus fruit cell walls that occurs during the fermentation process [33]. It is important to note that the complete conversion of naringin to naringenin may not occur during probiotic fermentation, likely due to the limited availability of β-glucosidase enzymes. Therefore, the increase in naringin levels following fermentation can be partially explained by the limited conversion of naringin to naringenin. The presence of naringin in the fermented juices contributes to the overall flavonoid content and may have potential implications for the sensory characteristics and health benefits of the final product. The naringin content in pomelo fruits can vary depending on the cultivar. Pink pomelo cultivars, such as Tubtim Siam, generally have higher levels of naringin compared to white varieties [34]. This difference in naringin content is likely due to genetic variations between the different pomelo cultivars. Therefore, when studying naringin levels in pomelo fruits, it is crucial to consider the specific cultivar being analyzed. Taking into account the specific cultivar of pomelo is important for accurately assessing and understanding the naringin content and its potential impact on the flavor and health benefits of the fermented juices.

3.4. Functional Properties of Fermented Pomelo Juices

The results from Figure 1 show how fermentation affects the functional properties of pomelo juice, specifically its ability to bind bile acids, inhibit cholesterol micellization, and inhibit lipase activity. The study found that all varieties of pomelo juice could bind primary (bile extract) and secondary bile acids (GDA), showing binding capacities ranging from 50.71% to 57.62% and 37.80% to 46.33%, respectively (Figure 1A,B). This ability is due to the presence of polyphenolic substances in pomelo juice that can interact with bile acids through ionic, hydrogen, and hydrophobic interactions, forming polyphenol–bile acid complexes [35]. Fermentation significantly improved the capacity of pomelo juice to bind bile acid, with a range of 60.43% to 67.51% for primary bile acid and 58.30% to 62.98% for secondary bile acid (GDA). The binding values of fermented pomelo juices were found to be similar to those of cholestyramine.

In addition to its remarkable bile acid binding capacity, fermented pomelo juice exhibited an enhanced capability to disrupt cholesterol micellization formation. The increase in cholesterol micellization formation inhibition was significantly higher in the fermented juice compared to the unfermented juice (Figure 1C). Based on the findings, we suggest that this inhibitory action can be attributed to the bile acid binding activity of pomelo juices and the presence of lactic acid bacteria. Citrus juices, containing soluble fiber like pectin, can bind cholesterol, thereby exhibiting a hypocholesterolemic effect [36]. Meanwhile, *L. paracasei*, belonging to the species of lactic acid bacteria, is recognized for assimilating cholesterol and binding it to their cell walls, leading to reduced solubility and disruption of cholesterol micellization [37]. Furthermore, the fermented pomelo juice showed a higher inhibition of pancreatic lipase activity (Figure 1D), indicating that the presence of polyphenols might be responsible for this effect. These compounds have been shown to form a complex with lipase through hydrophobic associations and hydrogen bonding, causing a conformational change in the enzyme that inhibits its binding to the substrate, leading to its inhibition [38]. Among the cultivars, Tubtim Siam generally exhibited the highest functional properties related to disrupting lipid absorption, which could be attributed to its higher phytochemical content and LAB count compared to the other cultivars. The increased functional properties observed in this study may be attributed to the rise in total phenolic and flavonoid content of the fermented pomelo juice. The inhibitory mechanisms of fermented pomelo juices may have the potential to improve lipid metabolism related to fat absorption, thereby lowering the risk of conditions such as hypercholesterolemia and obesity [39]. However, further experiments are warranted to investigate the effects of fermented pomelo juice on bile acid binding capacity in humans.

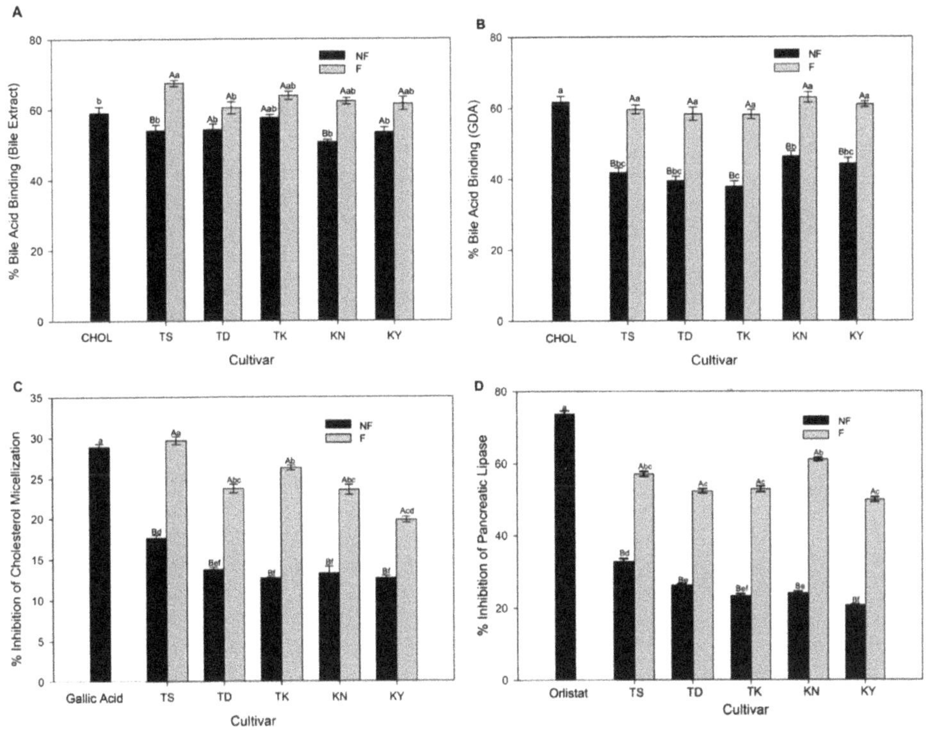

Figure 1. Effects of fermentation on bile acid binding (**A**,**B**), cholesterol micellization inhibition (**C**), and the pancreatic lipase inhibitory activity (**D**) of pomelo juices from different cultivars. Results are expressed as mean ± S.E.M. ($n = 3$). The means with different uppercase letters (fermentation effects) at the same cultivar are significantly different ($p < 0.05$). The means with different lowercase letters (treatment effects) are significantly different ($p < 0.05$). The abbreviations used for different cultivars are as follows: TS: Tubtim Siam; TD: Tong Dee; TK: Ta Koi; KN: Kao Numpueng; KY: Kao Yai; CHOL: Cholestyramine (3 mg/mL); Gallic acid at 0.5 mg/mL; Orlistat at 0.01 mg/mL; NF: Non-fermented juice; F: Fermented juice.

3.5. Volatile Compounds of Fermented Pomelo Juices

The tentative identification of volatile flavor compounds greatly influences the quality of citrus fruits, contributing to their complex aroma. The identification was carried out in both fermented and non-fermented pomelo cultivars using KI compared to the database (Table S2). Forty volatile compounds were tentatively identified in fermented and non-fermented pomelo juice samples. Among these, thirteen compounds are presented in Table 5, including nine compounds (1-Hexanol, 1-octen-3-ol, (*Z*)-3-hexenal, 2,4-decadienal, (*E*,*E*)-, ethyl acetate, myrcene, (+)-alpha-phellandrene, L-limonene, and indole) that were previously identified in pomelo juices [23,40,41].

Table 5. Volatile flavor compounds identified in pomelo juices.

Compound	Tentative Identification	Odor Description	Peak Area ($\times 10^2$)									
			KN		KY		TD		TK		TS	
			NF	F	NF	F	NF	F	NF	F	NF	F
						Alcohols						
1	1-Hexanol	Green, Fruity	N.D.	N.D.	N.D.	N.D.	N.D.	N.D.	2.84 ± 0.15 ^{A}	1.69 ± 0.09 ^{A}	4.08 ± 1.36 ^{A}	1.23 ± 0.1[illegible]
2	1-Octen-3-ol	Earthy, Herbaceous	1.68 ± 0.35 ^{A}	1.05 ± 0.06 ^{A}	N.D.	N.D.	1.17 ± 0.03 ^{A}	1.72 ± 0.03 ^{A}	N.D.	N.D.	N.D.	N.D.
						Aldehydes						
3	(Z)-3-Hexenal	Green, Leafy	N.D.	N.D.	3.09 ± 0.16 ^{A}	2.83 ± 0.15 ^{A}	N.D.	N.D.	7.73 ± 1.190	N.D.	15.67 ± 1.99	N.D.
4	2,4-Decadienal, (E,E)-	Citrus, Green	5.66 ± 0.81 ^{A}	3.45 ± 0.5 ^{B}	3.85 ± 0.93 ^{A}	1.46 ± 0.81 ^{B}	1.27 ± 0.97	N.D.	4.71 ± 0.52 ^{A}	3.54 ± 0.44 ^{A}	7.12 ± 1.92	N.D.
						Esters						
5	Ethyl acetate	Acidic, Sweet	5.91 ± 0.59 ^{B}	17.13 ±1.26 ^{A}	5.74 ± 0.08 ^{A}	5.69 ± 0.03 ^{A}	9.07 ± 1.71 ^{B}	30.66 ± 1.31 ^{A}	7.86 ± 0.04 ^{A}	7.73 ± 0.10 ^{A}	11.10 ± 0.10 ^{A}	11.14 ± 0.0[illegible] ^{A}
						Hydrocarbons						
6	Myrcene	Lemon, Musty	4.03 ± 0.57 ^{A}	1.22 ± 0.26 ^{B}	N.D.	N.D.	1.31 ± 0.43	N.D.	24.58 ± 2.70 ^{A}	18.38 ± 1.80 ^{B}	N.D.	N.D.
7	(+)-alpha-Phellandrene	Citrus, Green	8.94 ± 1.88 ^{A}	6.48 ± 1.25 ^{B}	N.D.	N.D.	N.D.	N.D.	N.D.	N.D.	N.D.	N.D.
8	L-Limonene	Citrus, Terpenic	37.69 ±1.75 ^{A}	27.91 ± 1.29 ^{B}	13.78 ± 0.11 ^{A}	4.12 ± 0.15 ^{B}	14.60 ± 0.10 ^{A}	2.56 ± 0.14 ^{B}	11.74 ± 0.13 ^{A}	1.19 ± 0.07 ^{B}	12.94 ± 1.01 ^{A}	9.07 ± 0.49 ^{B}
						Ketones						
9	Butan-2-one	Cheese, Ether-like	N.D.	62.75 ± 2.37	N.D.	121.52 ± 3.46	N.D.	92.70 ± 1.65	N.D.	66.80 ± 7.60	N.D.	165.92 ± 9.78
10	Acetoin	Creamy, Buttery	N.D.	26.98 ± 1.88	N.D.	2.82 ± 0.50	N.D.	1.98 ± 0.14	N.D.	2.37 ± 0.64	N.D.	3.38 ± 0.64
11	3,5-Octadien-2-one	Fruity, Mushroom	N.D.	28.02 ± 2.97	N.D.	23.09 ± 2.47	N.D.	5.04 ± 0.49	N.D.	21.02 ± 2.72	N.D.	11.47 ± 1.81
						Others						
12	Indole	Floral, Sweet	N.D.	N.D.	2.65 ± 0.38 ^{A}	3.87 ± 0.64 ^{A}	N.D.	N.D.	1.60 ± 0.32	N.D.	1.39 ± 0.34 ^{A}	1.34 ± 0.23 ^{A}
13	bis(2-furylmethyl) sulfide	Green, Earthy	13.57 ± 0.36 ^{A}	8.41 ± 0.14 ^{B}	14.58 ± 0.44 ^{A}	9.61 ± 1.53 ^{B}	12.87 ± 3.64 ^{A}	5.14 ± 0.94 ^{B}	15.88 ± 2.52 ^{A}	8.89 ± 0.27 ^{B}	19.43 ± 3.83 ^{A}	7.45 ± 1.79 ^{B}

The peak areas are obtained from the MXT-1701 column. Data are expressed as Mean ± SEM, $n = 3$. Means with different uppercase letters (fermentation effects) at the same cultivar are significantly different ($p < 0.05$). Abbreviations: KN: Kao Numpueng; KY: Kao Yai; TK: Ta Koi; TD: Tong Dee; TS: Tubtim Siam; NF: Non-fermented juice; F: Fermented juice.

The major compounds are associated with fruity flavors (ethyl acetate), citrus flavors (including L-limonene, myrcene, and 2,4-decadienal (*E,E*)-), and 'green' flavors (bis(2-furylmethyl) sulfide). As reported by Rosales and Suwonsichon [34], pomelo fruits contain a variety of volatile compounds, including hydrocarbons, alcohols, esters, aldehydes, and minor components that contribute to a fruity, citrusy, and herbaceous aroma (green).

Fermentation of pomelo juice resulted in an increase in total ketones, including butan-2-one, acetoin, and 3,5-octadien-2-one, which were not found in non-fermented pomelo juices. Lactic acid bacteria, such as *L. paracasei*, can metabolize citric acid and pyruvate, producing acetoin and butanediol, with butan-2-one being produced through the dehydration of 2,3-butanediol [22]. During the fermentation process of pomelo juice, some undesirable compounds (green odors), such as 2,4-decadienal, (*E,E*)-, (*Z*)-3-Hexenal, and bis(2-furylmethyl) sulfide, were significantly reduced ($p < 0.05$). Unsaturated aliphatic aldehydes such as 2,4-decadienal are known to be aroma-active compounds present in both white and pink pomelo varieties, contributing to their citrus, green aroma. The reduction in the intensity of these compounds is attributed to the reduction or oxidation of aldehydes by lactic acid bacteria during fermentation, resulting in the formation of alcohols or carboxylic acids [42]. After the fermentation process, there was a notable decrease in the presence of undesirable aromatic compounds, including l-limonene. L-limonene, a terpene hydrocarbon naturally present in citrus fruits, contributes to the citrusy aroma of pomelo juice, while trimethylamine and 1-butanamine are associated with strong unpleasant fishy and ammonia-like odors [43]. High concentrations of terpenes and terpenoids in citrus fruits have been linked to high degrees of bitterness, which can affect consumer preference.

Based on our findings, it can be inferred that the fermentation process involving *L. paracasei* has the potential to enhance the volatile aromatic profile of pomelo juice. This improvement is attributed to the reduction of undesirable compounds and the enhancement or synthesis of desirable flavor compounds. Moreover, our results indicate that different cultivars of pomelo exhibit varying aromatic compound profiles, which may contribute to differences in aroma, flavor, and other organoleptic characteristics.

An electronic nose system, utilizing ultrafast GC, has been employed for food quality control and to investigate the influence of specific factors on changes in volatile compounds in food [44,45]. This method entails the simultaneous analysis of chromatographic profiles from two GC columns with different polarities, providing a comprehensive exploration of the volatile profile in various foodstuffs. Notably, it offers a quicker and simpler alternative to other chromatographic techniques.

The tentative identification, achieved through Kovats relative retention indexes on the two columns of different polarities, is designed to enhance accuracy. However, the study has limitations as it relies solely on tentative identification through KI from a library database, lacking additional confirmation using MS spectra. Therefore, further research is imperative, utilizing GC-MS/MS for more precise identification. In particular, volatile compounds not found in pomelo juice but emerging after fermentation, such as butan-2-one, acetoin, and 3,5-octadien-2-one, should undergo further identification via GC-MS to confirm and elucidate how fermentation influences the volatile profile of the juices.

Furthermore, GC-MS/MS facilitates the quantification of volatile compounds using internal standards, enabling the examination of concentrations in pomelo juice and the assessment of how fermentation affects them. Additionally, to investigate how odor compounds impact the aroma of pomelo juice, further analyses are required. This includes odor threshold testing and the evaluation of odor activity values (OAV), determining the minimum detectable concentration of odor compounds by human olfactory receptors, and assessing their overall impact on the aroma profiles of pomelo juice.

3.6. Organoleptic Acceptability Scores of Fermented Pomelo Juices

The organoleptic test of pomelo juice before and after fermentation showed significant differences in acceptance scores (Figure 2). Although appearance and color remained unchanged during fermentation, the aroma scores decreased in some cultivars despite an improvement in the aromatic compound profile. This indicates that changes in the intensity of some desirable aromatic compounds may not have been noticeable to consumers. The organoleptic assessment of the fermented samples indicated slightly lower scores for sweetness and sourness acceptance compared to the non-fermented samples. This could be attributed to the decrease in total sugar content and the increase in acidity that occurs during the fermentation process. Interestingly, the fermented juice of KN showed more improvement in terms of bitterness and overall acceptability compared to the non-fermented juice. On the other hand, no significant change in bitterness and overall acceptability was observed for TD before and after fermentation. However, the overall acceptability of both the non-fermented and fermented juice of TS did not show a significant difference. In contrast, a decrease in the acceptability scores was observed for TK and KY following fermentation.

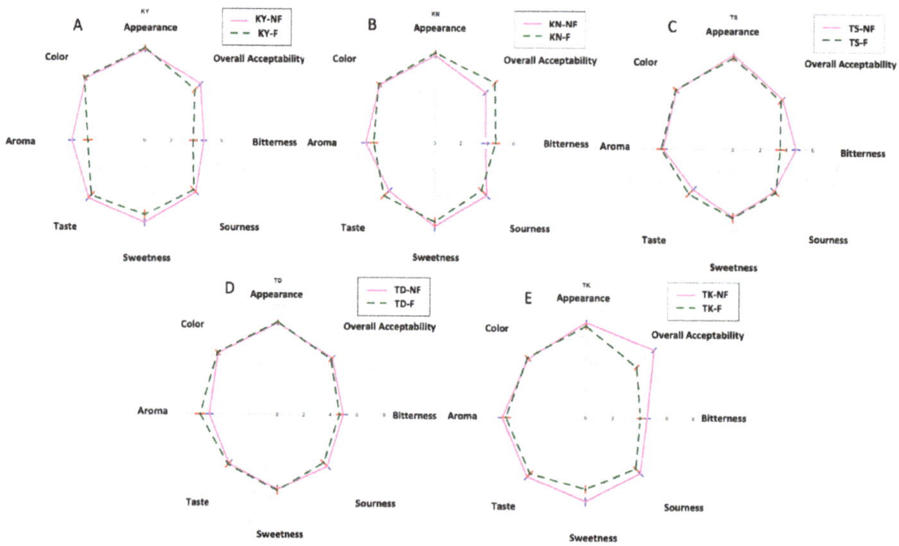

Figure 2. Acceptability scores of non-fermented and fermented pomelo juices from different cultivars (**A**–**E**). Data are presented as mean scores (n = 50). KY: Kao Yai (**A**); KN: Kao Numpueng (**B**); TS: Tubtim Siam(**C**); TD: Tong Dee (**D**); TK: Ta Koi (**E**); NF: Non-fermented; F: Fermented.

During the fermentation of pomelo juice, there was an increase in flavonoids, particularly naringin, which contributed to an elevation in bitterness, resulting in lower acceptance scores. Similar outcomes have been reported in previous studies on fermented mango and orange juices, where fermentation by lactobacilli bacteria resulted in decreased overall liking due to an increase in sourness, bitterness, and off-flavors in the juices [9,38]. Among the non-fermented samples, KY was the most preferred juice, while KN was preferred the most among the fermented samples. These findings highlight the need to address the bitterness and optimize sweetness and sourness to enhance the consumer acceptability of fermented pomelo juice.

3.7. Principal Component Analysis

The principal component analysis (PCA) depicted in Figure 3 demonstrated that PC1 and PC2 explained 73.2% and 10.7% of the variance, respectively. PC1 was strongly influenced by TFC, antioxidant properties, inhibition of cholesterol micellization, pancreatic lipase inhibition, and bile acid binding capacity, while TPC and acceptability scores were major factors influencing PC2. The samples were separated by PC1 due to the increased TFC, antioxidant capacity, and functional properties of the pomelo juices after fermentation. Additionally, there was a positive relationship between the phytochemical content and functional properties of pomelo juice samples, while there was an inverse relationship between functional properties and acceptability scores. Furthermore, the agglomerative hierarchical clustering of the pomelo juice samples revealed that the fermented Tubtim Siam juice (TSF) had the highest properties and was placed in the first cluster (A). The other fermented samples (TKF, TDF, KNF, and KYF) had lower properties and were placed in the second cluster (B). The non-fermented samples were grouped into two clusters: one composed of Tubtim Siam and Tong Dee (TS and TD) with higher properties than the other cluster, composed of Ta Koi, Kao Numpueng, and Kao Yai. These results suggest that Tubtim Siam exhibited the highest properties related to health benefits and functionality among all cultivars tested in both fermented and non-fermented samples.

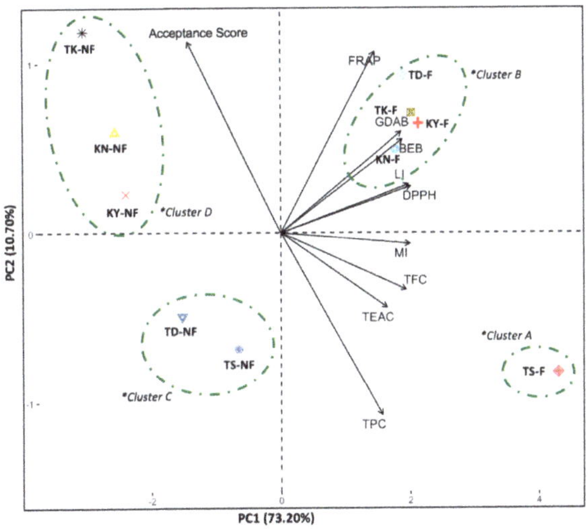

Figure 3. A biplot of the principal component analysis based on the phytochemical content, functional properties, and overall organoleptic acceptance scores of fermented and non-fermented pomelo juice cultivars. The data represent the average of three independent measurements. *Clusters were determined by Agglomerative Hierarchical Clustering. TS: Tubtim Siam; TD: Tong Dee; TK: Ta Koi; KN: Kao Numpueng; KY: Kao Yai; NF: Non-fermented; F: Fermented. TPC: Total Phenolic Content; TFC: Total Flavonoid Content; FRAP: ferric-reducing antioxidant power; DPPH: DPPH radical scavenging activity; TEAC: Trolox equivalent antioxidant capacity; MI: Cholesterol Micellization Inhibition; LI: Lipase activity inhibition; BEB: Bile Extract Binding; GDAB: Glycodeoxycholic acid Binding.

4. Conclusions

The current study provides insights into the use of different pomelo varieties for producing a probiotic beverage using *L. paracasei*. Fermentation was found to increase bioactive components and antioxidant capacity, leading to potential health benefits such as the reduction of lipid digestion which may result in a reduction of obesity and risks of cardiovascular diseases. Tubtim Siam was identified as the most suitable cultivar for producing fermented pomelo juice with desirable characteristics such as high *L. paracasei* survival, antioxidant properties, bile acid binding, lipase activity inhibition, and disruption of cholesterol micellization. This study also highlights the importance of considering cultivar variations when exploring the potential of pomelo in novel food applications.

Further studies should explore probiotic viability, phytochemical bioaccessibility during storage and digestion, and the effect on the gut microbiota to establish the functionality and health benefits of fermented pomelo beverages.

Supplementary Materials: The following supporting information can be downloaded at: https://www.mdpi.com/article/10.3390/foods12234278/s1, Table S1, Retention time (RT) and Kovats retention indices (KI) for alkane standards C6-C16. Table S2, Retention time (RT) and Kovats retention indices (KI) for identified volatile compounds.

Author Contributions: Conceptualization, Methodology, Investigation, Validation, Formal Analysis, Writing—Original Draft, V.B.; Investigation, M.M., C.C., K.K. (Kritmongkhon Kamonsuwan) and K.K. (Kasinee Katelakha); Supervision, Writing—Review and Editing, S.C.; Conceptualization, Methodology, Writing—Review and Editing, Supervision, Project Administration, and Funding Acquisition, S.A. All authors have read and agreed to the published version of the manuscript.

Funding: This study received support from the 90th Anniversary of Chulalongkorn University Scholarship, Ratchadaphiseksomphot Endowment Fund, Chulalongkorn University and the National Research Council of Thailand (NRCT) under Grant Number N42A640325.

Informed Consent Statement: Informed consent was obtained from all subjects involved in the study.

Data Availability Statement: Data are contained within the article or Supplementary Materials.

Acknowledgments: We extend our gratitude to the Halal Science Center at Chulalongkorn University, Thailand for their assistance with analytical instrument support.

Conflicts of Interest: The authors declare no conflict of interest.

References

1. Wojtunik-Kulesza, K.; Oniszczuk, A.; Oniszczuk, T.; Combrzyński, M.; Nowakowska, D.; Matwijczuk, A. Influence of in vitro digestion on composition, bioaccessibility and antioxidant activity of food polyphenols-a non-systematic review. *Nutrients* **2020**, *12*, 1401. [CrossRef] [PubMed]
2. Begum, P.; Madhavi, G.; Rajagopal, S.; Viswanath, B.; Razak, M.; Venkataratnamma, V. Probiotics as functional foods: Potential effects on human health and its impact on neurological diseases. *Int. J. Nutr. Pharmacol. Neurol. Dis.* **2017**, *7*, 23–33.
3. Abraham, B.P.; Quigley, E.M.M. Probiotics in inflammatory bowel disease. *Gastroenterol. Clin. N. Am.* **2017**, *46*, 769–782. [CrossRef]
4. Son, S.H.; Dae-Kyung, K.; Eun, B.; Hye-Lin, J.; Hyun-Dong, P.; Na-Kyoung, L.; Young-Seo, P. Potential probiotic *Lactobacillus plantarum* Ln4 from kimchi: Evaluation of β-galactosidase and antioxidant activities. *LWT-Food Sci. Technol.* **2017**, *85*, 181–186. [CrossRef]
5. Zoumpopoulou, G.; Pot, B.; Tsakalidou, E.; Papadimitriou, K. Dairy probiotics: Beyond the role of promoting gut and immune health. *Int. Dairy J.* **2016**, *67*, 46–60. [CrossRef]
6. Pothuraju, R.; Hussain, S.A. Probiotics: An important player in the obesity management alone? *Obes. Med.* **2017**, *8*, 13–14. [CrossRef]
7. Silva, A.R.A.; Silva, M.M.N.; Ribeiro, B.D. Health issues and technological aspects of plant-based alternative milk. *Food Res. Int.* **2020**, *131*, 108972. [CrossRef]
8. Leonard, W.; Liang, A.; Ranadheera, C.S.; Fang, Z.; Zhang, P. Fruit juices as a carrier of probiotics to modulate gut phenolics and microbiota. *Food Funct.* **2022**, *13*, 10333–10346. [CrossRef]
9. Quan, Q.; Liu, W.; Guo, J.; Ye, M.; Zhang, J. Effect of six lactic acid bacteria strains on physicochemical characteristics, antioxidant activities and sensory properties of fermented orange juices. *Foods* **2022**, *11*, 1920. [CrossRef]
10. Kwaw, E.; Ma, Y.; Tchabo, W.; Apaliya, M.T.; Xiao, L. Effect of lactic acid fermentation on the phytochemical, volatile profile and sensory attributes of mulberry juice. *J. Food Nutr. Res.* **2017**, *56*, 305–317.
11. Di Cagno, R.; Coda, R.; De Angelis, M.; Gobbetti, M. Exploitation of vegetables and fruits through lactic acid fermentation. *Food Microbiol.* **2013**, *33*, 1–10. [CrossRef]
12. Toh, J.; Khoo, H.; Azrina, A. Comparison of antioxidant properties of pomelo [*Citrus grandis* (L) Osbeck] varieties. *Int. Food Res. J.* **2013**, *20*, 1661–1668.
13. Mäkynen, K.; Jitsaardkul, S.; Tachasamran, P.; Sakai, N.; Puranachoti, S.; Nirojsinlapachai, N.; Chattapat, V.; Caengprasath, N.; Ngamukote, S.; Adisakwattana, S. Cultivar variations in antioxidant and antihyperlipidemic properties of pomelo pulp (*Citrus grandis* [L.] Osbeck) in Thailand. *Food Chem.* **2013**, *139*, 735–743. [CrossRef] [PubMed]
14. Marnpae, M.; Chusak, C.; Balmori, V.; Kamonsuwan, K.; Dahlan, W.; Nhujak, T.; Hamid, N.; Adisakwattana, S. Probiotic Gac fruit beverage fermented with *Lactobacillus paracasei*: Physiochemical properties, phytochemicals, antioxidant activities, functional properties, and volatile flavor compounds. *LWT-Food Sci. Technol.* **2022**, *169*, 113986. [CrossRef]

15. Nielsen, S.S. Phenol-Sulfuric Acid Method for Total Carbohydrates. In *Food Analysis Laboratory Manual*; Nielsen, S.S., Ed.; Springer: Boston, MA, USA, 2010; pp. 47–53.
16. Jain, A.; Jain, R.; Jain, S. Quantitative analysis of reducing sugars by 3,5-dinitrosalicylic acid (DNSA method). In *Basic Techniques in Biochemistry, Microbiology and Molecular Biology: Principles and Techniques*; Springer: New York, NY, USA, 2020; pp. 181–183.
17. Balmori, V.; Dizon, E.; Elegado, F. Effect of adjunct inoculation of *Lactobacillus plantarum* BS and *Pediococcus acidilactici* 3G3 on the microbiological, physicochemical, and sensory properties of fermented carabeef (*Pindang damulag*). *Food Res.* **2019**, *3*, 70–78. [CrossRef]
18. Chusak, C.; Chanbunyawat, P.; Chumnumduang, P.; Chantarasinlapin, P.; Suantawee, T.; Adisakwattana, S. Effect of gac fruit (*Momordica cochinchinensis*) powder on in vitro starch digestibility, nutritional quality, textural and sensory characteristics of pasta. *LWT-Food Sci. Technol.* **2020**, *118*, 108856. [CrossRef]
19. Chayaratanasin, P.; Barbieri, M.A.; Suanpairintr, N.; Adisakwattana, S. Inhibitory effect of *Clitoria ternatea* flower petal extract on fructose-induced protein glycation and oxidation-dependent damages to albumin in vitro. *BMC Complement. Altern. Med.* **2015**, *15*, 27. [CrossRef]
20. Adisakwattana, S.; Ruengsamran, T.; Kampa, P.; Sompong, W. In vitro inhibitory effects of plant-based foods and their combinations on intestinal α-glucosidase and pancreatic α-amylase. *BMC Complement. Altern. Med.* **2012**, *12*, 110. [CrossRef]
21. Mesquita, E.; Monteiro, M. Simultaneous HPLC determination of flavonoids and phenolic acids profile in Pêra-Rio orange juice. *Food Res. Int.* **2018**, *106*, 54–63. [CrossRef]
22. Wang, Y.; Wu, J.; Lv, M.; Shao, Z.; Hungwe, M.; Wang, J.; Bai, X.; Xie, J.; Wang, Y.; Geng, W. Metabolism Characteristics of Lactic Acid Bacteria and the Expanding Applications in Food Industry. *Front. Bioeng. Biotechnol.* **2021**, *9*, 612285. [CrossRef]
23. Chen, J.; Luo, W.; Cheng, L.; Wu, J.; Yu, Y.; Li, L.; Xu, Y. Influence of Cultivar and Turbidity on Physicochemical Properties, Functional Characteristics and Volatile Flavor Substances of Pomelo Juices. *Foods* **2023**, *12*, 1028. [CrossRef]
24. Makkumrai, W.; Huang, Y.; Qiang, X. Comparison of pomelo (*Citrus maxima*) grown in China and Thailand. *Front. Agric. Sci. Eng.* **2021**, *8*, 335–352.
25. Pinto, T.; Vilela, A.; Cosme, F. Chemical and sensory characteristics of fruit juice and fruit fermented beverages and their consumer acceptance. *Beverages* **2022**, *8*, 33. [CrossRef]
26. Fernández-Vázquez, R.; Hewson, L.; Fisk, I.; Vila, D.H.; Mira, F.J.H.; Vicario, I.M.; Hort, J. Colour influences sensory perception and liking of orange juice. *Flavour* **2014**, *3*, 1. [CrossRef]
27. Minaker, S.; Mason, R.; Chow, D. Optimizing color performance of the ngenuity 3-dimensional visualization system. *Ophthalmol. Sci.* **2021**, *1*, 100054. [CrossRef] [PubMed]
28. Anh, T.T.M.; Nguyen, T.B.; Duc, V.N.; Bujna, E.; Mai, D.; Nguyen, Q.D. Changes in bitterness, antioxidant activity and total phenolic content of grapefruit juice fermented by *Lactobacillus* and *Bifidobacterium* strains. *Acta Aliment.* **2020**, *49*, 103–110.
29. Zhao, D.; Shah, N.P. Changes in antioxidant capacity, isoflavone profile, phenolic and vitamin contents in soymilk during extended fermentation. *LWT-Food Sci. Technol.* **2014**, *58*, 454–462. [CrossRef]
30. Wu, C.; Li, T.; Qi, J.; Jiang, T.; Xu, H.; Lei, H. Effects of lactic acid fermentation-based biotransformation on phenolic profiles, antioxidant capacity and flavor volatiles of apple juice. *LWT-Food Sci. Technol.* **2020**, *122*, 109064. [CrossRef]
31. Šalić, A.; Šamec, D. Changes in the content of glucosinolates, polyphenols and carotenoids during lactic-acid fermentation of cruciferous vegetables: A mini review. *Food Chem. X* **2022**, *16*, 100457. [CrossRef] [PubMed]
32. Wdowiak, K.; Walkowiak, J.; Pietrzak, R.; Bazan-Woźniak, A.; Cielecka-Piontek, J. Bioavailability of hesperidin and its aglycone hesperetin-compounds found in citrus fruits as a parameter conditioning the pro-health potential (neuroprotective and antidiabetic activity)-mini-review. *Nutrients* **2022**, *14*, 2647. [CrossRef]
33. Addi, M.; Elbouzidi, A.; Abid, M.; Tungmunnithum, D.; Elamrani, A.; Hano, C. An overview of bioactive flavonoids from citrus fruits. *Appl. Sci.* **2022**, *12*, 29. [CrossRef]
34. Rosales, C.; Suwonsichon, S. Sensory lexicon of pomelo fruit over various cultivars and fresh-cut storage. *J. Sens. Stud.* **2015**, *30*, 12133. [CrossRef]
35. Naumann, S.; Haller, D.; Eisner, P.; Schweiggert-Weisz, U. Mechanisms of interactions between bile acids and plant compounds—A review. *Int. J. Mol. Sci.* **2020**, *21*, 6495. [CrossRef] [PubMed]
36. Oboh, G.; Bello, F.; Ademosun, A. Hypocholesterolemic properties of grapefruit (*Citrus paradisii*) and shaddock (*Citrus maxima*) juices and inhibition of angiotensin-1-converting enzyme activity. *J. Food Drug Anal.* **2014**, *4*, 477–484. [CrossRef] [PubMed]
37. Tomaro-Duchesneau, C.; Jones, M.L.; Shah, D.; Jain, P.; Saha, S.; Prakash, S. Cholesterol assimilation by *Lactobacillus* probiotic bacteria: An in vitro investigation. *BioMed Res. Int.* **2014**, *2014*, 380316. [CrossRef]
38. Buchholz, T.; Melzig, M.F. Polyphenolic compounds as pancreatic lipase inhibitors. *Planta Med.* **2015**, *81*, 771–783. [CrossRef]
39. Plaza-Díaz, J.; Ruiz-Ojeda, F.J.; Gil-Campos, M.; Gil, Á. Mechanisms of action of probiotics. *Adv. Nutr.* **2019**, *10*, S49–S66. [CrossRef]
40. Cheong, M.; Shao, Q.L.; Weibiao, Z.; Philip, C.; Bin, Y. Chemical Composition and Sensory Profile of Pomelo (*Citrus grandis* (L.) Osbeck) Juice. *Food Chem.* **2012**, *135*, 2505–2513. [CrossRef]
41. González-Mas, M.C.; Rambla, J.L.; Alamar, M.C.; Gutiérrez, A.; Granell, A. Comparative Analysis of the Volatile Fraction of Fruit Juice from Different Citrus Species. *PLoS ONE* **2011**, *6*, e22016. [CrossRef]
42. Mandha, J.; Shumoy, H.; Devaere, J.; Schouteten, J.J.; Gellynck, X.; De Winne, A.; Matemu, A.O.; Raes, K. Effect of lactic acid fermentation on volatile compounds and sensory characteristics of mango (*Mangifera indica*) juices. *Foods* **2022**, *11*, 383. [CrossRef]

43. Park, S.K.; Jo, D.M.; Yu, D.; Khan, F.; Lee, Y.B.; Kim, Y.M. Reduction of trimethylamine off-odor by lactic acid bacteria isolated from korean traditional fermented food and their in-situ application. *J. Microbiol. Biotechnol.* **2020**, *30*, 1510–1516. [CrossRef] [PubMed]
44. Strani, L.; D'Alessandro, A.; Ballestrieri, D.; Durante, C.; Cocchi, M. Fast Gc E-Nose and Chemometrics for the Rapid Assessment of Basil Aroma. *Chemosensors* **2022**, *10*, 105. [CrossRef]
45. Wu, X.; Fauconnier, M.L.; Bi, J. Characterization and Discrimination of Apples by Flash Gc E-Nose: Geographical Regions and Botanical Origins Studies in China. *Foods* **2022**, *11*, 1631. [CrossRef] [PubMed]

foods

MDPI

Article

The Effect of Co-Culture with Different *Pichia kluyveri* and *Saccharomyces cerevisiae* on Volatile Compound and Characteristic Fingerprints of Mulberry Wine

Bo Ding [1], Shutian Zhao [1], Wenxue Zhang [1,2,*], Ying Lin [1] and Ling Xiong [1]

[1] College of Biomass Science and Engineering, Sichuan University, Chengdu 610065, China; kelpbo@163.com (B.D.); zhaoshutian@stu.scu.edu.cn (S.Z.); 2022223080058@stu.scu.edu.cn (Y.L.); 2022223080061@stu.scu.edu.cn (L.X.)

[2] School of Liquor-Brewing Engineering, Sichuan University of Jinjiang College, Meishan 620860, China

* Correspondence: zhangwenxue@scu.edu.cn

Abstract: In this study, changes in volatile compounds co-fermented by different *Pichia kluyveri* with *Saccharomyces cerevisiae* were analyzed using GC-IMS and compared with *S. cerevisiae* fermentation, to investigate the production of aroma in mulberry wine during the fermentation process. A total of 61 compounds were accurately identified, including 21 esters, 10 alcohols, 8 aldehydes, 6 ketones, and 19 other volatiles. Compared with the single strain fermentation (*S. cerevisiae*), the content of 2-methylpropyl acetate, allyl Isothiocyanate, ethyl crotonate, isobutyl propanoate, and butyl 2-methylbutanoate, co-fermentation groups (*S. cerevisiae* with different *P. kluyveri*) showed a significant decrease. Alcohols, aldehydes, ketones, and organic acid were lower in both the F(S-P1) and F(S-P2) groups than in the F(S) group throughout fermentation. The 2-methylpentanoic acid only was contained in the F(S) group. The co-fermentation with different *P. kluyveri* could also be well distinguished. The content of Benzaldehyde and 4-methylphenol in the F(S-P1) group was significantly lower than that in the F(S-P2) group. The PCA results revealed effective differentiation of mulberry wine fermented by different fermentation strains from GC-IMS. The result showed that *P. kluyveri* could establish a new flavor system for mulberry wine, which plays a crucial role in enhancing the flavor of fruit wine.

Keywords: mulberry wine; *S. cerevisiae*; *P. kluyveri*; co-fermentation; GC-IMS

1. Introduction

Mulberry fruit (*Mores alba* L.) is one of the most popular fruits of the mulberry genus for their nutritional value, flavor, and ability to improve health [1,2]. Mulberry fruit is rich in nutrients including fatty acids, amino acids, vitamins, minerals, and bioactive compounds such as anthocyanins, rutin, quercetin, chlorogenic acid, and polysaccharides [3]. Due to the abundance of natural compounds it contains, previous studies have shown that mulberry fruit has anti-inflammatory, antioxidant, antibiotic, and hepatoprotective properties [4].

However, the storage and transport performance of fresh mulberries has not been satisfactory [5]. Plenty of mulberries are used for fresh consumption and primary processing, such as making juice, mulberry jam, and dried fruit [6]. To enrich mulberry products and adjust consumer needs, fruit wine, as a good value-added product, has gained increasing attention in recent years in China. Mulberry wine is one of the most popular fruit wines consumed worldwide [7,8]. As a traditional fermented alcoholic beverage with substantial territorial and socio-cultural significance, wine has a long history that is patterned by a tradition of innovation [9]. Presently, fruit wine, a different fruit-based fermented alcoholic beverage, has been accepted by people because of its low alcohol content and special flavor characteristics.

Generally, traditional fruit winemaking often involves fermenting a single yeast [10]. The *Saccharomyces cerevisiae* is a traditional ferment used to produce fruit wine because of its strong alcohol fermentation capacity [11]. However, single strain fermentation could make for a similar flavor and less characteristic profiles of the products. In recent years, non-*Saccharomyces* have received increasing attention. A cohort of scholars alternated to traditional fermentation performed by co-fermentation of *S. cerevisiae* and non-*Saccharomyces*, which lead to significant changes in the chemical composition and flavor characteristics of fruit wine [12]. Among the non-*Saccharomyces* yeasts, *Pichia kluyveri* has been considered a potential yeast in fruit wine processing because of its excellent aroma production ability. It has been reported that *P. kluyveri* increases the production of glycerol and esters [13]. And *P. kluyveri* could contribute to the production of aroma compounds, such as 2-phenylethyl acetate and 3-sulfanylhexan-1-ol acetate [14,15]. In addition, *Pichia fermentans*, either dead or alive, can make a positive contribution to promote the growth of *Oenococcus oeni*. And the co-fermentation with *Pichia fermentans*, improved the production of yeast-derived and grape-derived volatiles, which enhances the fresh fruity aroma of wines [16]. Therefore, it is believed that co-fermentation with *S. cerevisiae* and *P. kluyveri* could be a promising brewing method to introduce new flavors to mulberry wines. We intended to ferment mulberry juice by co-culturing of *S. cerevisiae* and *P. kluyveri*, thereby comparing the flavor differences of mulberry wine between single and co-culture fermentation.

For fruit wine, volatile compounds of flavor play a critical role in wine quality, which directly affects the quality and grade of wine. Typically, GC-MS (gas chromatography-mass spectrometry) can separate and identify volatile compounds from complex samples relatively accurately for the purpose of evaluating the flavor of products. However, the detection performance of GC-MS on low molecular weight (below 300 daltons) compounds is not ideal. Small molecules (C2–C10) with a low content of volatile components below 300 daltons, such as some ketones, acids, furans, and other compounds, can be key components in the modification of flavor. Previous studies have shown that these substances can be characteristic of different types of wine aroma. By monitoring these substances, the fermentation process can be precisely controlled. This will improve the quality of the product [17,18].

At present, GC-IMS (gas chromatography-ion mobility spectrometry) has a good detection effect on such small molecule volatile substances [19]. Ion mobility spectrometry (IMS) is an analytical technique to detect trace gases and to characterize chemical ionic substances based on the difference in the rate of migration of gas phase ions in an electric field [20]. The advantages of gas chromatography ion mobility spectrometry (GC-IMS) is that it requires no sample pretreatment, enables fast analysis timeframes, and has a low detection limit [21], which is a well-established technology that has been widely used in airport security, military security, and drug detection fields nowadays [22]. Especially, due to its lower testing and incubating temperature requirements, GC-IMS has promising prospects for food flavor analysis as it does not damage the original compounds of the food [23]. Previous research has confirmed that IMS is an effective method for the analysis and characterization of volatile compounds with different properties [24,25]. Many studies have also confirmed that GC-IMS technology minimizes loss of flavor substances, which is suitable for food volatile components analysis during storage [26] and processing [27].

Therefore, the aim of this study was to ferment mulberry juice by co-culturing *S. cerevisiae* and *P. kluyveri*, and monitor the release of VOCs in mulberry wine during fermentation; then, comparatively analyze the differences of the VOCs between single and co-culture fermentation; and finally, establish the characteristic volatile fingerprints of the VOCs in mulberry wine. On that basis, we can appraise the aroma influence of marker volatile composition in co-fermentation by difference strains so as to understand the immense potential of *P. kluyveri* in applications. At the same time, the creation of fingerprints of flavor compounds at different stages of fermentation will also contribute to the targeted regulation of flavor during the fermentation process.

2. Materials and Methods

2.1. Material

Healthy and ripe mulberries of "Yun No. 2" at commercial maturity were provided by Sixi Agricultural Development Co., Ltd. (Panzhihua, China). The strains *Saccharomyces cerevisia* (S), *Pichia kluyveri* (P1), and *Pichia kluyveri* (P2) were screened from the mulberry orchard (Panzhihua, China).

2.2. Mulberry Juice Fermentation

Squeezed and standardized mulberry juice (Brix, 25 °Bx). The strains *S. cerevisia* (S), *P. kluyveri* (P1), and *P. klüyveri* (P2) were separately pre-cultured in YPD medium at 28 °C with 170 rpm. After 40 h incubation, the yeast cells were separately collected, washed twice, and then resuspended in the 30 mL sterile saline solution. Finally, approximately 10^6 CFU/mL of each yeast strain was transferred into the 500 mL of pasteurized mulberry juice. Fermentation tests were set up as follows: (1) F(S): single inoculation with *S. cerevisiae*, (2) F(S-P1): co-inoculation of *S. cerevisiae* (S) and *P. kluyveri* (P1), (3) F(S-P2): co-inoculation of *S. cerevisiae* (S) and *P. kluyveri* (P2), and the inoculum ratio was 1:1 [10]. We set the natural fermentation group as the control (CK). The control group and each fermentation broth were fermented at 16 °C for 18 days. Then, the samples were injected into the centrifuge tube and stored at −80 °C in frozen storage every 6 days (0, 6, 12, 18 days) to prepare for the subsequent analyses.

2.3. Methods

All samples were analyzed for the presence of volatile compounds using a gas chromatography-ion mobility spectrometry (GC-IMS) instrument (FlavourSpec®, Gesellschaft für Analytische Sensorsysteme mbH, Dortmund, Germany), which connected with a MXT-WAX capillary column (30 m × 0.53 mm × 1.00 um, RESTEK, Bellefonte, PA, USA). The n-ketones C4–C9 (Sinopharm Chemical Reagent Beijing Co., Ltd., Beijing, China) were used as external references to calculate the retention index (RI) of volatile compounds. The VOCs were identified by comparing the RI and the drift time (in milliseconds) of the standard in the GC-IMS library.

A total of 3 mL of mulberry wine was absorbed, then placed into a 20 mL headspace vial. Subsequently, samples were incubated at 45 °C with 500 rpm for 10 min. After incubation, 500 μL of volatile composition was automatically injected into the injector by means of a heated syringe at 50 °C with splitless mode. The oven temperature was set as 60 °C with isothermal conditions. The samples were driven into the column by nitrogen at a programmed flow as follows: 2 mL/min for 10 min, 15 mL/min for 5 min, 50 mL/min for 5 min, 100 mL/min for 5 min. Then, they were ionized in the IMS ionization chamber of 45 °C. The drift gas flow was set at a constant flow of 150 mL/min.

All analyses were performed in triplicate. Identification of volatiles were identified by comparing retention indices (RI) and the drift time (in milliseconds) of standard in the GC-IMS library [19].

2.4. Statistical Analysis

The instrumental analysis software LAV version 2.0.0, G.A.S. (Laboratory Analytical Viewer) and three attached plug-ins were used for sample analysis. Particularly, the program GC-IMS Library Search was applied to analyze the characteristic flavor substances. Then, the Reporter and Gallery plug-ins (G.A.S., Dortmund, Germany) were used to characteristic fingerprint the volatile flavor compounds, and the multivariate data analysis (PCA) was performed using the Dynamic PCA plug-in (G.A.S., Dortmund, Germany). Correlation analysis among volatile compounds was analyzed using Origin 2023 b software (Origin-Lab Corporation, Northampton, MA, USA).

3. Results and Discussion

3.1. GC–IMS Topographic Plots in Mulberry Wine during Fermentation

The differences of volatile compounds in mulberry wine samples during the fermentation process were analyzed by GC-IMS. As shown in Figure 1, the GC-IMS analysis of sample volatiles resulted in a 3D-topographic plot, where *X*, *Y*, and *Z*-axes, respectively, represent the ion migration time (DT) for identification, the retention time (RT) of the gas chromatograph, and the peak intensity for quantification. The 2D-topographic spectra of volatile compounds during fermentation with treatment samples are shown in Figure 2.

In Figures 1 and 2, the overall blue background was the GC-IMS spectra, and the remarkable red vertical line at abscissa 1.0 was the reactive ion peak (RIP) after normalization. Each heterochromatic or homochromatic point on both sides of the RIP line represented a volatile compound, which indicated the concentration of compound by different levels of color saturation. In general, light color represents lower concentration and redness reflects higher concentration, which means, the darker color indicates the higher concentration of the volatile compound [28]. The effect of fermentation with different strains on the volatile compounds is a significant difference. And there are also varying degrees of differences in peak intensity.

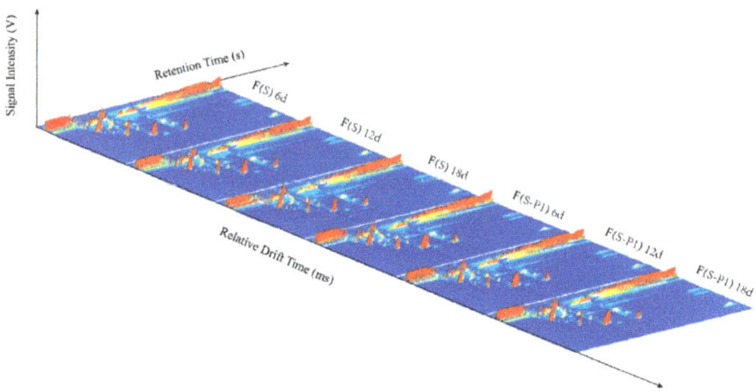

(A)

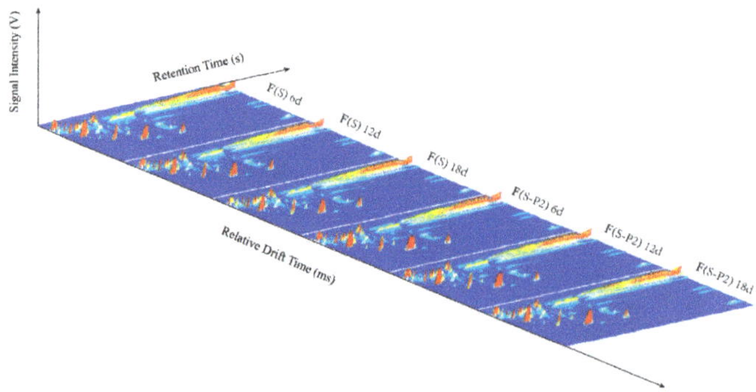

(B)

Figure 1. *Cont.*

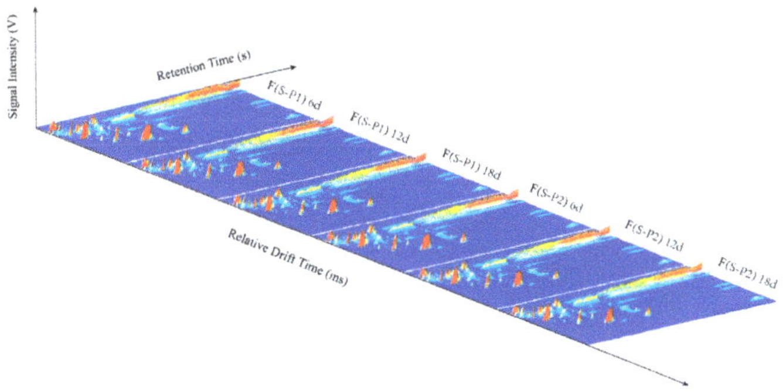

(C)

Figure 1. 3D-topographic plots in different mulberry wine samples during fermentation. (**A**) group F(S); (**B**) group F(S–P1); (**C**) group F(S–P2).

In Figure 2, all the mulberry wine samples are rich in volatile compounds. It can be seen that most of the peak signals appeared in the retention time of 100–900 s, and a few signals appeared in the retention time of 900–1400 s. The peak signals were observed in ranges of drift time of 1.0–2.0 ms.

The difference comparison model was applied to compare the aroma differences among the samples, conveniently. The topographical plot of F(S) 6d was taken as the reference and the topographical plots of other samples (including F(S) 12d, F(S) 18d, F(S-P1) 6d, F(S-P1) 12d, and F(S-P1) 18d) were deducted from the reference (Figure 3A). For each dot after deduction appearing in the figure, the red color means a higher concentration of a volatile compound than the reference, blue color indicates a lower concentration of a volatile compound than the reference, and the white dot implies the volatile compounds are consistent with the reference. Similarly, Figure 3B also displayed on this principle, and F(S) 6d was taken as the reference as well. As shown in Figure 3, significant differences between F(S) and F(S-P1) groups were observed, which indicated the difference in molecule structure and concentration of volatile compounds [29]. This appearance should be caused by the participation of non-*Saccharomyces* (P1 and P2). Notably, there were also significant differences among the fermentation broth in different periods. As shown in Figure 3, some volatiles showed significant color differences in the spectra at different stages of fermentation. It is able to show the changes in the content of the substance as the fermentation process takes place.

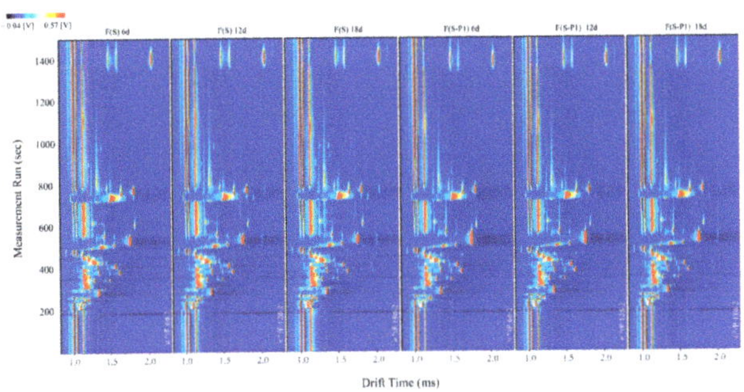

(A)

Figure 2. *Cont.*

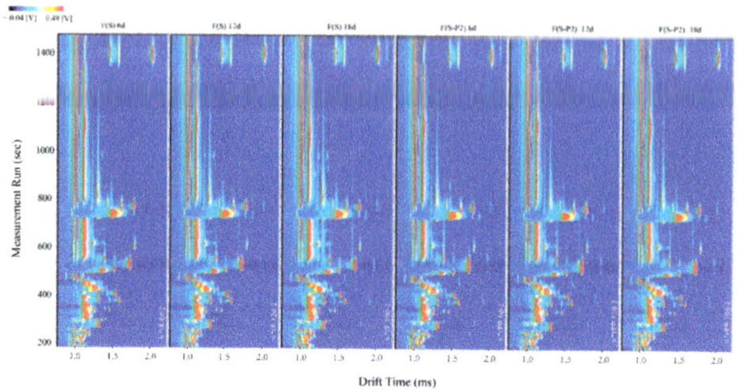

(**B**)

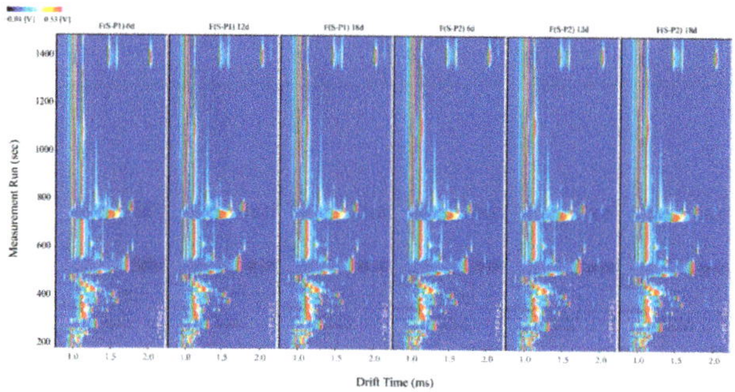

(**C**)

Figure 2. 2D-topographic plots in different mulberry wine samples during fermentation. (**A**) group F(S) vs. group F(S-P1); (**B**) group F(S) vs. group F(S-P2); (**C**) group F(S-P1) vs. group F(S-P2).

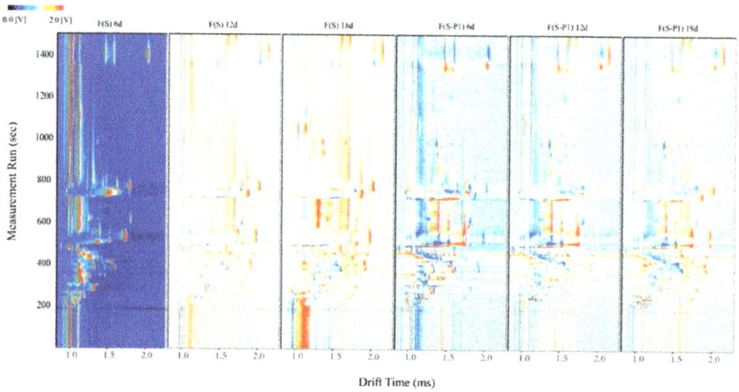

(**A**)

Figure 3. *Cont.*

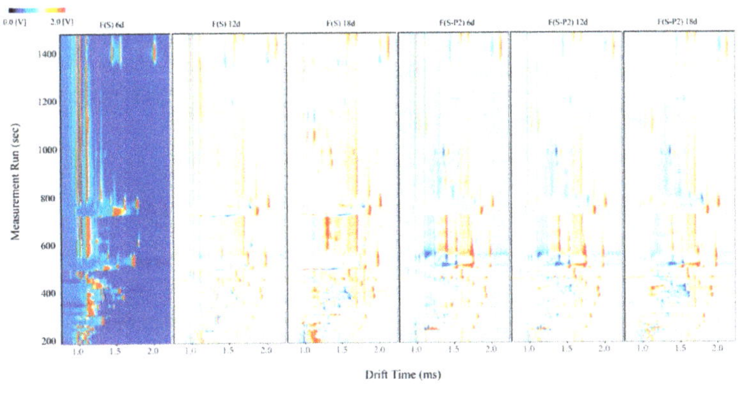

(B)

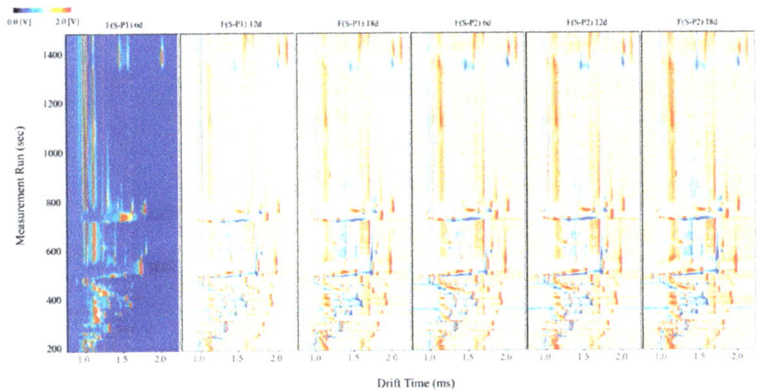

(C)

Figure 3. The comparison topographic plots in different mulberry wine samples during fermentation. (**A**) group F(S) vs. group F(S-P1); (**B**) group F(S) vs. group F(S-P2); (**C**) group F(S-P1) vs. group F(S-P2).

3.2. Fingerprint Analysis of Volatile Compounds during the Fermentation Process of Mulberry Wine with Different Strains

To further identify the differences of volatile compound profiles in different mulberry wines during fermentation, all information provided to qualitatively characterize by the fingerprint analysis technique. It means to give an overall profile of the aroma composition, not only based on the identification of one or two volatile compounds [30]. Therefore, the Gallery Plot plug-in could be used for previewing the volatile fingerprints by showing the color change trend of each point in the 2D plot, which included all volatile substances identified by the GC-IMS instrument (Figure 4). As shown in Figure 4, Gallery Plot listed volatile components together for intuitive comparison, all bright points of each row represented the entire signal peak of a sample, and each column showed the signal strength of the same substance detected in different periods of fermentation.

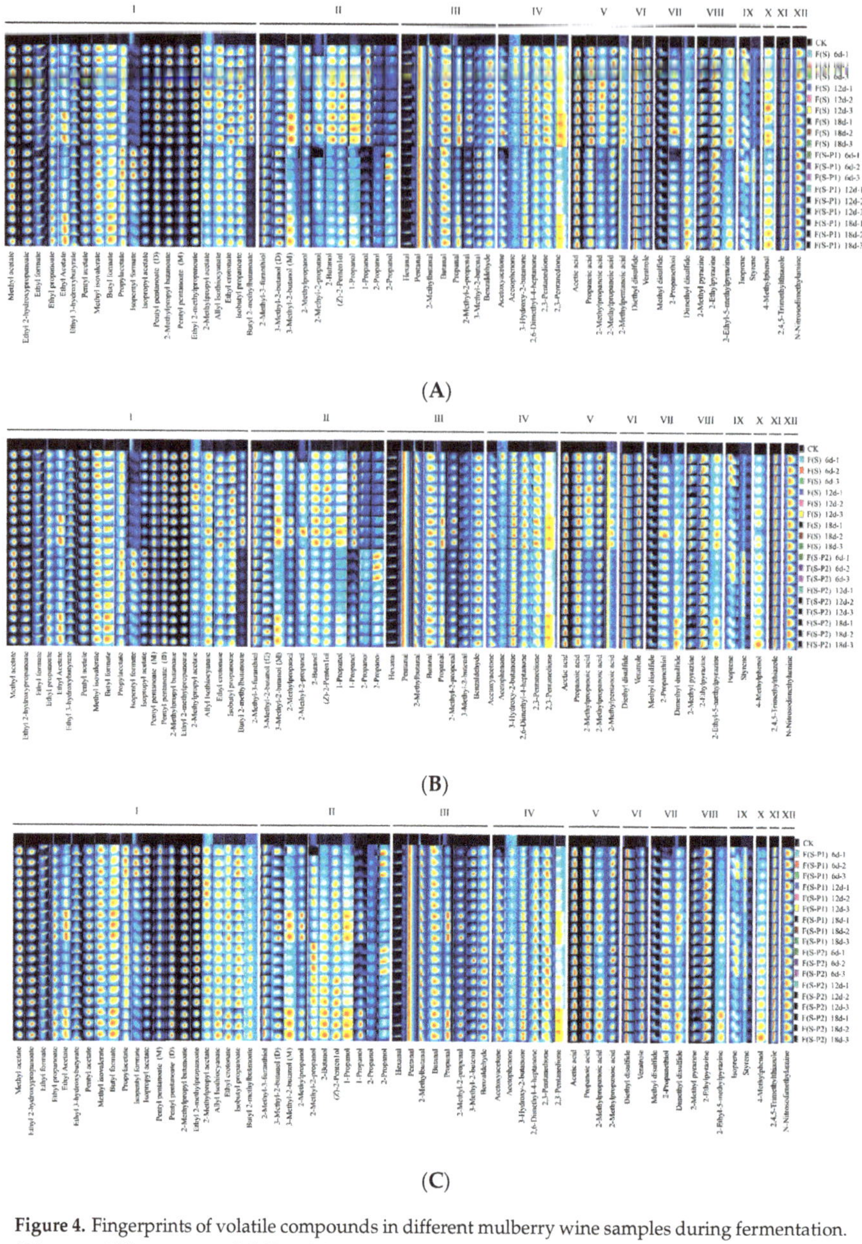

Figure 4. Fingerprints of volatile compounds in different mulberry wine samples during fermentation. (**A**) group F(S) vs. group F(S-P1); (**B**) group F(S) vs. group F(S-P2); (**C**) group F(S-P1) vs. group F(S-P2). I refers to esters, II refers to alcohols, III refers to aldehydes, IV refers to ketones, V refers to acids, VI refers to ethers, VII refers to sulpho compounds, VIII refers to pyrazine, IX refers to terpenes, X refers to phenol, XI refers to thiazole, and XII refers to nitrogen compound.

IMS drift times and retention indices were compared with authentic reference compounds to characterize these compounds. In Figure 4, generally, a total of 64 flavor substance signals were measured. Among these peaks, 61 compounds were accurately identified (based on the GC-IMS library, NIST database). Among these volatiles, as shown in

Table 1, included 21 esters, 10 alcohols, 8 aldehydes, 6 ketones, and 19 other volatiles such as terpenes and acids (Table 1). It should be noted that there were 3 compounds, which had monomer (M) and dimer (D) form, produced two peak signals by one volatile (Figure 4 and Table 1).

Table 1. GC–IMS integration parameters of volatile compounds obtained from three different mulberry wine samples.

No.	Compound	CAS	Molecule Formula	MW [a]	RI [b]	RT [c] [s]	DT [d] [ms]	Comment
I Esters								
1	Methyl acetate	79-20-9	$C_3H_6O_2$	74.1	545	250.986	1.2007	
2	Ethyl 2-hydroxYPPropanoate	97-64-3	$C_5H_{10}O_3$	118.1	811.2	505.891	1.552	
3	Ethyl formate	109-94-4	$C_3H_6O_2$	74.1	622.9	303.678	1.2087	
4	Ethyl propanoate	105-37-3	$C_5H_{10}O_2$	102.1	715.7	387.553	1.1441	
5	Ethyl Acetate	141-78-6	$C_4H_8O_2$	88.1	562	261.299	1.0895	
6	Ethyl 3-hydroxybutanoate	5405-41-4	$C_6H_{12}O_3$	132.2	950.1	762.614	1.6458	
7	Pentyl acetate	628-63-7	$C_7H_{14}O_2$	130.2	932.1	722.363	1.3195	
8	Methyl isovalerate	556-24-1	$C_6H_{12}O_2$	116.2	767	446.34	1.2081	
9	Butyl formate	592-84-7	$C_5H_{10}O_2$	102.1	746.3	421.426	1.2121	
10	Propylacetate	109-60-4	$C_5H_{10}O_2$	102.1	717.3	389.246	1.4832	
11	Isopentyl formate	110-45-2	$C_6H_{12}O_2$	116.2	775.9	457.729	1.6333	
12	Isopropyl acetate	108-21-4	$C_5H_{10}O_2$	102.1	670.6	343.527	1.489	
13	Pentyl pentanoate	2173-56-0	$C_{10}H_{20}O_2$	172.3	1146.9	1407.006	1.4827	Monomer
14	Pentyl pentanoate	2173-56-0	$C_{10}H_{20}O_2$	172.3	1145.7	1401.398	2.0471	Dimer
15	2-methylpropyl butanoate	539-90-2	$C_8H_{16}O_2$	144.2	954.5	773.001	1.8113	
16	Ethyl 2-methylpropanoate	97-62-1	$C_6H_{12}O_2$	116.2	736.4	410.037	1.5711	
17	2-Methylpropyl acetate	110-19-0	$C_6H_{12}O_2$	116.2	745.4	420.316	1.6156	
18	Allyl Isothiocyanate	1957/6/7	C_4H_5NS	99.2	878.7	615.893	1.3861	
19	Ethyl crotonate	623-70-1	$C_6H_{10}O_2$	114.1	846.4	560.062	1.5493	
20	Isobutyl propanoate	540-42-1	$C_7H_{14}O_2$	130.2	842.2	553.34	1.7146	
21	butyl 2-methylbutanoate	15706-73-7	$C_9H_{18}O_2$	158.2	1032	980.844	1.385	
II Alcohols								
22	2-Methyl-3-furanthiol	28588-74-1	C_5H_6OS	114.2	861.1	584.731	1.1402	
23	3-Methyl-2-butanol	598-75-4	$C_5H_{12}O$	88.1	677.8	350.056	1.238	Monomer
24	3-Methyl-2-butanol	598-75-4	$C_5H_{12}O$	88.1	673.3	345.996	1.4334	Dimer
25	2-methylpropanol	78-83-1	$C_4H_{10}O$	74.1	616	298.43	1.3799	
26	2-Methyl-2-propanol	75-65-0	$C_4H_{10}O$	74.1	509.6	231.177	1.1317	
27	2-Butanol	78-92-2	$C_4H_{10}O$	74.1	610.1	294.048	1.1509	
28	(Z)-2-Penten1ol	1576-95-0	$C_5H_{10}O$	86.1	751.2	427.121	1.4368	
29	1-Propanol	71-23-8	C_3H_8O	60.1	531.3	243.032	1.1178	
30	1-Propanol	71-23-8	C_3H_8O	60.1	558.9	259.421	1.1108	
31	2-Propanol	67-63-0	C_3H_8O	60.1	531.6	243.228	1.1779	Monomer
32	2-Propanol	67-63-0	C_3H_8O	60.1	536.4	245.931	1.2186	Dimer
III Aldehydes								
33	Hexanal	66-25-1	$C_6H_{12}O$	100.2	805.1	497.264	1.2544	
34	Pentanal	110-62-3	$C_5H_{10}O$	86.1	689.2	360.844	1.1745	
35	2-Methylbutanal	96-17-3	$C_5H_{10}O$	86.1	678.4	350.696	1.4006	
36	Butanal	123-72-8	C_4H_8O	72.1	610.1	294.048	1.1133	
37	Propanal	123-38-6	C_3H_6O	58.1	468.6	211.079	1.1393	
38	2-Methyl-2-propenal	78-85-3	C_4H_6O	70.1	572.1	267.793	1.2092	
39	3-Methyl-2-butenal	107-86-8	C_5H_8O	84.1	775.9	457.729	1.3505	
40	Benzaldehyde	100-52-7	C_7H_6O	106.1	960.5	787.283	1.4573	
IV Ketones								
41	Acetoxyacetone	592-20-1	$C_5H_8O_3$	116.1	488.5	220.495	1.0945	
42	Acetophenone	98-86-2	C_8H_8O	120.2	1095.1	1195.186	1.5748	
43	3-Hydroxy-2-butanone	513-86-0	$C_4H_8O_2$	88.1	737	410.749	1.3244	
44	2,6-Dimethyl-4-heptanone	108-83-8	$C_9H_{18}O$	142.2	972.2	815.848	1.3218	
45	2,3-Pentanedione	600-14-6	$C_5H_8O_2$	100.1	689.8	361.451	1.2115	
46	2,3-Pentanedione	600-14-6	$C_5H_8O_2$	100.1	710.9	382.528	1.217	

Table 1. *Cont.*

No	Compound	CAS	Molecule Formula	MW [a]	RI [b]	RT [c] [s]	DT [d] [ms]	Comment
V Acids								
47	Acetic acid	64-19-7	$C_2H_4O_2$	60.1	600.8	287.345	1.341	
48	Propanoic acid	79-09-4	$C_3H_6O_2$	74.1	711.4	382.987	1.3465	
49	2-Methylpropanoic acid	79-31-2	$C_4H_8O_2$	88.1	769.8	449.899	1.3746	
50	2-Methylpropanoic acid	79-31-2	$C_4H_8O_2$	88.1	746.9	422.138	1.3906	
51	2-Methylpentanoic acid	97-61-0	$C_6H_{12}O_2$	116.2	1031.6	979.77	1.2668	
VI Ethers								
52	Diethyl disulfide	110-81-6	$C_4H_{10}S_2$	122.2	916.8	689.903	1.1448	
53	Veratrole	91-16-7	$C_8H_{10}O_2$	138.2	1146.3	1404.202	1.5962	
VII Sulpho compound								
54	Methyl disulfide	624-92-0	$C_2H_6S_2$	94.2	771.5	452.035	1.1459	
55	2-Propanethiol	75-33-2	C_3H_8S	76.2	562.5	261.649	1.1434	
56	Dimethyl disulfide	624-92-0	$C_2H_6S_2$	94.2	741.7	416.084	1.1397	
VIII Pyrazine								
57	2-Methyl pyrazine	109-08-0	$C_5H_6N_2$	94.1	808.2	501.576	1.3943	
58	2-Ethylpyrazine	13925-00-3	$C_6H_8N_2$	108.1	938.5	736.295	1.5252	
59	2-Ethyl-5-methylpyrazine	13360-64-0	$C_7H_{10}N_2$	122.2	1004.9	901.848	1.2119	
IX Terpenes								
60	Isoprene	78-79-5	C_5H_8	68.1	524.7	239.328	1.2107	
61	Styrene	100-42-5	C_8H_8	104.2	905.9	667.742	1.431	
X Phenol								
62	4-Methylphenol	106-44-5	C_7H_8O	108.1	1069.9	1104.14	1.142	
XI Thiazole								
63	2,4,5-Trimethylthiazole	13623-11-5	C_6H_9NS	127.2	1002.6	895.425	1.1343	
XII Nitrogen compound								
64	N-Nitrosodimethylamine	62-75-9	$C_2H_6N_2O$	74.1	755.5	432.25	1.2705	

[a] MW: molecular mass; [b] RI: retention index; [c] RT: retention time; [d] Dt: drift time.

The content of some ester substances in all experimental groups was relatively high in the early stage of fermentation, but showed an overall decreasing trend as fermentation progressed. That was probably because the rapid proliferation of yeast during fermentation contributes to the formation of esters [28,31]. Compared to the F(S) group, the F(S-P1) and F(S-P2) groups showed a significant decrease in this study. However, no significant difference was found between the F(S-P1) and F(S-P2) groups. As shown in Figure 4A, there were five identified ester substances, including 2-methylpropyl acetate, allyl Isothiocyanate, ethyl crotonate, isobutyl propanoate, and butyl 2-methylbutanoate. The F(S-P1) group substances were lower than those of the F(S) group during fermentation. It was also observed that the incorporation of *Pichia* (P1 and P2) had a greater effect on the formation of small molecule alcohols (Figure 4A,B). High levels of alcohol were observed in the F(S) group at different stages of fermentation. Alcohols were lower in both the F(S-P1) and F(S-P2) groups than in the F(S) group throughout fermentation. In this study, the levels of ester substances such as butyl 2-methylbutytanoate were negatively correlated with alcoholic substances. This phenomenon might be due to the fact that *Pichia* could significantly enhance the enzymatic reaction between higher alcohols and acetyl-CoA, thereby producing more esters [32]. Overall, the results of this study suggested that small molecule alcohols may also influence the formation of the corresponding esters. This was in line with the previous reports [28].

There were relatively few types of aldehydes and ketones in this study. Two types of carbonyl compounds gradually accumulated during the fermentation process. The F(S-P1) and F(S-P2) groups were significantly lower than the F(S) group. Benzaldehyde had a strong bitter almond and nutty aroma with a slight fruity sweetness [33]. There was no significant difference in the content of benzaldehyde between the F(S) group and the F(S-P2) group (Figure 4B), but the production of the F(S P1) group was lower than that of the other groups throughout the fermentation process (Figure 4A,C). The results suggested that during the process of metabolism of *Pichia* (P1), the yield of benzaldehyde was consistently lower than that of the other two types of strain. In addition, a certain amount of benzaldehyde was produced during the metabolism of strain *S. cerevisiae* (S) itself (Figure 4A,B). The lower overall levels in the F(S-P1) group could also be due to the fact that the process of metabolism of *P. kluyveri* (P1) consumes a large amount of benzaldehyde, resulting in the absence of corresponding odor.

The organic acid substances F(S-P1) and F(S-P2) groups were significantly lower than the F(S) group, and there was no significant difference between the F(S-P1) and F(S-P2) groups. This could prevent the wine from developing an overly sharp acidic taste. But improper control could also make the wine taste monotonous. As shown in Figure 4C, most of the acids accumulated gradually with the fermentation process, while 2-methylpropanoic acid decreased gradually with the fermentation process. The content of 2-methylpropanoic acid in the F(S) group was consistently higher than that in the F(S-P1) and F(S-P2) groups. And the F(S-P1) group showed a significant decrease in content after 6 days, whereas the F(S-P2) group showed the same situation after 12 days. Notably, 2-methylpentanoic acid appeared only in the F(S) group and accumulated with fermentation. It did not appear in the F(S-P1) and F(S-P2) groups. It should be noted that the F(S) group has a distinct grassy taste compared to the other two groups in this study. The 2-methylpentanoic acid was probably partly responsible for this significant difference. This result is also consistent with previous reports [34].

The content of other substances, including ethers, pyrazines and others, was not significantly different among the experimental groups, and there was little change during the fermentation process. These substances were mostly formed in the early stages of fermentation and their levels did not change significantly as the fermentation process progresses (Figure 4A–C). It is noteworthy that both the F(S-P1) and the F(S-P2) groups showed a certain amount of dimethyl disulfide at the late stage of the fermentation process. Sulfur-containing substances, such as dimethyl disulfide, have an odor of their own. It would bring the special bouquet that the sulfur-containing substances had the right amount. However, its high levels could be very detrimental to fruit wines [35].

In addition, the only phenolic substance 4-Methylphenol is described as caramel flavor (threshold, 20 μg/L), which may become a potential hazardous substance [36]. In this study, 4-methylphenol levels remained consistently low in the F(S-P1) group (Figure 4A,C). It was higher in the F(S-P2) group than in the F(S) and F(S-P1) groups (Figure 4B,C). The results indicated that *Pichia* (P1) had a better inhibitory effect on the production of 4-methylphenol.

3.3. PCA Analysis of Different Mulberry Wine during Fermentation

In order to distinguish and present the differences among the different mulberry wine samples during fermentation, the principal component analysis (PCA) technique was used to evaluate the differences. In this study, the signal intensity of different volatile compounds identified by GC-IMS was used to structure the PCA model. The PCA of volatile compounds in different mulberry wine samples during fermentation is presented in Figure 5. The distribution map indicated that the total contribution ratio of the first two principal components was 64.6%, which collected 47.5% (PC 1) and 17.1% (PC 2) of the cumulative variance contribution rate. The PCA results exactly indicated that the fermentation with single and co-culture of yeast would be distinguished in the distribution map. The same sample also showed significant stage differences at different stages of fermentation. As show in Figure 5, all samples of F(S) could be separated by the positive score values of PC1, and the F(S-P1) group could be distinguished according to the negative scores of PC1. On the other hand, the score values of PC2 could be used to separate the different fermentation stages of fruit wine. On this foundation, together with biplot of PCA analysis, the result showed that the non-*Saccharomyces Pichia kluyveri* P1 and P2 could enhance the characteristic flavor of mulberry wine by improving the composition and content of characteristic aroma compounds, leading to the creation of new aromatic mulberry wines. As a result of differential metabolism of non-Saccharomyces, it would provide flavors with different characteristics. This also results in significant differences in wine flavor type.

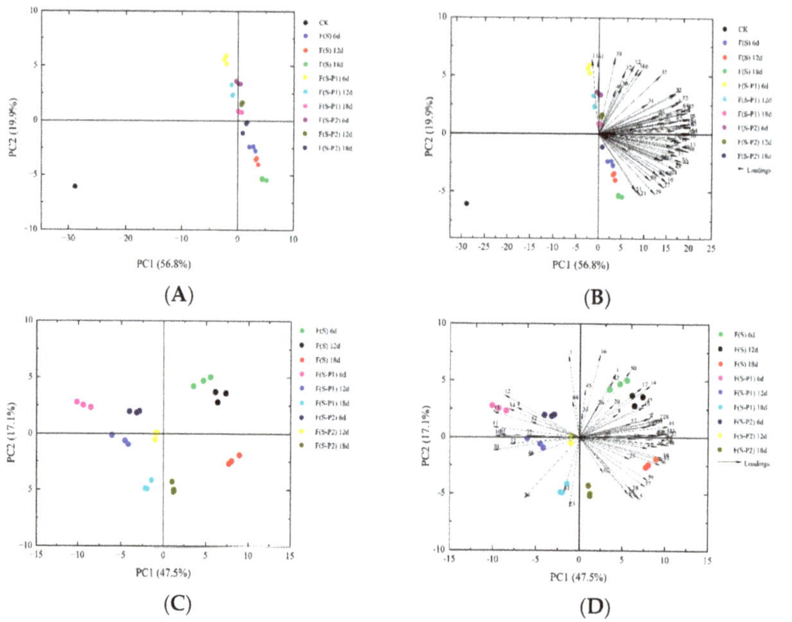

Figure 5. PCA analysis of flavor compounds in mulberry wine with different strains during the fermentation process. (**A**,**B**) The score plot and biplot of PCA analysis (with CK). (**C**,**D**) The score plot and biplot of PCA analysis (without CK).

4. Conclusions

Aroma is one of the most important means of evaluating and identifying a fruit wine. Fruit wines are frequently made with a particular type of *Saccharomyces cerevisiae*, which means that the raw materials significantly impact the flavor of the product. It should be noted that traditional methods are advantageous in molding product flavour, but the extension of fruit wine varieties is significantly restricted. Appropriate application of non-*Saccharomyces* could help fruit wines develop new and more attractive flavors. In this study, a total of 64 signal peaks from topographic plots were identified in different mulberry wine samples. The specificity of *P. kluyveri* (P1, P2) in regulating the production of 2-methylpentanoic acid, benzaldehyde and, conversely, 4-methylphenol is reported for the first time in this article, and deserves further investigation in future research. Additionally, the VOCs fingerprint also demonstrates the process of change in the content of different volatile substances at different stages of fermentation. In addition, the results of PCA showed that the different mulberry wines were well distinguished. Therefore, non-*Saccharomyces* will provide more options for creating fruit wine flavors and has great research potential for producing new types of fruit wine.

By constructing effective fingerprints, key flavor compounds can be identified. This allows rapid differentiation of the flavor characteristics of wines fermented by different non-*Saccharomyces*. This feature can be used to create fingerprint maps during fermentation to monitor changes in the levels of the key flavor compounds. It also means that by regulating the fermentation process based on changes in key flavor compounds, the quality of wine can be improved.

Author Contributions: B.D.: investigation, writing—original draft, formal analysis, visualization. S.Z., L.X. and Y.L.: investigation, resources. W.Z.: project administration, reviewing, funding acquisition. All authors have read and agreed to the published version of the manuscript.

Funding: This research was funded by the Sichuan Provincial Science and Technology Plan Project (2021YFN0012).

Institutional Review Board Statement: Not applicable.

Informed Consent Statement: Not applicable.

Data Availability Statement: Data is contained within the article.

Conflicts of Interest: The author declares no conflict of interest.

References

1. Zhang, H.X.; Ma, Z.F.; Luo, X.Q.; Li, X.L. Effects of Mulberry Fruit (*Morus alba* L.) Consumption on Health Outcomes: A Mini-Review. *Antioxidants* **2018**, *7*, 69. [CrossRef] [PubMed]
2. Sangteerakij, D.; Rodyu, S.; Junden, S. Nutrition composition and anthocyanin content of mulberry fruits (*Mores alba* L.) by different ripening stage grown in the south of Thailand, applicants in sorbet and sherbet ice-cream and analysis of physical, chemical and evaluation. *J. Fac. Sci.* **2023**, *35*, 130–138.
3. Yuan, Q.X.; Zhao, L.Y. The Mulberry (*Morus alba* L.) Fruit-A Review of Characteristic Components and Health Benefits. *J. Agric. Food Chem.* **2017**, *65*, 10383–10394. [CrossRef] [PubMed]
4. Manach, C.; Mazur, A.; Scalbert, A. Polyphenols and prevention of cardiovascular diseases. *Curr. Opin. Lipidol.* **2005**, *16*, 77–84. [CrossRef] [PubMed]
5. Dimitrijevic, D.; Kostic, D.; Paunovic, D.; Mitic, M.; Krstic, J.; Misic, I.R.; Arsic, B. Antimicrobial Activity and The Quantitative Analyses of Phenolic Compounds and Heavy Metals of Red Mulberry Extracts (*Morus rubra* L.) From Serbia. *Stud. Univ. Babes-Bolyai Chem.* **2022**, *67*, 195–207. [CrossRef]
6. Gahukar, R.T. Plant-derived products in crop protection: Effects of various application methods on pests and diseases. *Phytoparasitica* **2016**, *44*, 379–391. [CrossRef]
7. Tao, Y.; Wang, Y.L.; Yang, J.; Wang, Q.; Jiang, N.; Chu, D.T.; Han, Y.B.; Zhou, J.Z. Chemical composition and sensory profiles of mulberry wines as fermented with different Saccharomyces cerevisiae strains. *Int. J. Food Prop.* **2017**, *20*, 2006–2021.
8. Zhang, S.R.; Xing, X.; Chu, Q.; Sun, S.Y.; Wang, P. Impact of co-culture of Lactobacillus plantarum and Oenococcus oeni at different ratios on malolactic fermentation, volatile and sensory characteristics of mulberry wine. *LWT—Food Sci. Technol.* **2022**, *169*, 113995. [CrossRef]
9. Pretorius, I.S. Tasting the terroir of wine yeast innovation. *FEMS Yeast Res.* **2020**, *20*, foz084. [CrossRef]
10. van Wyk, N.; Grossmann, M.; Wendland, J.; von Wallbrunn, C.; Pretorius, I.S. The Whiff of Wine Yeast Innovation: Strategies for Enhancing Aroma Production by Yeast during Wine Fermentation. *J. Agric. Food Chem.* **2019**, *67*, 13496–13505. [CrossRef]
11. Gonzalez, R.; Morales, P. Truth in wine yeast. *Microb. Biotechnol.* **2021**, *15*, 1339–1356. [CrossRef]
12. Wei, J.P.; Zhang, Y.X.; Wang, Y.W.; Ju, H.M.; Niu, C.; Song, Z.H.; Yuan, Y.H.; Yue, T.L. Assessment of chemical composition and sensorial properties of ciders fermented with different non-*Saccharomyces* yeasts in pure and mixed fermentations. *Int. J. Food Microbiol.* **2020**, *318*, 108471. [CrossRef]
13. Gao, M.M.; Hu, J.N.; Wang, X.J.; Zhang, H.Y.; Du, Z.P.; Ma, L.J.; Du, L.P.; Zhang, H.L.; Tian, X.J.; Yang, W.M. Effects of *Pichia kluyveri* on the flavor characteristics of wine by co-fermentation with *Saccharomyces cerevisiae*. *Eur. Food Res. Technol.* **2023**, *249*, 1449–1460. [CrossRef]
14. Dutraive, O.; Benito, S.; Fritsch, S.; Beisert, B.; Patz, C.D.; Rauhut, D. Effect of Sequential Inoculation with Non-*Saccharomyces* and *Saccharomyces* Yeasts on Riesling Wine Chemical Composition. *Fermentation* **2019**, *5*, 79. [CrossRef]
15. Liu, Y.J.; Lu, Y.Y.; Liu, S.Q. The potential of spent coffee grounds hydrolysates fermented with *Torulaspora delbrueckii* and *Pichia kluyveri* for developing an alcoholic beverage: The yeasts growth and chemical compounds modulation by yeast extracts. *Curr. Res. Food Sci.* **2021**, *4*, 489–498. [CrossRef]
16. Zhao, H.Y.; Li, Y.Q.; Liu, L.X.; Zheng, M.N.; Feng, Z.J.; Hu, K.; Tao, Y.S. Effects of inoculation timing and mixed fermentation with *Pichia fermentans* on *Oenococcus oeni* viability, fermentation duration and aroma production during wine malolactic fermentation. *Food Res. Int.* **2022**, *159*, 111604. [CrossRef]
17. Cuadros-Inostroza, A.; Giavalisco, P.; Hummel, J.; Eckardt, A.; Willmitzer, L.; Peña-Cortés, H. Discrimination of Wine Attributes by Metabolome Analysis. *Anal. Chem.* **2010**, *82*, 3573–3580. [CrossRef]
18. Mierczynska-Vasilev, A.; Bindon, K.; Gawel, R.; Smith, P.; Vasilev, K.; Butt, H.J.; Koynov, K. Fluorescence correlation spectroscopy to unravel the interactions between macromolecules in wine. *Food Chem.* **2021**, *352*, 129343. [CrossRef]
19. Yang, X.L.; Zhang, T.X.; Yang, D.D.; Xie, J.C. Application of gas chromatography-ion mobility spectrometry in the analysis of food volatile components. *Acta Chromatogr.* **2023**, *35*, 35–45. [CrossRef]
20. Shvartsburg, A.A.; Tang, K.Q.; Smith, R.D. Differential Ion Mobility Separations of Peptides with Resolving Power Exceeding 50. *Anal. Chem.* **2010**, *82*, 32–35. [CrossRef]

21. Rodríguez-Maecker, R.; Vyhmeister, E.; Meisen, S.; Martinez, A.R.; Kuklya, A.; Telgheder, U. Identification of terpenes and essential oils by means of static headspace gas chromatography-ion mobility spectrometry. *Anal. Bioanal. Chem.* **2017**, *409*, 6595–6603. [CrossRef] [PubMed]
22. Perl, T.; Jünger, M.; Vautz, W.; Nolte, J.; Kuhns, M.; Zepelin, M.B.V.; Quintel, M. Detection of characteristic metabolites of *Aspergillus fumigatus* and *Candida* species using ion mobility spectrometry–metabolic profiling by volatile organic compounds. *Mycoses* **2011**, *54*, 828–837. [CrossRef] [PubMed]
23. Othman, A.; Goggin, K.A.; Tahir, N.I.; Brodrick, E.; Singh, R.; Sambanthamurthi, R.; Parveez, G.K.A.; Davies, A.N.; Murad, A.J.; Muhammad, N.H.; et al. Use of headspace-gas chromatography-ion mobility spectrometry to detect volatile fingerprints of palm fibre oil and sludge palm oil in samples of crude palm oil. *BMC Res. Notes* **2019**, *12*, 229. [CrossRef] [PubMed]
24. Arroyo-Manzanares, N.; Martín-Gómez, A.; Jurado-Campos, N.; Garrido-Delgado, R.; Arce, C.; Arce, L. Target vs spectral fingerprint data analysis of Iberian ham samples for avoiding labelling fraud using headspace–gas chromatography-ion mobility spectrometry. *Food Chem.* **2018**, *246*, 65–73. [CrossRef] [PubMed]
25. Gu, S.; Chen, W.; Wang, Z.H.; Wang, J. Rapid determination of potential aflatoxigenic fungi contamination on peanut kernels during storage by data fusion of HS-GC-IMS and fluorescence spectroscopy. *Postharvest Biol. Technol.* **2020**, *171*, 111361. [CrossRef]
26. Jia, S.L.; Li, Y.; Zhuang, S.; Sun, X.H.; Zhang, L.T.; Shi, J.; Hong, H.; Luo, Y.K. Biochemical changes induced by dominant bacteria in chill-stored silver carp (*Hypophthalmichthys molitrix*) and GC-IMS identification of volatile organic compounds. *Food Microbiol.* **2019**, *84*, 103248. [CrossRef] [PubMed]
27. Gallegos, J.; Arce, C.; Jordano, R.; Arce, L.; Medina, L.M. Target identification of volatile metabolites to allow the differentiation of lactic acid bacteria by gas chromatography-ion mobility spectrometry. *Food Chem.* **2017**, *220*, 362–370. [CrossRef]
28. Chen, Y.; Li, P.; Liao, L.Y.; Qin, Y.Y.; Jiang, L.W.; Liu, Y. Characteristic fingerprints and volatile flavor compound variations in Liuyang Douchi during fermentation via HS-GC-IMS and HS-SPME-GC-MS. *Food Chem.* **2021**, *361*, 130055. [CrossRef]
29. Feng, T.; Sun, J.Q.; Song, S.Q.; Wang, H.T.; Yao, L.Y.; Sun, M.; Wang, K.; Chen, D. Geographical differentiation of Molixiang table grapes grown in China based on volatile compounds analysis by HS-GC-IMS coupled with PCA and sensory evaluation of the grapes. *Food Chem. X* **2022**, *15*, 100423. [CrossRef]
30. López-Feria, S.; Cárdenas, S.; Valcárcel, M. Simplifying chromatographic analysis of the volatile fraction of foods. *TrAC Trends Anal. Chem.* **2008**, *27*, 794–803. [CrossRef]
31. He, G.Q.; Huang, J.; Liang, R.; Wu, C.D.; Zhou, R.Q. Comparing the differences of characteristic flavour between natural maturation and starter culture for *Mucor-type Douchi*. *Int. J. Food Sci. Technol.* **2016**, *51*, 1252–1259. [CrossRef]
32. Xu, E.B.; Long, J.; Wu, Z.Z.; Li, H.Y.; Wang, F.; Xu, X.M.; Jin, Z.Y.; Jiao, A.Q. Characterization of Volatile Flavor Compounds in Chinese Rice Wine Fermented from Enzymatic Extruded Rice. *J. Food Sci.* **2015**, *80*, C1476–C1489. [CrossRef] [PubMed]
33. García-Carpintero, E.G.; Sánchez-Palomo, E.; Gallego, M.A.G.; González-Viñas, M.A. Characterization of impact odorants and sensory profile of Bobal red wines from Spain's La Mancha region. *Flavour Fragr. J.* **2012**, *27*, 60–68. [CrossRef]
34. Gracia-Moreno, E.; Lopez, R.; Ferreira, V. Quantitative determination of five hydroxy acids, precursors of relevant wine aroma compounds in wine and other alcoholic beverages. *Anal. Bioanal. Chem.* **2015**, *407*, 7925–7934. [CrossRef]
35. Ugliano, M.; Dieval, J.B.; Siebert, T.E.; Kwiatkowski, M.; Aagaard, O.; Vidal, S.; Waters, E.J. Oxygen Consumption and Development of Volatile Sulfur Compounds during Bottle Aging of Two Shiraz Wines. Influence of Pre- and Postbottling Controlled Oxygen Exposure. *J. Agric. Food Chem.* **2012**, *60*, 8561–8570. [CrossRef]
36. Fedrizzi, B.; Magno, F.; Badocco, D.; Nicolini, G.; Versini, G. Aging effects and grape variety dependence on the content of sulfur volatiles in wine. *J. Agric. Food Chem.* **2007**, *55*, 10880–10887. [CrossRef]

foods

MDPI

Article

Characterization and Preliminary Safety Evaluation of *Akkermansia muciniphila* PROBIO

Xin Ma [1,†], Meng Tian [2,†], Xueping Yu [1], Ming Liu [3], Bin Li [3], Dayong Ren [2,*] and Wei Wang [1,*]

1 State Key Laboratory of Bioreactor Engineering, East China University of Science and Technology, Shanghai 200237, China; pkartest@yeah.net (X.M.); yuxueping@thankcome.com (X.Y.)
2 College of Food Science and Engineering, Jilin Agricultural University, Changchun 130118, China; jodie0730@163.com
3 China National Research Institute of Food and Fermentation Industries, Beijing 100015, China; lm_bob@163.com (M.L.); libin_2957@163.com (B.L.)
* Correspondence: rendayong@jlau.edu.cn (D.R.); weiwang@ecust.edu.cn (W.W.); Tel.: +86-15104409354 (D.R.); +86-21-64253707 (W.W.)
† These authors contributed equally to this work.

Abstract: In addition to providing certain health advantages to the host, a bacterial strain must possess a clearly defined safety profile to be regarded as a probiotic. In this study, we present a thorough and methodical assessment of the safety of a novel strain of bacteria, *Akkermansia muciniphila* PROBIO, which was isolated from human feces. Firstly, we examined the strain's overall features, such as its gastrointestinal tolerance and its physiological and biochemical traits. Next, we verified its genotoxic properties through bacterial reverse mutation and in vitro mammalian cell micronucleus assays. The drug sensitivity of *A. muciniphila* PROBIO was subsequently examined through an analysis of its antibiotic resistance genes. Additionally, the toxicological impact was verified through acute and sub-chronic toxicity studies. A genome-based safety assessment was conducted to gain further insights into gene function, including potential virulence factors and pathogenic properties. Finally, we assessed whether moxifloxacin resistance in *A. muciniphila* PROBIO is transferred using in vitro conjugation experiments. *A. muciniphila* PROBIO exhibited superior gastrointestinal tolerance, with no observed hematological or histopathological abnormalities. Moreover, the outcomes pertaining to mutagenic, clastogenic, or toxic impacts were found to be negative, even at exceedingly high dosages. Moreover, no adverse effects associated with the test substance were observed during the examination of acute and sub-chronic toxicity. Consequently, it was plausible to estimate the no-observed-adverse-effect level (NOAEL) to be 6.4×10^{11} viable bacteria for an average individual weighing 70 kg. Additionally, only three potential drug resistance genes and one virulence factor gene were annotated. *A. muciniphila* PROBIO is naturally resistant to moxifloxacin, and resistance does not transfer. Collectively, the data presented herein substantiate the presumed safety of *A. muciniphila* PROBIO for its application in food.

Keywords: *Akkermansia muciniphila*; safety; toxicity; genome; antibiotic

1. Introduction

Akkermansia muciniphila, a member of the phylum *Verrucomicrobia*, is an oval-shaped, Gram-negative, and strictly anaerobic bacterium that resides as a commensal in the human intestinal tract. This type of bacterium is immobile and lacks the ability to form spores. *A. muciniphila* was first isolated from the human intestine in 2004 [1] and has been found in the small and large intestines, with a particularly high abundance in the cecum, where most of the mucin is produced [2]. The proportion of *A. muciniphila* in the intestinal microbiota of humans and other mammal species ranges from 3% to 5% [3]. Additionally, *A. muciniphila* has been detected in the breast milk and breast tissue of lactating women, utilizing breast milk oligosaccharides as a source of energy, carbon, and nitrogen [4]. This bacterium is vertically transmitted from breast milk to infants and plays a role in the development of the intestinal microbiota during early infancy (around 1 month of age), with a significant increase in abundance during adulthood [3]. A recent comprehensive analysis of a large global dataset revealed that *A. muciniphila* was present in 77.73% of the global evaluated cohorts and in 81.81% of the western population cohorts [5].

Although *A. muciniphila* has only been isolated for a short time, it has been extensively researched for its impact on human health. *A. muciniphila* produces a group of enzymes that specialize in the degradation of mucins on the surface of the human gastrointestinal tract [6]. This degradation of human mucus proteins by *A. muciniphila* serves as a competitive exclusion mechanism against pathogenic mucus degradants and may have gatekeeper and signaling functions [7,8]. The decomposition of mucus leads to the production of beneficial metabolites, such as acetic acid, propionic acid, and oligosaccharides, contributing to improved metabolism and reduced inflammation [9,10]. *A. muciniphila* also influences mucus production by regulating the differentiation of intestinal epithelial cells to improve the integrity of the intestinal barrier and mucosal homeostasis [11]. As *A. muciniphila* has been found to be abundant in the intestines of healthy individuals and reduced in patients with inflammatory bowel disease, obesity, and other disorders, it has been proposed as a potential biomarker for intestinal health [9,12]. A large number of studies have shown that changes in the abundance of *A. muciniphila* are related to metabolic disorders [13,14], immune diseases [15,16], and cancer [17]. Notably, oral administration of live or pasteurized *A. muciniphila* can regulate the metabolism of the host, resulting in a decrease in serum triglyceride levels and an improvement in insulin sensitivity [18]. This intervention was shown to offer a certain degree of protection against diet-induced obesity, diabetes, and intestinal barrier dysfunction in mice. Furthermore, studies have demonstrated an elevation in mucins among individuals who showed positive reactions to immune checkpoint inhibitors [19]. Additionally, the presence of *A. muciniphila* has been found to augment the anti-tumor efficacy of Programmed Cell Death Protein 1 (PD-1) blockers [20]. Given the diverse advantageous biological properties exhibited by *A. muciniphila* in relation to human health, it is widely acknowledged as a promising candidate for novel pharmaceutical development or as a next-generation probiotic [21].

Despite the availability of data supporting the positive effects of *A. muciniphila*, the lack of thorough safety research is mostly to blame for the strain's absence from the list of strains that have been certified by governmental agencies in many nations across the world. Several preclinical and clinical studies have indicated that *A. muciniphila* is predominantly eliminated through fecal excretion following oral administration, and supplementation of *A. muciniphila* would not cause substantial changes in the overall composition of the gut microbiome [13,22]. In 2021, the European Food Safety Authority (EFSA) issued a scientific opinion on the safety of pasteurized *A. muciniphila* as a novel food product [23]. This viewpoint led to the conclusion that the daily consumption of 3.4×10^{10} cells of pasteurized *A. muciniphila* was safe for the target population. Nevertheless, these findings alone do not provide adequate evidence to support the safety of *A. muciniphila* for use as a probiotic. In accordance with the guidelines set forth by the Food and Agriculture Organization (FAO) and the World Health Organization (WHO), probiotic strains must meet specific safety criteria. While probiotic strains derived from species typically present in fermented foods are generally deemed safe, strains originating from nontraditional species may elicit heightened concerns regarding potential adverse effects [24]. Therefore, further investigation is needed to ensure their safety. In our previous study, we successfully isolated a novel strain of *A. muciniphila* PROBIO (AKK PROBIO, AKK. pro) from human fecal samples. Our preliminary unpublished data showed that this strain exhibited promising health advantages, including the ability to reduce colitis symptoms, boost PD-1 antibodies, and produce short-chain fatty acids. The current study aimed to evaluate the initial safety of the novel *A. muciniphila* PROBIO at the strain level. The assessment encompassed an examination of various aspects of the strain, including the physicochemical characteristics of the cell surface, biofilm formation, resistance to antibiotics, tolerance to simulated gastrointestinal fluids, mutagenicity, acute toxicity, and sub-chronic toxicity.

2. Materials and Methods

2.1. Bacterial Strains and Culture Conditions

A. muciniphila PROBIO was isolated from the feces of a healthy adult and identified using 16S rRNA gene sequence analysis. This strain is currently stored in the China General Microbiological Culture Collection Center (Beijing, China) under the preservation number CGMCC No. 20955. AKK PROBIO was cultured in Brain Heart Infusion (BHI) broth medium (BD, Sparks, MD, USA) for 48 to 72 h at 37 °C under anaerobic conditions in a Basic1 anaerobic workstation (ChongQing Jiangxue Science and Technology, Chongqing, China) for oxygen deprivation (80% N_2, 15% CO_2, 5% H_2). *Enterococcus faecalis* ATCC 29212 was used as a recipient, which was grown in Mueller-Hinton (MH) broth medium (BD, Sparks, MD, USA) at 37 °C for 24 h under anaerobic conditions.

2.2. Tolerance to Artificial Gastric and Intestinal Fluids

Gastrointestinal tolerance analysis refers to the method of Ulsemer [25]. The gastric juice (5 M HCl, 2 g/L NaCl, 3.2 g/L pepsin, pH 2.0) and intestinal juice (6.8 g/L KH_2PO_4, 10 g/L pancreatin, 0.1 M NaOH, pH 8.0) were prepared, followed by filtration (0.22 μm) and sterilization. The AKK PROBIO was activated twice in a 5 mL BHI liquid culture medium, then centrifuged at 3337× *g* for 10 min. The resulting bacterial precipitation was washed twice with sterile physiological saline and resuspended in 5 mL of physiological saline. The bacterial liquid was combined with sterile artificial gastric juice or intestinal juice (1:1, *v*/*v*). The mixture was thoroughly shaken and subsequently placed in a constant-temperature incubator (37 °C). The viable bacteria count was determined at 0 h and 4 h. The survival rate of AKK PROBIO in artificial gastric juice or intestinal juice was calculated using Formula (1).

$$Survival\ rate = \frac{4\ \mathrm{h}\ viable\ count\ (\mathrm{CFU/mL})}{0\ \mathrm{h}\ viable\ count\ (\mathrm{CFU/mL})} \quad (1)$$

2.3. Evaluation of Hydrophobicity, Auto-Aggregation, and Biofilm Formation

Bacterial adhesion to hydrocarbons (BATH) was used to evaluate the cell surface hydrophobicity of AKK PROBIO [26]. Specifically, *A. muciniphila* PROBIO was cultured in BHI medium and incubated overnight at 37 °C. Cell precipitates were obtained through centrifugation (4 °C; 7104× *g*; 10 min), washed twice, and resuspended in phosphate buffer (PBS, pH 7). The absorbance (OD_{600} nm) was adjusted to 0.5 in order to standardize the bacterial count. The biphasic mixture was thoroughly homogenized by adding an equivalent amount of xylene and vortexing it for 5 min. The aqueous phase was then extracted, and its absorbance at 600 nm was measured after an incubation time at room temperature at intervals of 15, 30, and 60 min. The hydrophobicity of AKK PROBIO was calculated using Formula (2).

$$Hydrophobicity\ \% = \frac{p0 - p1}{P1} \times 100 \quad (2)$$

where $p0$ and $p1$ are the absorbance values before and after extraction with hydrocarbons, respectively.

For auto-aggregation [26], the cell suspension of AKK PROBIO was prepared as described above and incubated at 37 °C. The absorbance was adjusted (OD_{600} nm) to 0.5. The bacterial suspension was cultivated at 37 °C, and the absorbance at 600 nm was read after intervals of 4, 8, 12, 16, 20, and 24 h. The auto-aggregation of AKK PROBIO was calculated using the following Formula (3):

$$Auto-aggregation\ \% = 1 - (\frac{p1}{P0}) \times 100 \quad (3)$$

where $p0$ is the absorbance at time 0, and $p1$ is the absorbance detected after 4, 8, 12, 16, 20, and 24 h.

For biofilm formation [27], AKK PROBIO was cultured overnight in tryptone soy broth (TSB) (BD, Sparks, MD, USA) at 37 °C. The bacterial cells were then harvested through centrifugation at 7104× *g* for 10 min at 4 °C, followed by two washes with PBS. The cells were subsequently resuspended in different media, including TSB medium without glucose and TSB medium with varying concentrations of glucose (0.25%, 1%, and 2.5%, *w*/*v*). Each bacterial suspension (200 µL) was aliquoted into a 96-well plate, with the uninoculated culture medium serving as the negative control. The plate was incubated at 37 °C for 24 h. The culture medium was aspirated from each well and subsequently washed three times with sterile physiological saline in order to eliminate any unattached cells. The attached cells were then fixed with methanol (200 µL per well) and allowed to dry for 15 min. Subsequently, the cells were subjected to staining with 2% crystal violet (200 µL per well) for a duration of 5 min. Following this, the wells were washed three times with sterile physiological saline to remove any excess dye. Once the plate had air dried, the attached cells were resuspended in 160 µL of glacial acetic acid (33%, *v*/*v*), and their absorbance at 600 nm was measured. The critical value, denoted as OD_0, was established as the mean optical density (OD) value of the negative control. Based on the OD measurement, the strain was categorized into four groups: non-biofilm producers ($OD \leq OD_0$), weak biofilm producers ($OD_0 < OD \leq 2 \times OD_0$), medium biofilm producers ($2 \times OD_0 < OD \leq 4 \times OD_0$), or strong biofilm producers ($4 \times OD_0 < OD$) [27].

2.4. Bacterial Reverse Mutation Test (Ames Test)

The bacterial reverse mutation test was performed according to the methods of plate incorporation and preincubation [28,29]. The bacterial strains employed in this study included *Salmonella Typhimurium* TA97a, TA98, TA100, and TA1535, and *Escherichia coli* WP2 uvrA. Distilled water was utilized as the carrier to create a bacterial suspension (100 mg/mL). *A. muciniphila* PROBIO, as the test item, had the potential to influence the test outcomes in various ways, thereby compromising the validity of the study. Consequently, the bacteria present in the test sample were eliminated through mechanical methods. During the preparation process, the soluble components of the test items are dissolved in the carrier and subsequently subjected to centrifugation at 4 °C and 8000 rpm for 10 min to eliminate any remaining bacteria. Following centrifugation, the supernatant was aseptically filtered using a 0.22 µm filter. Subsequently, a lower concentration preparation was prepared and utilized for the treatment of the test bacterial strain.

The initial mutation study (plate incorporation method) and the confirmed mutation study (pre-culture procedure) were conducted based on the findings from preliminary experiments. Confirmatory mutation studies were conducted at concentrations of 15.81, 50, 158.1, 500, 1581, and 5000 µg/plate, adhering to the highest concentration advised by the prevailing regulatory guidelines. The positive controls encompassed Dixon, sodium azide, methyl methane sulfonate, and 2-aminofluorene, which were dissolved in either distilled water or dimethyl sulfoxide. The survival rate of the test bacterial cells was checked by the plate experiment in each test. If there was an increase in the number of responses dependent on the dosage, it could be inferred that the test item exhibited mutagenic properties. This inference holds true when at least one dose group demonstrates reproducible positive reactions that are biologically related, regardless of the presence or absence of metabolic activation. Additionally, if the number of reversals observed in *S. typhimurium* TA98, TA100, and *E. coli* WP2 uvrA strains was more than twice the number observed in the negative control, it was considered to be biologically significant. In the case of the *S. typhimurium* TA97a and TA1535 strains, if the number of reversals was more than three times the number observed in the negative control, it was also considered to be biologically significant.

2.5. *In Vitro Mammalian Cell Micronucleus Test*

The genotoxic potential of AKK PROBIO in inducing micronucleus formation in mouse erythrocytes was assessed using the cytokinesis-block method [30]. The Kunming (Aneta, #192) mice used in this study were obtained from the animal experimental center of Jilin Agricultural University and were approximately 7 weeks old. A 5-day adaptation period was provided for the animals, and only those in acceptable health were included in the study. The animals had access to distilled water. They were housed either in groups of five animals per cage or two animals per cage and were exposed to a 12-h light cycle.

In each experimental group, a total of five male mice were administered bacterial suspension at doses of 222.2, 666.7, and 2000 mg/kg body weight. The negative control group received distilled water, while the positive control group was administered cyclophosphamide at a dose of 60 mg/kg body weight. Following euthanasia, bone marrow samples were promptly collected from the femur of each mouse and prepared for glass slides, which were subsequently stained and evaluated for scoring purposes.

2.6. *Antibiotic Susceptibility Test*

The agar dilution method recommended by the Clinical and Laboratory Standards Institute (CLSI) was employed to assess the susceptibility of AKK PROBIO to clindamycin, meropenem, tetracycline, moxifloxacin, chloramphenicol, and ceftriaxone [31]. Similarly, the microbroth dilution method recommended by CLSI was utilized to determine the quality control strain, *Bacteroides fragilis* ATCC 25285 or *Staphylococcus aureus* ATCC 29213. Additionally, two databases, the CARD (The Comprehensive Antibiotic Resistance Database, https://card.mcmaster.ca/ accessed on 6 April 2023) and ResFinder (https://cge.cbs.dtu.dk/services/ResFinder/, accessed on 6 April 2023), were employed to predict potential drug resistance-related genes in *A. muciniphila* PROBIO.

2.7. *Oral Acute and 90-Day Sub-Chronic Toxicity Studies*

In order to evaluate the acute toxicity of AKK PROBIO, KM mice under specific pathogen-free (SPF) conditions were randomly divided into four groups (n = 8). Each group was given a suspension of AKK PROBIO (1×10^9, 5×10^{10}, and 5×10^{11} Colony-Forming Units (CFU)/day, respectively) or saline (0.5 mL/day) by oral gavage for 1 day, and their general health and weight were observed every day for 7–10 days. At the end of the experimental period, blood samples were collected for hematology and serum biochemical analysis. All animals in each group were examined for histopathology of major organs and tissues. The protocols were approved by the ethics committee of the Laboratory Animal Center of Jilin Agricultural University (permit number: 20220711002).

To assess the potential systemic toxicity of AKK PROBIO, a sub-chronic oral toxicity study was conducted in vivo following the guidelines provided by the OECD test guidelines [32]. A total of 64 KM mice (32 males and 32 females) were obtained from the animal experimental center of Jilin Agricultural University. The administration of AKK PROBIO began when the animals were approximately 6–7 weeks old. The experimental environment was maintained at a room temperature of 22 ± 3 °C and a relative humidity of 55% ± 15%. The animals were divided into four groups in a random manner: group 1 only received saline; group 2 received a bacterial suspension at a dosage of 9.2×10^8 CFU/kg body weight per day; group 3 received a bacterial suspension at a dosage of 27.6×10^8 CFU/kg body weight per day; and group 4 served as a control. The animals were administered the suspension orally using a gavage technique for a period of 90 days. Daily records of clinical signs were maintained, and comprehensive clinical observation was conducted once a week. Additionally, the animals' weight, food consumption, and water consumption were recorded on a weekly basis.

2.8. *Genome Sequencing, Assembly, and Analysis*

Genomic DNA extraction, determination of the quality and quantity of extracted DNA, DNA fragmentation, and sequencing with the PacBio Sequel II platform (Frasergen

Bioinformatics Co., Ltd., Wuhan, China) were carried out. The raw sequencing reads (also called polymerase reads) generated from the PacBio platforms were processed using SMRTlink v8.0. Error correction of the reads was performed as part of the Canu (version 2.0) assembly process. The genome was assembled by Canu v2.0. Canu detects and annotates that the final assembly is circular when the best overlap graph (BOG) is circular. As an alternative method, we confirmed genome circularization by identifying repeated sequences at the two ends using BLASTN (version 2.7.1). The NCBI Prokaryotic Genome Automatic Annotation Pipeline (PGAP) was applied to complete genome annotation. The virulence factors and pathogenicity of AKK PROBIO were predicted using the VFD database (virulence factors of pathological bacteria, http://www.mgc.ac.cn/VFs/main.htm, accessed on 6 April 2023), VirulenceFinder database (https://cge.cbs.dtu.dk/services/VirulenceFinder, accessed on 6 April 2023), and PathogenFinder database (https://cge.cbs.dtu.dk/services/PathogenFinder, accessed on 6 April 2023).

2.9. In Vitro Conjugation Experiments

The in vitro conjugation was carried out following the methods of Ashwini [33] to determine the transferability of the *adeF* gene of moxifloxacin resistance detected in AKK PROBIO (donor). *E. faecalis* ATCC 29212 (recipient) was used as a recipient to perform the conjugation in vitro. Briefly, from the overnight-grown cultures of the donor and the recipient, the cultures (1 mL) were centrifuged at 5000× *g* for 5 min to collect the bacterial cells. They were washed three times with PBS and resuspended to prepare a bacterial suspension of 0.5 on the McFarland scale. AKK PROBIO was thoroughly mixed with ATCC 29212 in volume ratios of 1:1 and 10:1, respectively.

We inoculated the mixtures (donor/recipient ratios of 1:1 and 10:1) into MH liquid culture medium at a ratio of 2% (v/v), which were then incubated at 37 °C for 24 h.

We took 0.1 mL of the mixtures (donor/recipient ratios of 1:1 and 10:1), spread them on to MH agar medium, and incubated them at 37 °C for 24 h. We then collected all colonies and resuspended them with PBS for a 10-fold gradient dilution.

The mixtures (donor/recipient ratios of 1:1 and 10:1) were filtrated through a sterile filter (0.22 μm). After the collection of donor and recipient cells on the filter, PBS (5 mL) was passed through the filter. We placed the filter membrane containing bacteria on MH agar medium and cultured them at 37 °C until colonies grew on the surface of the membrane. The bacterial colonies on the filter membrane were washed with PBS, and a dilution gradient was performed.

The transferability was detected on selective MH agar plates containing different concentrations of moxifloxacin. The bacterial solution of *E. faecalis* ATCC 29212 without AKK PROBIO was used as a negative control. Three parallel tests were run in each case.

2.10. Statistical Analysis

All experiments were performed at least in triplicate using independent assays, and values were expressed as the mean ± standard error. Statistical analysis was performed through the analysis of variance (ANOVA), followed by Tukey's multiple comparison. A p-value of <0.05 was considered statistically significant. The software Graph Pad Prism (version 8.0) was used for the analysis.

3. Results

3.1. Gastrointestinal Tract (GI) Tolerance and Surface Properties

We determined the survival of *A. muciniphila* PROBIO after exposure to artificial gastric and intestinal fluids in vitro. After 240 min of exposure, the survival rates of AKK PROBIO in gastric and intestinal fluids were 84% (Figure 1a) and 82% (Figure 1b), respectively, suggesting that AKK PROBIO has good GI tolerance. The maximum values of hydrophobicity and auto-aggregation capacity of AKK PROBIO were 31% (60 min, Figure 1c) and 52% (24 h, Figure 1d), respectively. The ability of AKK PROBIO to form biofilms was weak regardless of the amount of glucose used. The results of these in vitro tests suggested that AKK PROBIO has a low ability to colonize the gut.

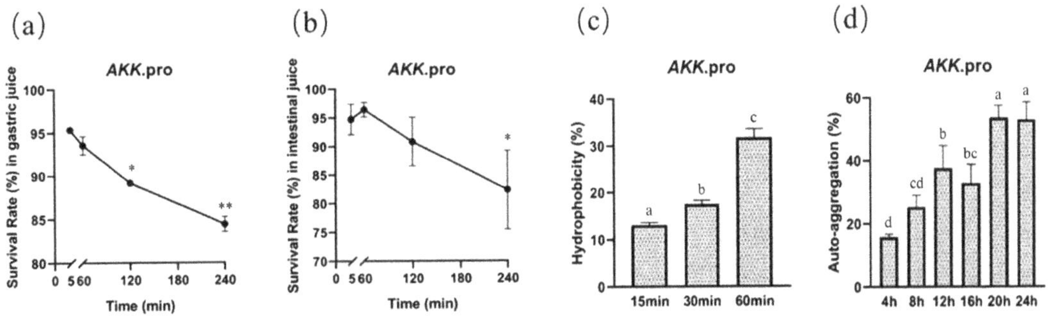

Figure 1. General characteristics of *A. muciniphila* PROBIO. (**a**) Strain tolerance to artificial gastric fluid. (**b**) Strain tolerance to artificial intestinal fluid. (**c**) Adhesion of AKK PROBIO to hydrocarbons. (**d**) Auto-aggregation percentages of AKK PROBIO. The asterisk (*/**) indicates a significant ($p < 0.05$) or highly significant ($p < 0.01$) difference. Different letters indicate significant differences ($p < 0.05$).

3.2. Bacterial Reverse Mutation Test

As shown in Table 1, within the confirmatory mutation study employing the plate incorporation technique, *E. coli* WP2 uvrA exhibited the highest rate of reversion when subjected to metabolic activation S9. Nevertheless, no discernible dose–response relationship was observed, and the quantities of revertant colonies fell within the range of historical controls. The mean values of revertant colony numbers observed in the untreated, negative, and positive control plates were found to fall within the historical control range for all strains. The reference mutagens exhibited a noticeable increase in the number of induced revertant colonies in each strain, regardless of the presence or absence of metabolic activation. The results indicated that, under the specific experimental conditions employed, the test item did not induce gene mutations through alterations in base pairs or frameshifts within the genomes of the strains utilized. AKK PROBIO showed no mutagenic activity in the examined bacterial strains under the present study conditions.

Table 1. Summary of reverted colonies of *Salmonella typhimurium* (Ames test) with and without metabolic activation with S9.

Concentrations	*S. typhimurium*								*E. coli*	
	TA97a		TA98		TA100		TA1535		WP2 uvrA	
	−S9	+S9	−S9	+S9	−S9	+S9	−S9	+S9	−S9	+S9
Distilled water	122.0 ± 30.7	126.2 ± 22.1	41.4 ± 8.1	35.0 ± 6.4	144.6 ± 25.5	109.0 ± 43.8	12.0 ± 4.4	16.0 ± 6.5	122.4 ± 19.4	162.0 ± 32.4
AKK PROBIO	98.2 ± 7.7	101.8 ± 7.7	36.8 ± 7.5	31.6 ± 6.8	61.6 ± 12.9	52.0 ± 9.6	10.0 ± 3.9	12.4 ± 2.9	113.6 ± 5.8	130.8 ± 8.4
Positive control 1	1169.4 ± 181.5	-	1525.0 ± 271.7	-	-	-	-	-	-	-
Positive control 2	-	-	-	-	1347.0 ± 202.3	-	-	-	613.4 ± 61.9	-
Positive control 3	-	-	-	-	-	-	141.8 ± 26.6	-	-	-
Positive control 4	-	1288.4 ± 296.3	-	1398.6 ± 309.1	-	1247.4 ± 59.7	-	163.2 ± 9.9	-	548.6 ± 111.8

Data are presented as revertant colonies per plate (mean ± standard deviation). The positive controls 1–4 were: Dixon, methyl methanesulfonate, sodium azide, and 2-aminofluorene, respectively.

3.3. Micronucleus Analysis

The findings of the in vivo mammalian erythrocyte micronucleus study indicated that exposure to AKK PROBIO in mice did not result in any treatment-related impacts on body weight, mortality, systemic toxicity, or observable symptoms. The examination of bone marrow slides in all experimental groups involved the quantification of polychromatic erythrocytes (PCEs) and mature red blood cells in order to evaluate the incidence of micronuclei. The findings are presented in Table 2, indicating that there was no significant disparity in the [illegible] of micronuclei in PCEs between the negative control and test groups ($p > 0.05$). However, the positive control treatment exhibited a substantial and statistically significant elevation in the frequency of micronuclei in PCEs ($p < 0.001$). The occurrence of micro-nucleated PCEs among all groups showed no significant differences. The results suggested that AKK PROBIO did not display genotoxic activity.

Table 2. In vitro mammalian cell micronucleus test with *A. muciniphila* PROBIO.

Group	PCE (%)	NCE Ratio	PCE Ratio
Negative control	36.34 ± 4.63	0.06 ± 0.09	0.09 ± 0.04
Low dose	36.28 ± 7.83	0.03 ± 0.04	0.11 ± 0.06
Mid dose	34.83 ± 4.73	0.02 ± 0.05	0.12 ± 0.06
High dose	32.73 ± 4.46	0.04 ± 0.03	0.12 ± 0.08
Positive control	24.6 ± 5.90	0.33 ± 0.10 ***	4.76 ± 2.13 ***

PCE, polychromatic erythrocyte; NCE, polychromatic erythrocyte; ***, $p < 0.001$.

3.4. Antibiotic Resistance

The antibiotic resistance analysis results are presented in Table 3. AKK PROBIO exhibited susceptibility to clindamycin, meropenem, tetracycline, chloramphenicol, and ceftriaxone. However, it was resistant to moxifloxacin. The predicted results of drug resistance-related genes are shown in Table S1. In the genome of AKK PROBIO, two genes associated with antibiotic resistance were identified, specifically pertaining to quinolones and aminoglycosides. One of these genes, *adeF*, is part of the gene cluster *adeFGH*. When the genes *adeF*, *adeG*, and *adeH* within the gene cluster are simultaneously expressed, they facilitate the encoding and production of the multidrug-resistant efflux pump *AdeFGH*, thereby conferring resistance to fluoroquinolones and other pharmaceutical agents upon the strain. In the ResFinder database, only one putative antibiotic resistance-related gene, *aadA15*, was identified, which was linked to aminoglycoside antibiotic resistance (Table S2).

Table 3. Antibiotic resistance of *A. muciniphila* PROBIO to different antibiotics.

Antibiotics	Judging Rule			Tested Strain		Quality Control Strain		
				A. muciniphila PROBIO		*B. fragilis* ATCC 25285		
	S	I	R	MIC (μg/mL)	Result	Quality Control Range	MIC (μg/mL)	Within the Quality Control Range
Clindamycin	≤2	4	≥8	0.5	Sensitivity	0.5–2	2	Yes
Meropenem	≤4	8	≥16	0.06	Sensitivity	0.03–0.25	0.25	Yes
Tetracycline	≤4	8	≥16	4	Sensitivity	0.125–0.5	0.5	Yes
Moxifloxacin	≤2	4	≥8	>32	Resistance	0.125–0.5	0.5	Yes
Chloramphenicol	≤8	16	≥32	2	Sensitivity	2–8	8	Yes
Antibiotics	**Judging Rule**			**Tested Strain**		**Quality Control Strain**		
				***A. muciniphila* PROBIO**		***S. aureus* ATCC 29213**		
	S	**I**	**R**	**MIC (μg/mL)**	**Result**	**Quality Control Range**	**MIC (μg/mL)**	**Within the Quality Control Range**
Ceftriaxone	≤16	32	≥64	2	Sensitivity	1–8	4	Yes

S, susceptible; I, intermediate; R, resistance; MIC, minimal inhibitory concentration.

3.5. Acute and Sub-Chronic Toxicities

We found that none of the indices of the mice fed with AKK PROBIO at any given concentration were significantly different from the healthy control group in either the acute toxicity test or the sub-chronic toxicity test. Specifically, no treatment-related deaths or behavioral abnormalities were observed in the mice during the experiments. Figure 2a–d displays body weights and food consumption during the entire administration period. There was no significant difference in weight gain or food intake in all groups ($p > 0.05$) (Figure 2a–d). There were no significant changes in the blood glucose level (Figure 2e,f), glucose tolerance (Figure 2g), or insulin tolerance (Figure 2h) in all groups after 90 days of oral administration of AKK PROBIO ($p > 0.05$). There were no significant differences observed in red blood cells (RBCs), hemoglobin (HGB), platelets (PLTs), or other routine blood parameters among the groups ($p > 0.05$) (Table 4). Similarly, no significant differences were found in the biochemical markers associated with glucose and lipid metabolism in either the serum or liver across the groups ($p > 0.05$) (Table 5). The weights of organs in all groups showed no significant differences ($p > 0.05$) (Table 6). Furthermore, the structural integrity of the liver and renal tissues was unaffected by AKK PROBIO (Figure 3). All of these findings indicated that AKK PROBIO is safe for long-term usage at high doses (9.2×10^8–2.76×10^9 CFU/kg bw/day).

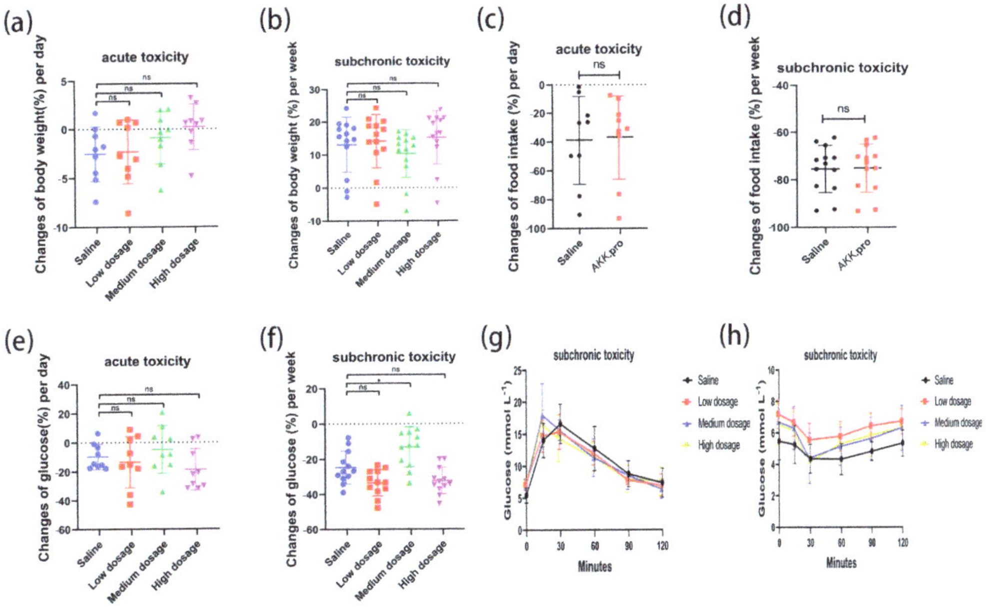

Figure 2. Changes in glucose metabolism in acute and sub-chronic toxicities of *A. muciniphila* PROBIO in normal mice. (**a**,**b**) Body weight; (**c**,**d**) food consumption; (**e**,**f**) blood sugar; (**g**) glucose tolerance; (**h**) insulin tolerance. "ns" indicates no significant difference between the two groups ($p > 0.05$); asterisk (*) indicates $p < 0.05$.

Table 4. Summary of hematological values in the study of acute and sub-chronic toxicity in mice ($n = 8$).

Groups	Blood Routine Indexes	Saline	Low Dose	Medium Dose	High Dose	Unit
Acute toxicity	white blood cells	0.7 ± 0.2	0.9 ± 0.4	1.1 ± 0.5	0.7 ± 0.3	10^9/L
	red blood cells	1.3 ± 0.6	2.1 ± 0.6	2.2 ± 0.8	1.4 ± 0.5	10^{12}/L
	hemoglobin content	85.8 ± 15.3	87.5 ± 16.1	96.0 ± 14.1	110.0 ± 11.3	g/L
	platelets	1028.8 ± 408.3	921.0 ± 360.8	891.5 ± 316.9	1012.5 ± 344.3	10^9/L
	lymphocytes	0.4 ± 0.2	0.8 ± 0.5	0.9 ± 0.4	0.5 ± 0.3	10^9/L
	neutrophils	0 ± 0	0.1 ± 0.07	0.1 ± 0.08	0 ± 0	10^9/L

Table 4. *Cont.*

Groups	Blood Routine Indexes	Saline	Low Dose	Medium Dose	High Dose	Unit
Acute toxicity	platelets–large cells	314.5 ± 107.4	273.5 ± 95.2	258.5 ± 86.9	228.5 ± 64.3	10^9/L
	hematocrit value	7.3 ± 2.31	11.7 ± 2.7	11.7 ± 3.6	7.4 ± 2.6	%
	mean corpuscular hemoglobin concentration	1667.5 ± 303.1	802.0 ± 427.0	892.0 ± 422.8	1579.0 ± 438.4	g/L
Sub-chronic toxicity	white blood cells	1.8 ± 0.2	1.5 ± 0.4	1.6 ± 0.5	1.3 ± 0.3	10^9/L
	red blood cells	7.4 ± 0.5	7.3 ± 0.3	7.6 ± 0.3	7.6 ± 0.5	10^{12}/L
	hemoglobin content	107.5 ± 4.3	105.0 ± 3.2	110.6 ± 4.1	107.2 ± 3.4	g/L
	platelets	500.1 ± 62.5	476.3 ± 75.4	475.6 ± 50.6	508.2 ± 71.7	10^9/L
	lymphocytes	1.7 ± 0.6	1.4 ± 0.4	1.5 ± 0.7	1.1 ± 0.2	10^9/L
	neutrophils	0.1 ± 0.01	0.1 ± 0.05	0.1 ± 0.03	0.2 ± 0.02	10^9/L
	platelets–large cells	61.2 ± 30.2	101.8 ± 41.3	98.3 ± 55.3	71.7 ± 43.1	10^9/L
	hematocrit value	37.9 ± 2.1	36.6 ± 1.2	38.1 ± 0.9	37.6 ± 1.0	%
	mean corpuscular hemoglobin concentration	284.3 ± 12.3	286.5 ± 9.1	290.3 ± 17.5	285.2 ± 13.4	g/L

Table 5. Biochemical indicators in the serum and liver in acute and sub-chronic toxicity experiments (n = 8).

Groups	Indicators	Saline		Low Dose		Medium Dose		High Dose	
		Serum	Liver	Serum	Liver	Serum	Liver	Serum	Liver
Acute toxicity	BG (mmol/L) *	8.5 ± 1.6	9.1 ± 2.1	8.0 ± 1.2	8.6 ± 1.5	8.3 ± 0.7	8.7 ± 1.6	8.3 ± 1.3	9.1 ± 1.1
	TG (mmol/L)	1.5 ± 0.6	0.4 ± 0.05	1.4 ± 0.6	0.3 ± 0.03	1.2 ± 0.3	0.3 ± 0.03	1.4 ± 0.4	0.3 ± 0.03
	TC (mmol/L)	4.9 ± 2.0	0.05 ± 0.03	4.7 ± 1.6	0.04 ± 0.01	5.2 ± 1.7	0.04 ± 0.01	4.2 ± 0.8	0.03 ± 0.01
	BA (μmol/L)	3.3 ± 0.7	56.2 ± 35.6	3.3 ± 0.6	49.8 ± 35.3	2.9 ± 0.4	45.7 ± 25.7	2.8 ± 0.3	52.7 ± 35.2
	TP (μg/mL)	241.4 ± 28.5	781.9 ± 71.0	260.1 ± 80.3	806.2 ± 47.9	217.6 ± 59.0	812.1 ± 37.3	240.9 ± 70.8	813.1 ± 69.4
	ALT (U/L)	15.3 ± 5.5	9.2 ± 2.7	17.3 ± 4.8	9.9 ± 2.6	15.8 ± 3.8	9.1 ± 3.0	18.0 ± 4.9	8.9 ± 3.4
	AST (U/L)	40.7 ± 9.0	10.1 ± 1.9	42.5 ± 3.8	9.6 ± 2.0	43.5 ± 8.7	10.1 ± 2.2	48.3 ± 7.5	9.3 ± 2.1
	Cr (μmol/L)	15.7 ± 4.9	-	20.7 ± 8.1	-	14.2 ± 4.2	-	15.8 ± 5.0	-
	BUN (mmol/L)	5.7 ± 1.8	-	5.3 ± 1.4	-	5.7 ± 2.0	-	5.7 ± 1.7	-
Sub-chronic toxicity	BG (mmol/L)	8.4 ± 1.5	11.8 ± 2.3	7.9 ± 1.3	11.2 ± 2.1	8.5 ± 1.6	13.0 ± 1.4	8.3 ± 1.2	11.5 ± 2.0
	TG (mmol/L)	1.2 ± 0.4	1.7 ± 0.6	1.1 ± 0.2	2.6 ± 1.2	1.3 ± 0.5	2.4 ± 1.0	1.8 ± 0.9	1.9 ± 0.7
	TC (mmol/L)	1.8 ± 0.7	0.3 ± 0.1	1.6 ± 0.7	0.2 ± 0.1	1.8 ± 0.8	0.2 ± 0.1	2.2 ± 0.5	0.2 ± 0.1
	BA (μmol/L)	5.0 ± 1.8	2.6 ± 0.7	5.7 ± 1.2	2.1 ± 0.5	4.1 ± 1.0	2.4 ± 0.7	5.8 ± 2.0	2.8 ± 0.5
	TP (μg/mL)	1629 ± 271.1	595.0 ± 123.1	1512 ± 261.7	621.5 ± 90.2	1684 ± 225.1	654.4 ± 86.9	1516 ± 209.9	628.2 ± 89.2
	ALT (U/L)	10.8 ± 4.1	9.2 ± 2.7	10.7 ± 3.8	9.9 ± 2.6	11.1 ± 4.1	9.1 ± 2.8	10.8 ± 4.0	9.0 ± 2.3
	AST (U/L)	44.3 ± 10.1	268.6 ± 29.5	52.6 ± 8.7	240.9 ± 50.4	48.7 ± 12.5	232.2 ± 25.1	47.2 ± 7.5	280.4 ± 27.5
	Cr (μmol/L)	13.9 ± 2.9	-	14.1 ± 3.1	-	13.9 ± 1.6	-	13.9 ± 2.1	-
	BUN (mmol/L)	3.9 ± 0.7	-	4.5 ± 1.6	-	4.3 ± 1.6	-	4.0 ± 0.8	-

Abbreviations: BG, blood glucose; TG, triglycerides; TC, total cholesterol; BA, bile acid; TP, total protein; ALT, alanine aminotransferase; AST, aspartate aminotransferase; Cr, creatinine; BUN, blood urea nitrogen; *, Units: BG, TG, and TC (mmol/L serum, or mmol/gprot liver); -, untested.

Table 6. Summary of relative organ weights (g/kg bw) in acute and sub-chronic toxicity experiments ($n = 8$).

		Heart	Liver	Spleen	Kidney	Thymus	Brain	Testicle	Lung	Stomach	Intestines
Acute toxicity	Saline	5.8 ± 1.0	43.7 ± 7.5	2.4 ± 0.5	13.6 ± 2.5	2.7 ± 0.5	9.9 ± 1.6	6.2 ± 0.5	6.1 ± 0.8	17.3 ± 5.6	101.4 ± 11.3
	Low dosage	5.6 ± 0.8	44.3 ± 6.6	3.1 ± 0.5	14.2 ± 2.6	2.5 ± 0.3	10.7 ± 2.1	6.5 ± 0.4	6.3 ± 0.9	16.5 ± 7.3	97.7 ± 7.5
	Medium dosage	4.8 ± 1.6	47.4 ± 7.7	2.5 ± 0.2	13.1 ± 1.9	3.6 ± 0.6	11.9 ± 2.2	6.1 ± 0.8	5.9 ± 0.6	15.9 ± 5.1	98.5 ± 6.5
	High dosage	4.7 ± 1.7	48.0 ± 3.9	2.8 ± 0.8	12.4 ± 1.3	3.5 ± 0.8	10.6 ± 4.8	7.2 ± 0.7	6.7 ± 1.1	15.8 ± 6.3	113.3 ± 9.3
Subchronic toxicity	Saline	4.8 ± 0.5	38.7 ± 4.7	2.1 ± 0.6	11.6 ± 2.0	2.6 ± 0.9	9.1 ± 3.3	6.0 ± 0.9	6.2 ± 0.9	16.1 ± 6.6	80.3 ± 12.7
	Low dosage	4.7 ± 0.7	37.8 ± 2.3	2.5 ± 0.8	12.6 ± 1.9	2.6 ± 0.9	10.2 ± 2.1	6.2 ± 1.3	5.9 ± 0.6	15.8 ± 7.2	79.6 ± 16.0
	Medium dosage	4.7 ± 0.5	36.2 ± 4.3	2.6 ± 0.5	12.2 ± 1.8	2.4 ± 0.2	9.0 ± 2.0	6.4 ± 0.7	5.2 ± 0.8	14.4 ± 4.3	80.2 ± 10.8
	High dosage	5.0 ± 0.3	38.3 ± 6.4	2.8 ± 0.7	12.0 ± 1.2	3.6 ± 1.4	9.2 ± 1.7	5.8 ± 0.6	5.6 ± 0.4	15.7 ± 2.5	87.6 ± 8.5

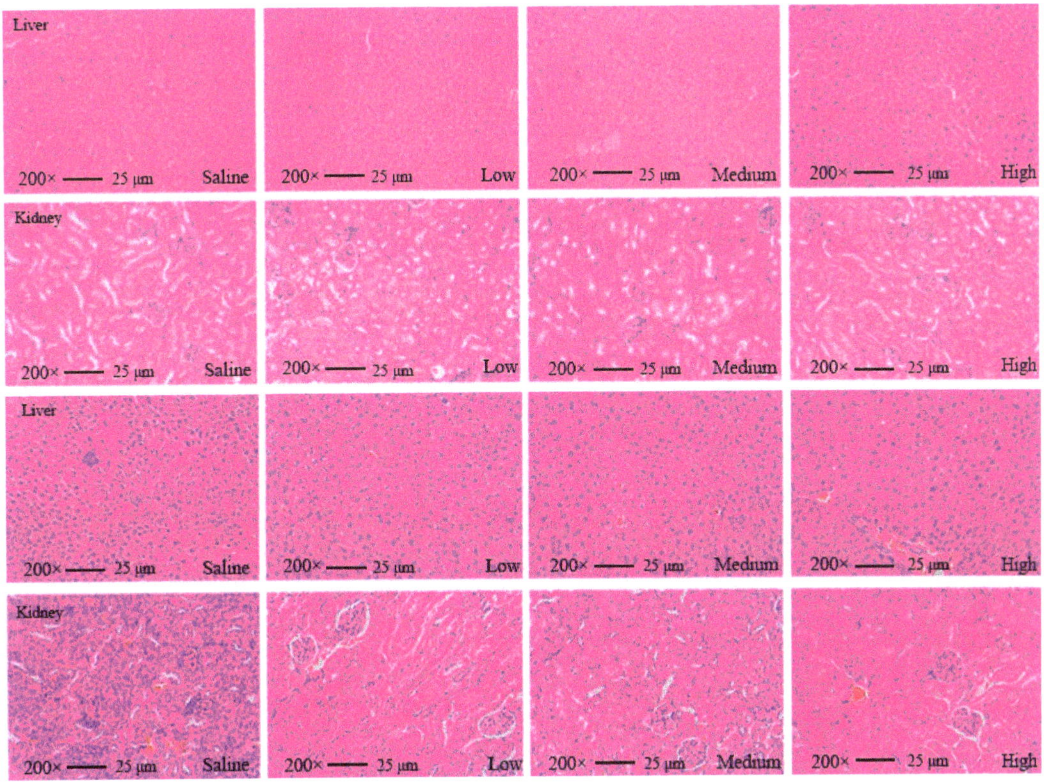

Figure 3. Tissue slice images of the liver and kidney from mice in the acute toxicity experiment (top two) and sub-chronic toxicity experiment (bottom two) at different concentrations of *A. muciniphila* PROBIO.

3.6. Genetic Characteristics

The complete genome sequences of *Akkermansia muciniphila* PROBIO are available at NCBI SRA accession number PRJNA1065580. The genome of AKK PROBIO consists of a single chromosome measuring 2,863,710 base pairs, with a GC content of approximately 55.32% (Table S3). Utilizing the integrated data from genome sequencing, gene prediction, and non-coding RNA prediction, a gene circle diagram was constructed (Figure 4). To gain further insights into the functional annotation of genes, searches were conducted in the COG, KEGG, and GO databases. Out of the total predicted genes, 59.10% were successfully assigned to COG categories (Figure 5a). Among these assignments, 210 genes were associated with general function prediction, 195 genes were linked to cell wall, cell membrane, and envelope synthesis, 182 genes were involved in translation, ribosome structure, and biosynthesis, and 168 genes were associated with amino acid transport and metabolism. Additionally, 38.91% of the genes could be assigned to KEGG annotation (Figure 5b), with a total of 666 genes found to be related to metabolism. Specifically, 126 genes were involved in carbohydrate metabolism, 125 genes were associated with amino acid metabolism, and 102 genes were linked to functional factors and vitamin metabolism. The GO analysis revealed that there were 1322 genes associated with molecular function, 1332 genes associated with cellular components, and 1439 genes involved in biological processes (Figure 5c). Only one virulence factor gene, *tufA*, was identified in the VFD database (Table S3). According to the PathogenFinder database, AKK PROBIO was predicted to be a non-human pathogen (Figure S1). These results indicated that AKK PROBIO was non-toxigenic at the genetic level.

3.7. In Vitro Conjugal Transfer of Moxifloxacin Resistance

The results of moxifloxacin resistance gene transfer under different conditions for two donor/receptor ratios (1:1 and 10:1) are shown in Table 7. The MIC value of receptor strain *E. faecalis* 29212 for moxifloxacin did not change under different conditions of conjugation, indicating that the recipient strain did not obtain the moxifloxacin resistance gene from *A. muciniphila* PROBIO. Therefore, these results suggested that the moxifloxacin resistance gene from the donor strain AKK PROBIO did not transfer.

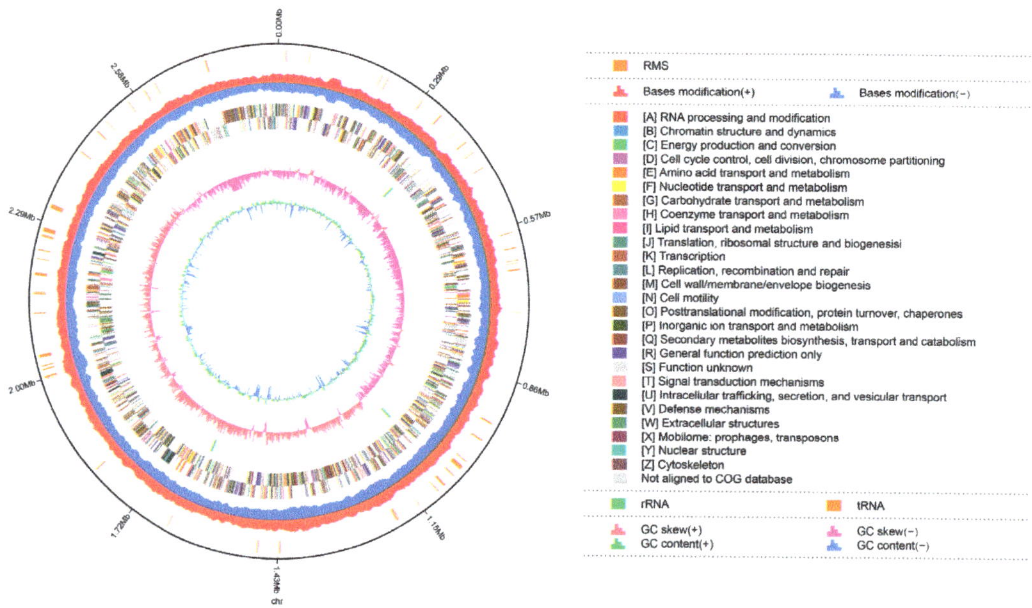

Figure 4. Genome map of *A. muciniphila* PROBIO.

Table 7. Transfer of the moxifloxacin resistance gene.

Growth in Moxifloxacin Medium (μg/mL)		0	0.05	0.125	0.5	2	4	8
1:1 Liquid	*A. muciniphila* and *E. faecalis*	+	+	+	−	−	−	−
	E. faecalis alone	+	+	+	−	−	−	−
10:1 Liquid	*A. muciniphila* and *E. faecalis*	+	+	+	−	−	−	−
	E. faecalis alone	+	+	+	−	−	−	−
1:1 Agar	*A. muciniphila* and *E. faecalis*	+	+	+	−	−	−	−
	E. faecalis alone	+	+	+	−	−	−	−
10:1 Agar	*A. muciniphila* and *E. faecalis*	+	+	+	−	−	−	−
	E. faecalis alone	+	+	+	−	−	−	−
1:1 0.22 μm Filter	*A. muciniphila* and *E. faecalis*	+	+	+	−	−	−	−
	E. faecalis alone	+	+	+	−	−	−	−
10:1 0.22 μm Filter	*A. muciniphila* and *E. faecalis*	+	+	+	−	−	−	−
	E. faecalis alone	+	+	+	−	−	−	−

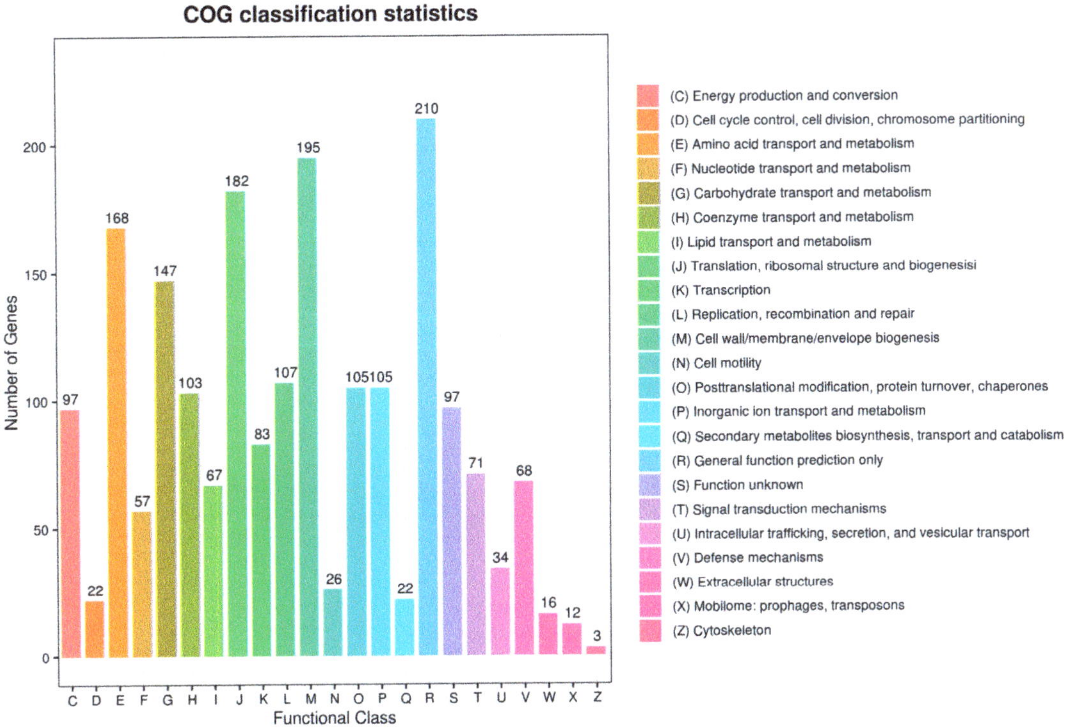

(**a**)

Figure 5. *Cont.*

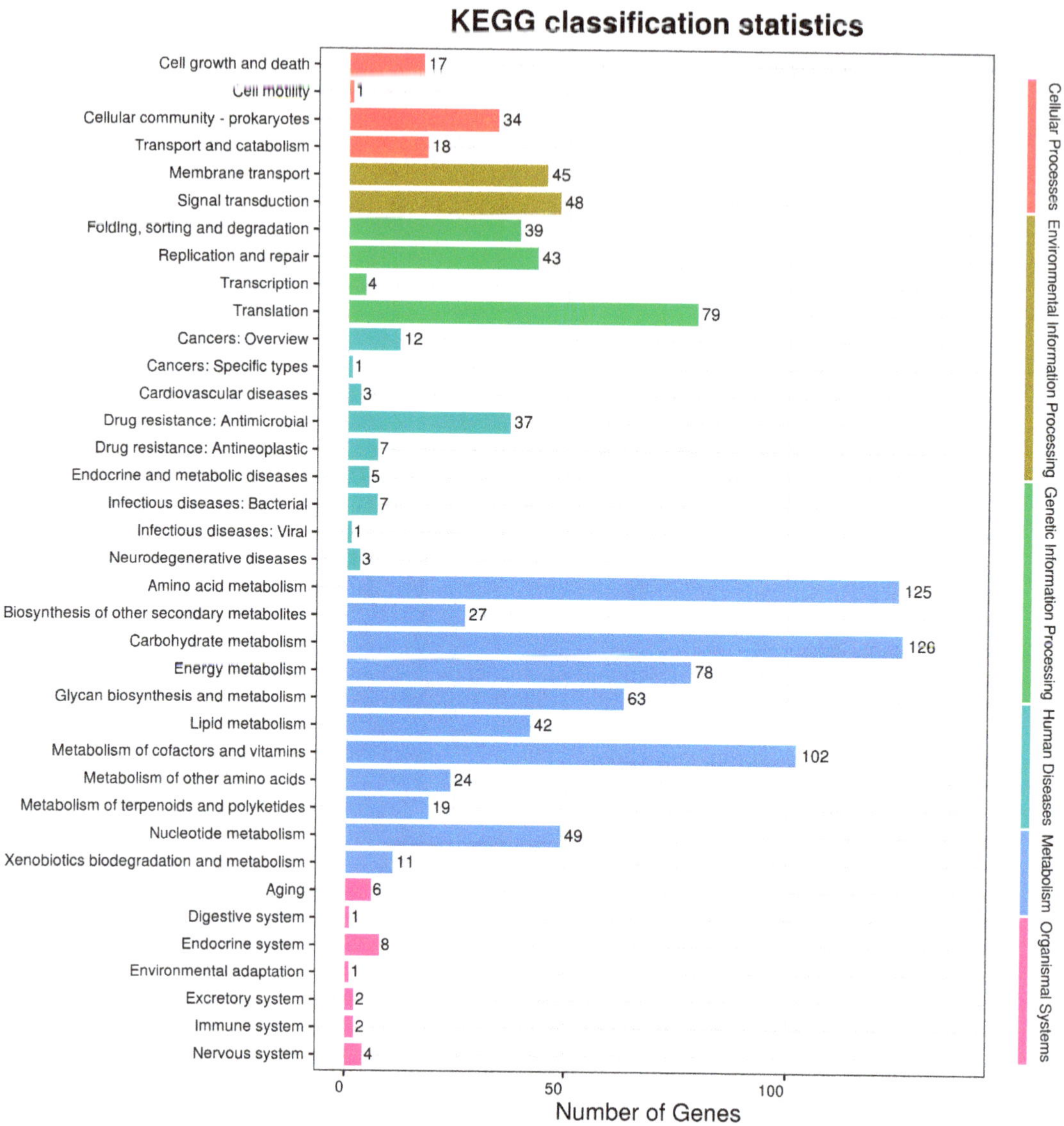

(b)

Figure 5. *Cont.*

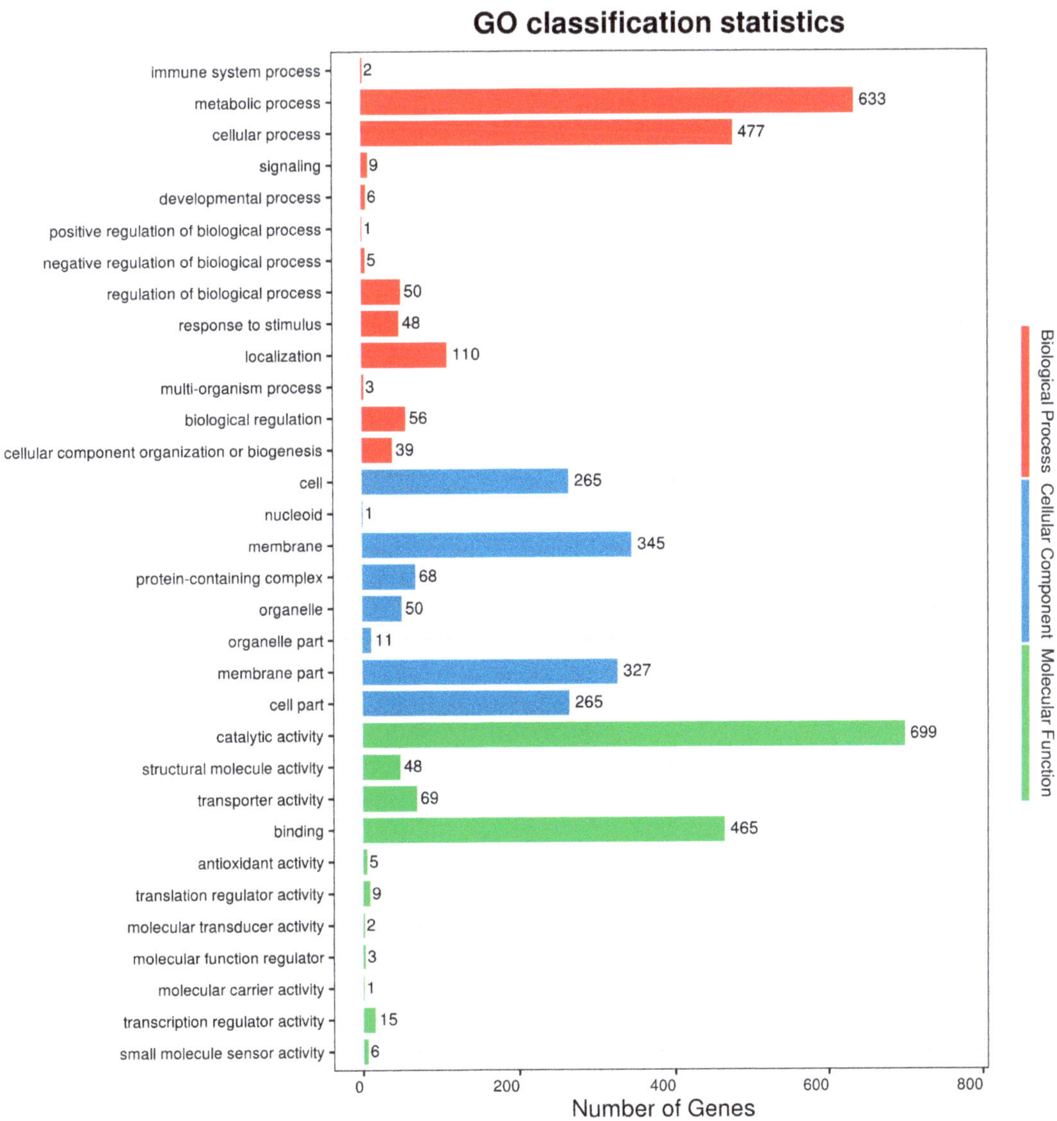

(c)

Figure 5. Functional distribution of COG annotation (**a**), KEGG annotation (**b**), and GO annotation (**c**). The X axis represents the number of unigenes. Unigenes were grouped into 23 KOG categories in the COG functional classification.

4. Discussion

Clear safety is a prerequisite for bacterial strains to be classified as probiotic strains. Consequently, probiotics such as *Bifidobacterium* and *Lactobacillus*, which are widely acknowledged as safe for human consumption, have garnered significant interest in both probiotic research and industry [34]. *A. muciniphila* is abundantly found in the human intestine and breast milk. Numerous studies have demonstrated a correlation between the absence of or reduction in *A. muciniphila* and various diseases, including obesity [35,36], diabetes [37], hepatic steatosis [38], inflammation [15], and the efficacy of immunotherapy [39]. Following the administration of *A. muciniphila*, a partial alleviation of symptoms associated with these diseases has been observed, thereby suggesting its potential as a next-generation probiotic. However, the safety assessment of *A. muciniphila* remained limited; only *A. muciniphila* ATCC BAA-835, DSM 22959, and AM02/06 have been evaluated and confirmed in animals [4,22,26]. A randomized controlled trial that applied up to 10^{10} CFU/day of live or pasteurized *A. muciniphila* ATCC BAA-835 found no adverse events or significant changes in clinical safety parameters for individuals with excess body weight [22]. However, the EFSA did not support this claim due to the limited subjects, disease scope, and short test period of the research [23]. The limited safety assessment of *A. muciniphila* in comparison to conventional probiotics hampers its utilization in clinical investigations and its potential for probiotic development. Therefore, it is imperative to thoroughly consider and evaluate the safety of other *A. muciniphila* strains.

A new *A. muciniphila* strain (*A. muciniphila* PROBIO) with potential probiotic properties was isolated from human feces. Given that *A. muciniphila* is not categorized as an opportunistic pathogen, this newly isolated strain has promise in terms of safety. Furthermore, the species demonstrates the ability to secrete a diverse range of enzymes, thereby augmenting its purported capacity for promoting health [40]. With the aim of carrying out a preliminary safety evaluation of a new strain that has no history of use in the food industry, we systematically evaluated and considered the safety of AKK PROBIO at the strain level, including the genetic information and physiological and biochemical characteristics of the strain, and also carried out an extensive toxicological evaluation.

The potential pathogenicity of *A. muciniphila* is related to adherence, which is related to the initial pathogenic behavior and the degradation of mucosal layers [41]. In this study, we evaluated the adhesion potential of AKK PROBIO to the intestinal mucosa. AKK PROBIO showed weak adhesion compared to other conventional probiotics, which indicates that it can colonize the intestine for only a short period of time. Notably, certain health effects of probiotics are thought to be influenced by their ability to adhere to the gut epithelium. However, as long as regular bacterial consumption takes place, probiotic colonization may not be required to achieve probiotic benefits. Therefore, the probiotic potential of the strain AKK PROBIO should not be affected by its weak adherence. We used a classical probiotic colonization evaluation method; however, as *A. muciniphila* does not use glucose as a limiting carbon source, there may be some limitations in the method.

Antibiotic resistance is a significant concern when assessing potential probiotic candidates [42]. The EFSA has introduced the concept of the qualified presumption of safety, which emphasizes the necessity of thoroughly evaluating the antibiotic resistance genes and horizontal transfer capabilities of biological agents [43]. Thus far, no transferable antibiotic resistance genes for *A. muciniphila* have been documented. We chose seven widely used antibiotics, guided by the predictions of CARD and ResFinder, to comprehensively assess antibiotic resistance. Our results diverged from those reported by Cozzolino et al., as they observed resistance of *A. muciniphila* to chloramphenicol [26]. In our study, *A. muciniphila* PROBIO exhibited sensitivity to most of the common antibiotics, except moxifloxacin. These results underscore the strain-specific nature of antibiotic resistance. The CARD database and ResFinder databases were used to explore the drug resistance of *A. muciniphila* PROBIO across a broader spectrum. The prediction yielded only three resistance genes belonging to two categories, indicating that AKK PROBIO is susceptible to the majority of antibiotics. According to the study by Filardi et al., all strains of *A. muciniphila* harbor the gene *adeF*, encoding for a subunit of the resistance-nodulation-cell division efflux pump system [44].

However, it is important to acknowledge that relying solely on the antibiotic resistance database may lead to erroneous conclusions, necessitating more antibiotic sensitivity tests to assess the safety of AKK PROBIO in subsequent investigations. Based on the antibiotic susceptibility evaluation tests, it was found that AKK PROBIO was indeed resistant to moxifloxacin. All *A. muciniphila* strains, including the type strain, showed low sensitivity to ciprofloxacin [44]. Both moxifloxacin and ciprofloxacin are quinolone antibiotics, so *A. muciniphila* may be innately resistant to quinolone antibiotics. The *adeF* gene of the strain was analyzed by bioinformatics as a possible cause of moxifloxacin resistance, and experimental validation was carried out to explore the non-transferability of resistance genes in *A. muciniphila*. *E. faecalis* is an opportunistic pathogen with a propensity to acquire resistance determinants through horizontal gene transfer [45] and is often used as a receptor for antibiotic transfer experiments [46]. *E. faecalis* ATCC 29212, which is sensitive to moxifloxacin, was used as a receptor strain, and the change in MIC value of this strain was observed by a gradient MH dish with moxifloxacin resistance to determine whether this receptor strain could obtain the moxifloxacin resistance gene from *A. muciniphila*. No change in the MIC value of *E. faecalis* ATCC 29212 against moxifloxacin was observed under different binding environments. The moxifloxacin resistance gene *adeF* of the donor strain AKK PROBIO was not transferable.

The findings from the in vitro genotoxicity assessments indicated that AKK PROBIO does not exhibit mutagenic properties, as determined by the bacterial reverse mutation test, nor does it demonstrate clastogenic or aneugenic effects, as determined by the in vitro mammalian cell micronucleus test. In both a 7-day acute toxicity experiment and a 90-day sub-chronic toxicity experiment, no adverse effects were observed in relation to the test items in clinical observations: body weight, food intake, water intake, routine blood parameters, organ weight, histopathology, glucose, and clinical chemistry parameters. Therefore, the no-observed-adverse-effect level (NOAEL) for the 90-day study was concluded to be 9.2×10^9 cells/kg body weight/day (the highest dose tested), corresponding to doses of 6.4×10^{11} viable bacteria for an average 70 kg human being. The value obtained for viable AKK PROBIO cells was within the range of typical values for probiotics.

Genome-based safety assessments of probiotics are gaining popularity due to their cost-efficiency and speed. Whole-genome sequencing (WGS) of individual probiotic strains provides more information about gene function, such as putative virulence factors and antibiotic resistance genes [47]. In the present study, the gene function prediction results showed that the majority of these genes exhibit multifunctionality, primarily encompassing the synthesis of structural proteins and proteins involved in physiological activities, such as the general metabolic processes of the cell. None of these genes have previously been documented in relation to the pathogenesis of *A. muciniphila*. The virulence factor and pathogenicity were the main concerns when considering *A. muciniphila* for probiotic definition. Only one putative virulence factor gene, *tufA*, was annotated in the genome of AKK PROBIO through comparative analysis with the VFD database. The gene *tufA* encodes the translation elongation factor Tu 1, which primarily participates in protein translation and is widely present in prokaryotes [48]. There has been no report about the role of *tufA* in causing disease. No other strain-specific virulence factors of *A. muciniphila* PROBIO were detected, and no virulence-related factors were found to be associated with the production and secretion systems of toxins.

5. Conclusions

In summary, we evaluated the strain *A. muciniphila* PROBIO in terms of the physicochemical characteristics of the cell surface, biofilm formation, resistance to antibiotics, tolerance to simulated gastrointestinal fluids, mutagenicity, acute toxicity, sub-chronic toxicity, and conjugal transfer of moxifloxacin resistance. Overall, *A. muciniphila* PROBIO in its viable form was shown to (i) have a better GI tolerance, suggesting its desirable probiotic properties; (ii) have no mutagenic or clastogenic properties; (iii) have no risk of spreading antibiotic resistance genes; (iv) have no toxicological effect, as confirmed by acute and sub-chronic toxicity studies; (v) have no potential virulence factors or any pathogenic properties; and (vi) its moxifloxacin resistance does not metastasize. Considering these findings together, *A. muciniphila* PROBIO is most likely safe for use as a probiotic candidate. This study supplements the initial safety assessment work already carried out for *A. muciniphila* and contributes to the development of next-generation probiotics.

Supplementary Materials: The following supporting information can be downloaded at: https://www.mdpi.com/article/10.3390/foods13030442/s1. Figure S1: Prediction of pathogenicity of *A. muciniphila PROBIO*; Table S1: Two genes related to potential drug resistance of *A. muciniphila* PROBIO detected based on the CARD database; Table S2: One gene related to potential drug resistance of *A. muciniphila* PROBIO detected based on the ResFinder database; Table S3: Genome sequence characteristics of *A. muciniphila* PROBIO; Table S4: A gene related to virulence factors of *A. muciniphila* PROBIO detected based on the VFD database.

Author Contributions: Methodology, investigation, formal analysis, X.M.; software, M.T. and X.Y.; validation and visualization: X.M., M.T. and X.Y.; data curation, M.L. and B.L.; writing—original draft preparation, X.M. and M.T.; conceptualization, supervision, resources, writing—review and editing: W.W. and D.R.; project administration, D.R.; funding acquisition: W.W. All authors have read and agreed to the published version of the manuscript.

Funding: This work was supported by the Suzhou Social Development Science and Technology Innovation Project (No. 2022SS35).

Institutional Review Board Statement: The protocols were approved by the ethics committee of the Laboratory Animal Center of Jilin Agricultural University (permit number: 20220711002).

Informed Consent Statement: Not applicable.

Data Availability Statement: The data used in the current study are available upon reasonable request to the corresponding author. The data are not publicly available due to privacy restrictions.

Acknowledgments: Many thanks to Yongli Zhang for her help in the cultivation of *Akkermansia muciniphila.*

Conflicts of Interest: The authors declare no conflicts of interest.

References

1. Derrien, M.; Vaughan, E.E.; Plugge, C.M.; de Vos, W.M. *Akkermansia muciniphila* gen. nov., sp. nov., a human intestinal mucin-degrading bacterium. *Int. J. Syst. Evol. Microbiol.* **2004**, *54 Pt 5*, 1469–1476. [CrossRef]
2. Yan, J.; Sheng, L.; Li, H. *Akkermansia muciniphila*: Is it the Holy Grail for ameliorating metabolic diseases? *Gut Microbes* **2021**, *13*, 1984104. [CrossRef]
3. Sharon, G.; Ioannis, K.; Willem, D.V.; Clara, B. *Akkermansia muciniphila* in the Human Gastrointestinal Tract: When, Where, and How? *Microorganisms* **2018**, *6*, 75.
4. Hou, F.; Tang, J.; Liu, Y.; Tan, Y.; Wang, Y.; Zheng, L.; Liang, D.; Lin, Y.; Wang, L.; Pan, Z.; et al. Safety Evaluation and Probiotic Potency Screening of *Akkermansia muciniphila* Strains Isolated from Human Feces and Breast Milk. *Microbiol. Spectr.* **2023**, *11*, e0336122. [CrossRef] [PubMed]
5. Falony, G.; Joossens, M.; Vieira-Silva, S.; Wang, J.; Darzi, Y.; Faust, K.; Kurilshikov, A.; Bonder, M.J.; Valles-Colomer, M.; Vandeputte, D.; et al. Population-level analysis of gut microbiome variation. *Science* **2016**, *352*, 560–564. [CrossRef]
6. Ghaffari, S.; Abbasi, A.; Somi, M.H.; Moaddab, S.Y.; Leylabadlo, H.E. *Akkermansia muciniphila*: From its critical role in human health to strategies for promoting its abundance in human gut microbiome. *Crit. Rev. Food Sci. Nutr.* **2022**, *63*, 7357–7377. [CrossRef]
7. Belzer, C.; de Vos, W.M. Microbes inside—From diversity to function: The case of *Akkermansia*. *ISME J.* **2012**, *6*, 1449. [CrossRef]
8. Earley, H.; Lennon, G.; Balfe, I.; Coffey, J.C.; O'Connell, P.R. The abundance of *Akkermansia muciniphila* and its relationship with sulphated colonic mucins in health and ulcerative colitis. *Sci. Rep.* **2019**, *9*, 15683. [CrossRef]
9. Gao, X.; Yue, C.; Tian, R.; Yu, L.; Tian, F.; Zhao, J.; Chen, W.; Zhai, Q. *Akkermansia muciniphila*-directed polyphenol chlorogenic acid intervention for obesity in mice. *Food Sci. Hum. Wellness* **2024**, *13*, 90–100. [CrossRef]
10. Shuoker, B.; Pichler, M.J.; Jin, C.; Sakanaka, H.; Wu, H.; Gascuea, A.M.; Liu, J.; Nielsen, T.S.; Holgersson, J.; Karlsson, E.N. Sialidases and fucosidases of *Akkermansia muciniphila* are crucial for growth on mucin and nutrient sharing with mucus-associated gut bacteria. *Nat. Commun.* **2023**, *14*, 1833. [CrossRef]
11. Sugihara, K.; Kitamoto, S.; Saraithong, P.; Nagao-Kitamoto, H.; Hoostal, M.; McCarthy, C.; Rosevelt, A.; Muraleedharan, C.K.; Gillilland, M.G., 3rd; Imai, J.; et al. Mucolytic bacteria license pathobionts to acquire host-derived nutrients during dietary nutrient restriction. *Cell Rep.* **2022**, *40*, 111093. [CrossRef] [PubMed]
12. Hasani, A.; Ebrahimzadeh, S.; Hemmati, F.; Khabbaz, A.; Hasani, A.; Gholizadeh, P. The role of *Akkermansia muciniphila* in obesity, diabetes and atherosclerosis. *J. Med. Microbiol.* **2021**, *70*, 001435. [CrossRef] [PubMed]
13. Everard, A.; Belzer, C.; Geurts, L.; Ouwerkerk, J.P.; Cani, P.D. Cross-talk between *Akkermansia muciniphila* and intestinal epithelium controls diet-induced obesity. *Proc. Natl. Acad. Sci. USA* **2013**, *110*, 9066–9071. [CrossRef] [PubMed]

14. Plovier, H.; Everard, A.; Druart, C.; Depommier, C.; Van Hul, M.; Geurts, L.; Chilloux, J.; Ottman, N.; Duparc, T.; Lichtenstein, L. A purified membrane protein from *Akkermansia muciniphila* or the pasteurized bacterium improves metabolism in obese and diabetic mice. *Nat. Med.* **2016**, *23*, 107. [CrossRef] [PubMed]
15. Li, J.; Lin, S.; Vanhoutte, P.M.; Woo, C.W.; Xu, A. *Akkermansia muciniphila* Protects Against Atherosclerosis by Preventing Metabolic Endotoxemia-Induced Inflammation in Apoe$^{(-/-)}$ Mice. *Circulation* **2016**, *133*, 2434–2446. [CrossRef] [PubMed]
16. Zheng, M.; Han, R.; Yuan, Y.; Xing, Y.; Zhang, W.; Sun, Z.; Liu, Y.; Li, J.; Mao, T. The role of *Akkermansia muciniphila* in inflammatory bowel disease: Current knowledge and perspectives. *Front. Immunol.* **2022**, *13*, 1089600. [CrossRef] [PubMed]
17. Wang, F.; Cai, K.; Xiao, Q.; He, L.; Xie, L.; Liu, Z. *Akkermansia muciniphila* administration exacerbated the development of colitis-associated colorectal cancer in mice. *J. Cancer* **2022**, *13*, 124–133. [CrossRef] [PubMed]
18. Chelakkot, C.; Choi, Y.; Kim, D.K.; Park, H.T.; Ghim, J.; Kwon, Y.; Jeon, J.; Kim, M.S.; Jee, Y.K.; Gho, Y.S. *Akkermansia muciniphila*-derived extracellular vesicles influence gut permeability through the regulation of tight junctions. *Exp. Mol. Med.* **2018**, *50*, e450. [CrossRef]
19. Matson, V.; Fessler, J.; Bao, R.; Chongsuwat, T.; Zha, Y.; Alegre, M.L.; Luke, J.J.; Gajewski, T.F. The commensal microbiome is associated with anti-PD-1 efficacy in metastatic melanoma patients. *Science* **2018**, *359*, 104–108. [CrossRef]
20. Derosa, L.; Routy, B.; Thomas, A.M.; Iebba, V.; Zalcman, G.; Friard, S.; Mazieres, J.; Audigier-Valette, C.; Moro-Sibilot, D.; Goldwasser, F.; et al. Intestinal *Akkermansia muciniphila* predicts clinical response to PD-1 blockade in patients with advanced non-small-cell lung cancer. *Nat. Med.* **2022**, *28*, 315–324. [CrossRef]
21. Cani, P.D.; Depommier, C.; Derrien, M.; Everard, A.; Vos, W.M.D. *Akkermansia muciniphila*: Paradigm for next-generation beneficial microorganisms. *Nat. Rev. Gastroenterol. Hepatol.* **2022**, *19*, 625–637. [CrossRef]
22. Depommier, C.; Everard, A.; Druart, C.; Plovier, H.; Van Hul, M.; Vieira-Silva, S.; Falony, G.; Raes, J.; Maiter, D.; Delzenne, N.M. Supplementation with *Akkermansia muciniphila* in overweight and obese human volunteers: A proof-of-concept exploratory study. *Nat. Med.* **2013**, *25*, 1096–1103. [CrossRef]
23. Turck, D.; Bohn, T.; Castenmiller, J.; De Henauw, S.; Hirsch-Ernst, K.I.; Maciuk, A.; Mangelsdorf, I.; McArdle, H.J.; Naska, A.; Pelaez, C.; et al. Safety of pasteurised *Akkermansia muciniphila* as a novel food pursuant to Regulation (EU) 2015/2283. *EFSA J.* **2021**, *19*, e06780.
24. Lund, B.; Edlund, C. Probiotic *Enterococcus faecium* strain is a possible recipient of the vanA gene cluster. *Clin. Infect. Dis.* **2001**, *32*, 1384–1385. [CrossRef]
25. Ulsemer, P.; Toutounian, K.; Schmidt, J.; Karsten, U.; Goletz, S. Preliminary safety evaluation of a new *Bacteroides xylanisolvens* isolate. *Appl. Environ. Microbiol.* **2012**, *78*, 528–535. [CrossRef]
26. Cozzolino, A.; Vergalito, F.; Tremonte, P.; Iorizzo, M.; Lombardi, S.J.; Sorrentino, E.; Luongo, D.; Coppola, R.; Di Marco, R.; Succi, M. Preliminary Evaluation of the Safety and Probiotic Potential of *Akkermansia muciniphila* DSM 22959 in Comparison with Lactobacillus rhamnosus GG. *Microorganisms* **2020**, *8*, 189. [CrossRef] [PubMed]
27. Dey, D.K.; Koo, B.G.; Sharma, C.; Kang, S.C. Characterization of *Weissella confusa* DD_A7 isolated from kimchi. *LWT* **2019**, *111*, 663–672. [CrossRef]
28. Ames, B.N.; Mccann, J.; Yamasaki, E. Methods for detecting carcinogens and mutagens with the *Salmonella*/mammalian-microsome mutagenicity test. *Mutat. Res.* **1975**, *31*, 347–363. [CrossRef] [PubMed]
29. Maron, D.M. Revised methods for the Salmonella mutagenicity test. *Mutat. Res.* **1983**, *113*, 173–215. [CrossRef] [PubMed]
30. Fenech, M. The cytokinesis-block micronucleus technique and its application to genotoxicity studies in human populations. *Environ. Health Perspect. Suppl.* **1993**, *101*, 101.
31. Clinical and Laboratory Standards Institute. *Performance Standards for Antimicrobial Susceptibility Testing*; Clinical and Laboratory Standards Institute: Wayne, PA, USA, 2007; pp. 106–112.
32. Trial, C.F. OECD Guideline for the Testing of Chemicals. *Mortality* **1998**, *601*, 858.
33. Ashwini, M.; Ray, M.; Sumana, K.; Halami, P.M. Prevalence of macrolide-lincosamide-streptogramin resistant lactic acid bacteria isolated from food samples. *J. Food Sci. Technol.* **2023**, *60*, 630–642. [CrossRef] [PubMed]
34. De-Chun, W.; Su, J. Probiotics Application in Food Industry. *Modern Food Sci. Technol.* **2005**, *4*, 124–125.
35. Davis, J.A.; Collier, F.; Mohebbi, M.; Stuart, A.L.; Loughman, A.; Pasco, J.A.; Jacka, F.N. Obesity, *Akkermansia muciniphila*, and Proton Pump Inhibitors: Is there a Link? *Obes. Res. Clin. Pract.* **2020**, *14*, 524–530. [CrossRef]
36. Ostadrahimi, K.A. *Akkermansia muciniphila* bacteria: A new perspective on the management of obesity: An updated review. *Rev. Med. Microbiol.* **2019**, *30*, 83–89.
37. Zhang, J.; Ni, Y.; Qian, L.; Fang, Q.; Zheng, T.; Zhang, M.; Gao, Q.; Zhang, Y.; Ni, J.; Hou, X. Decreased Abundance of *Akkermansia muciniphila* Leads to the Impairment of Insulin Secretion and Glucose Homeostasis in Lean Type 2 Diabetes (Adv. Sci. 16/2021). *Adv. Sci.* **2021**, *8*, 2170102. [CrossRef]
38. Grander, C.; Grabherr, F.; Enrich, B.; Tilg, H. *Akkermansia muciniphila* tend to influence experimental non-alcoholic steatohepatitis. *Z. Für Gastroenterol.* **2017**, *55*, e28–e56. [CrossRef]
39. Jiang, Y.; Xu, Y.; Zheng, C.; Ye, L.; Jiang, P.; Malik, S.; Xu, G.; Zhou, Q.; Zhang, M. Acetyltransferase from *Akkermansia muciniphila* blunts colorectal tumourigenesis by reprogramming tumour microenvironment. *Gut J. Br. Soc. Gastroenterol.* **2023**, *72*, 1308–1318.
40. Zou, Y.; Chen, T. Engineered *Akkermansia muciniphila*: A promising agent against diseases (Review). *Exp. Ther. Med.* **2020**, *20*, 285. [CrossRef]

41. Kosciow, K.; Deppenmeier, U. Characterization of three novel β-galactosidases from *Akkermansia muciniphila* involved in mucin degradation. *Int. J. Biol. Macromol.* **2020**, *149*, 331–340. [CrossRef]
42. Liu, C.; Zhang, Z.-Y.; Dong, K.; Yuan, J.-P.; Guo, X.-K. Antibiotic Resistance of Probiotic Strains of Lactic Acid Bacteria Isolated from Marketed Foods and Drugs. *Biomed. Environ. Sci.* **2009**, *22*, 401–412. [CrossRef] [PubMed]
43. Theodor, B.; Akihito, E.; Miguel, G.; Gabriel, V.; Wolfgang, K.; De, V.W.M.; Seppo, S.; Carlos, G.-G. Safety of Novel Microbes for Human Consumption: Practical Examples of Assessment in the European Union. *Front. Microbiol.* **2017**, *8*, 1725.
44. Filardi, R.; Gargari, G.; Mora, D.; Arioli, S. Characterization of antibiotic-resistance traits in *Akkermansia muciniphila* strains of human origin. *Sci. Rep.* **2022**, *12*, 19426. [CrossRef] [PubMed]
45. Dale, J.L.; Cagnazzo, J.; Phan, C.Q.; Barnes, A.M.; Dunny, G.M. Multiple Roles for Enterococcus faecalis Glycosyltransferases in Biofilm-Associated An-tibiotic Resistance, Cell Envelope Integrity, and Conjugative Transfer. *Antimicrob. Agents Chemother.* **2015**, *59*, 4094–4105. [CrossRef] [PubMed]
46. Preethi, C.; Thumu, S.C.R.; Halami, P.M. Occurrence and distribution of multiple antibiotic-resistant *Enterococcus* and *Lactobacillus* spp. from Indian poultry: In vivo transferability of their erythromycin, tetracycline and vancomycin resistance. *Ann. Microbiol.* **2017**, *67*, 395–404. [CrossRef]
47. Peng, X.; Ed-Dra, A.; Yue, M. Whole genome sequencing for the risk assessment of probiotic lactic acid bacteria. *Crit. Rev. Food Sci. Nutr.* **2022**, *63*, 11244–11262. [CrossRef]
48. Furano, A.V. The elongation factor Tu coded by the tufA gene of *Escherichia coli* K-12 is almost identical to that coded by the tufB gene. *J. Biol. Chem.* **1977**, *252*, 2154–2157. [CrossRef]

foods

MDPI

Article

Supplementation with *Lactiplantibacillus brevis* GKEX Combined with Resistance Exercise Training Improves Muscle Mass, Strength Performance, and Body Fat Condition in Healthy Humans

Mon-Chien Lee [1,2], Yi-Ju Hsu [1], Chin-Shan Ho [1], You-Shan Tsai [3], Chin-Chu Chen [3,4] and Chi-Chang Huang [1,5,*]

1 Graduate Institute of Sports Science, National Taiwan Sport University, Taoyuan 333325, Taiwan; 1061304@ntsu.edu.tw (M.-C.L.); ruby780202@ntsu.edu.tw (Y.-J.H.); kilmur23@ntsu.edu.tw (C.-S.H.)
2 Center for General Education, Taipei Medical University, Taipei 110301, Taiwan
3 Biotech Research Institute, Grape King Bio Ltd., Taoyuan 325002, Taiwan; youshan.tsai@grapeking.com.tw (Y.-S.T.); gkbioeng@grapeking.com.tw (C.-C.C.)
4 Institute of Food Science and Technology, National Taiwan University, Taipei 106319, Taiwan
5 Tajen University, Pingtung 907101, Taiwan
* Correspondence: john5523@ntsu.edu.tw; Tel.: +886-3-3283201 (ext. 2409)

Abstract: In addition to maintaining good exercise and dietary habits, recent studies have shown that probiotics may have potential benefits for muscle mass and strength. It is worth noting that the effects may vary depending on the specific strains used. To date, no studies have analyzed the effects of *Lactiplantibacillus brevis* in this context. Here, we combine the *L. brevis* strain GKEX with resistance training to further understand its effects on muscle mass, thickness, performance, and fat loss. In a six-week intervention for a double-blind randomized trial, 52 healthy subjects were divided into two groups (10 male and 16 female participants in each group): a placebo group (two capsules/day, containing 0 CFU of GKEX per capsule) and a GKEX group (two capsules/day, containing 1×10^{10} CFU of GKEX per capsule). Before the intervention, no differences were observed between the two groups in any of the tests (body composition, muscle thickness, exercise performance, and blood parameters). However, supplementation with GKEX significantly improved muscle mass and thickness, as well as grip strength, muscle strength, and explosive performance, when compared to the associated parameters before the intervention. Additionally, GKEX supplementation promoted a reduction in the body fat percentage ($p < 0.05$). Through analysis of the change amount, we observed that GKEX supplementation yielded significantly improved benefits when compared to the placebo group ($p < 0.05$). In summary, our findings support the notion that a six-week resistance exercise training program combined with *L. brevis* GKEX supplementation has superior additive effects that enhance muscle mass and strength performance, while also reducing body fat percentage. This intervention can promote muscle gain and fat loss.

Keywords: *Lactiplantibacillus brevis*; probiotic; resistance training; muscle mass; strength

1. Introduction

Appropriate and regular exercise training is considered to be effective in preventing injuries, improving muscle mass and strength performance, improving the physiological metabolism, and so on [1]. In addition to exercise training, appropriate nutritional supplementation also plays a considerable role in exercise performance and promoting muscle growth and synthesis [2]. In the past, common sports nutrition supplements were nothing more than antioxidant or protein supplements; however, in recent years, more and more studies have shown that probiotics are supplements with the potential to increase muscle mass and exercise performance. In a past meta-trial combining the results of 24 studies, it was found that probiotic supplementation significantly improved skeletal muscle mass and strength performance, although it did not help total lean mass [3]. In addition, another study has also shown that supplementation with *Lactiplantibacillus plantarum* TWK10 for six consecutive weeks without exercise training significantly improved the muscle mass as well as reducing the body fat rate and exercise fatigue indicators of non-athletes [4].

Probiotics are live micro-organisms that, when taken orally for several weeks and ingested in sufficient amounts, increase the number of beneficial bacteria in the gut, thereby providing health benefits to the host [5,6]. More and more evidence has demonstrated that the gut microbiota can stimulate the IGF-1/p70s6k/mTOR pathway in muscle cells through the synthesis of amino acids, thereby promoting muscle synthesis [7]. The exact mechanism by which probiotics stimulate increases in muscle mass or strength is currently unknown, but the most commonly accepted factor may be through improved protein digestion and amino acid absorption into the bloodstream [8], thereby promoting skeletal muscle anabolism and improving muscle mass and functional performance [9]. Some past studies have confirmed that probiotic supplementation can modulate the gut microbiota to enhance protein assimilation, upregulate mTOR activation and mitochondrial function, and reduce inflammatory cytokine activity, thereby improving muscle anabolism and reducing muscle loss [10,11]. A meta-analysis of the overall effects reported in seven studies has also shown that probiotic supplementation improved muscle mass and strength [12].

Lactobacillus brevis is a heterofermentative Gram-positive organism which is widely present in fermented foods of plant and animal origin, highly adhesive to low-pH bile salts and pancreatic juice, competitive against *Salmonella* and *Escherichia coli*, and promotes primary bile acid biosynthesis, amino acids, and other metabolic pathways [13,14]. A previous study has pointed out that daily supplementation with *L. brevis* FZU0713 for 8 weeks significantly regulated liver genes (including Acat2, Acox1, Hmgcr, Cd36, Srebp-1c, and Cyp7a1), thereby promoting the lipid metabolism and maintaining bile acid stability. In addition, they also found that the levels of short-chain fatty acids (SCFAs) in the feces of HFD-fed rats were significantly increased, which are considered to be able to effectively change the composition and metabolic function of gut micro-organisms [15]. Another previous study has shown that SCFAs play an important role in skeletal muscle function and exercise capacity [16], possibly through activating multiple regulatory pathways (e.g., UCP2-AMPK-ACC and PGC1-α) to increase ATP production and muscle fiber metabolic efficiency to affect skeletal muscle performance [17]. In addition, they act by inhibiting histone deacetylase, thereby preventing apoptosis and preventing muscle protein catabolism [18].

While there are no studies directly confirming the relationship between *L. brevis* and muscle and exercise performance, past studies have shown that *L. brevis* can increase the production of gamma-aminobutyric acid (GABA), the main inhibitory neurotransmitter in the mammalian central nervous system [19]. It can increase plasma growth hormone concentrations, promoting amino acid transport, cellular amino acid uptake, and skeletal muscle growth [20]. A previous study demonstrated that when whey protein was combined with GABA supplementation and resistance exercise training for 12 weeks, it significantly increased growth hormone levels and total body lean tissue weight [21]. Therefore, this study aims to investigate whether supplementation with *L. brevis* combined with resistance exercise training can provide benefits in increasing muscle mass and strength performance, and further explore its adaptability effects on the human body. We hypothesize that the synergistic effect of *L. brevis* combined with resistance exercise training can promote muscle mass and strength performance. This will be the first study to explore the effects of *L. brevis* on muscle and exercise performance, serving not only as a foundation for understanding its mechanisms and applications but also as a potential choice for sports nutrition supplementation.

2. Materials and Methods

2.1. Sample Preparation

L. brevis GKEX was isolated from douchi (fermented black soybeans) and preserved at the National Institute of Technology and Evaluation, Biological Resource Center (NBRC, Chiba, Japan), with collection number NITE BP-03696. The bacterial strain was cultured in MRS broth (BD, Franklin Lakes, NJ, USA) for 16 h at 32 °C with agitation at 100 rpm. Subsequently, the seed culture was 0.01% inoculated in a 15T bioreactor containing 80% culture medium as the working volume. The scaled-up culture was maintained at 32 °C for 16 h with pH control set to 6.0. The bioreactor medium comprised 5% glucose, 2% yeast extract, 0.05% $MgSO_4$, 0.1% K_2HPO_4, and 0.1% Tween 80. Following harvest, *L. brevis* GKEX was centrifuged (MBPX810, Alfa Laval AB, Lund, Sweden) at 7000 rpm and the bacterial pellet was mixed with 10% skim milk for lyophilization (FD24, Kingmech Scientific, Taoyuan, Taiwan) under 30 °C for 5 days. The resulting freeze-dried GKEX powder was then filled into No. 0 vegetable capsules, with each capsule containing 1.0×10^{10} CFU of live bacteria. Since the bacterial powder was mixed with microcrystalline α-cellulose and then filled into capsules, we used microcrystalline α-cellulose as a placebo substitute and filled it into the capsules accordingly. All the manufacturing processes were conducted at Grape King Bio Ltd. (Taoyuan, Taiwan) in compliance with the ISO22000 system [22].

2.2. Participants

The Harvard calculator (http://hedwig.mgh.harvard.edu/sample_size/size.html, accessed on 20 June 2022) was used to determine the sample size, assuming a parallel design with a significance level of 0.05, a power of 0.8, and a standard deviation of 0.8 for the difference. The trial included a total of 52 subjects (including 20 males and 36 females) from the National Sports University, aged 20–39 years old, who were not athletes. None of them met our exclusion criteria, including (1) BMI ≥ 27; (2) a history of metabolic disease, asthma, cancer, cardiovascular disease, or hypertension; (3) limb and neuromuscular movement disorders occurring within the last 6 months, indicating that the subject is unable to exercise; (4) had taken anti-inflammatory, analgesic, acute, or chronic disease or other drugs in the past month; (5) previous hepatobiliary and gastrointestinal surgery (except hernia and polypectomy); (6) those who smoke or drink; (7) having taken probiotic powder, capsules, or lozenges (including yogurt and other related foods) within two weeks; (8) allergy to food or lactic acid bacteria products; (9) students or stakeholders of the principal investigator. All eligible participants gave written informed consent before starting the experiment. The experiment was performed after receiving approval from the Institutional Review Board of Landseed International Hospital (Taoyuan, Taiwan; LSHIRB No. 22-022-A2) and was registered at clinicaltrials.gov under registration number NCT05909475. This study was conducted according to the guidelines of the Declaration of Helsinki.

2.3. Experimental Design

After all subjects were recruited, numbers were assigned according to the order of registration. Then, when the gender ratio of the two groups was the same, we randomly divided them into two groups through numbered sampling and conducted a double-blind experiment (with 10 male and 16 female participants in each group): a placebo group (2 capsules/day, containing 0 CFU of GKEX per capsule) and a GKEX group (2 capsules/day, containing 1×10^{10} CFU of GKEX per capsule). There were no significant differences between the placebo and GKEX groups in terms of age (placebo: 22.2 ± 1.9; GKEX: 22.6 ± 2.4 years old) and height (placebo: 166.2 ± 8.9; GKEX: 168.6 ± 8.7 cm). The trial lasted for 6 weeks, during which all subjects were required to supplement with two capsules daily and perform resistance exercise training three times a week. Before and after the intervention, all subjects underwent tests for dietary record, an exercise performance test, and blood samples were collected to analyze and measure body composition indicators. Among these, only body composition was required to be measured every two weeks.

Furthermore, all subjects were required to maintain their original lifestyle and habits. The nutrient intake was analyzed by a nutritionist. Table 1 presents the daily nutritional intake of participants, including carbohydrates, proteins, fats, and total calories, both before the experiment and six weeks after the intervention. No notable distinctions were observed between the groups, and there was no statistically significant alteration when comparing the pre- and post-intervention values within each group. The experimental design is shown in Figure 1.

Table 1. Dietary intake of subjects before and after the 6-week GKEX intervention.

Dietary Intake	Before		After	
	Placebo	GKEX	Placebo	GKEX
Carbohydrates (g/day)	164 ± 38	164 ± 42	163 ± 44	165 ± 43
Proteins (g/day)	124 ± 547	126 ± 39	125 ± 55	125 ± 42
Fats (g/day)	70 ± 26	71 ± 19	70 ± 25	72 ± 19
Total calories (kcal/day)	1785 ± 521	1803 ± 376	1784 ± 522	1805 ± 378

Data are presented as mean ± SD.

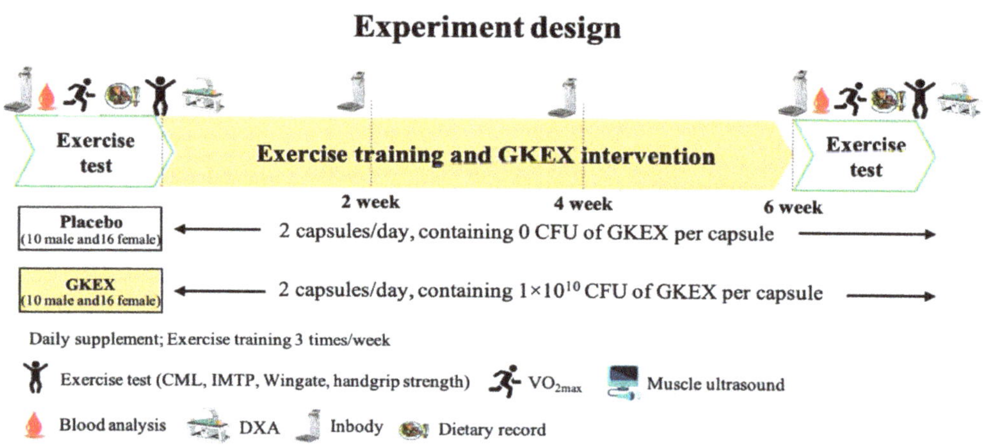

Figure 1. Experimental procedure.

2.4. Resistance Exercise Training

All subjects were instructed to engage in resistance training sessions three times per week, utilizing pneumatic resistance training equipment. The equipment comprised abdominal/back 5310, abduction/adduction 3520-HI5, push-up/pull-down 3120, leg extension/curl 3530, twist rehab 5340, and leg press 5540 (AB Hur Oy, Kokkola, Finland). This equipment selection was designed to target the enhancement of strength in specific muscle groups, including the upper limbs, back, abdomen, and legs. The incorporation of pneumatic resistance training equipment provided a standardized platform for participants to engage in targeted exercises, fostering comprehensive strength development in the upper limbs, back, abdomen, and legs. Before initiation of the training regimen, each participant underwent an assessment of their 3RM (repetition maximum) and, subsequently, their 1RM (repetition maximum) was calculated. This calculation involved the application of a coefficient formula derived from the prior literature, ensuring the determination of an appropriate resistance training intensity [23]. The training protocol spanned four weeks, with distinct progressions in intensity. In the initial week, participants executed a set of 15 repetitions at an intensity equivalent to 60% of their 1RM. Progressing to the second week, the intensity was elevated to 70% of 1RM, with a set comprising 12 repetitions. In the subsequent week, there was a further increase in intensity to 75% of 1RM, maintaining a set of 12 repetitions. Until the last week, subjects consistently maintained an intensity of 75% of 1RM, completing 2 sets of 10 repetitions. Researchers adjusted the resistance equipment weight and number of sets based on each subject's individual 1RM in accordance with the exercise protocol. Subjects were required to complete the designated movements and sets during each training session, with multiple professionals present to monitor the integrity and safety of their movements.

2.5. Maximal Oxygen Uptake (VO2max)

Assessment of the maximum oxygen consumption and exercise performance in the subjects involved the utilization of a treadmill (Pulsar, h/p/cosmos, Traunstein, Bavaria, Germany) and an automated breathing analyzer (Vmax 29c, Sensor Medics, Yorba Linda, CA, USA). Following the manufacturer's instructions, the flow sensor, as well as the O_2 and CO_2 sensors, underwent calibration before each test to guarantee accurate measurements. An automated calibration function was employed for the O_2 and CO_2 sensors. Heart rate (HR) monitoring was performed using a Polar heart rate device. Following the Bruce protocol, the treadmill was initiated at a speed of 7.2 km/h, progressively increasing by 1.8 km/h every 2 min until subjects reached a point of fatigue, as outlined in the Bruce protocol [24]. The determination of maximum O_2 consumption occurred when the respiratory exchange rate (VCO_2/VO_2, the volume ratio of CO_2 produced to O_2) exceeded 1.10 and coincided with reaching the maximum heart rate (maximum HR = 220 − age). The average of the three highest peaks in VO_{2max} was computed to derive individual VO_{2max} values.

2.6. Handgrip Strength Test

Handgrip strength was quantified in kilograms using a Takei digital grip strength meter (T. K. K. 5401, Takei Scientific Instruments Co., Ltd., Kamo, Niigata, Japan). Prior to the formal testing, subjects were instructed to apply minimal force to the gripper as a preliminary measure, in order to ensure their comprehension of the operating procedure and gripping distance. The researchers randomly assigned either the dominant or non-dominant hand as the starting point for the test. During the examination, participants were directed to exert maximal effort while squeezing the gripper with one hand, sustaining the squeeze for a minimum of 5 s. To mitigate the effects of fatigue, subjects alternated hands at 60 s intervals and repeated the test. This exchange method was repeated three times, and the maximum grip strengths of both the dominant and non-dominant hands were individually recorded [25].

2.7. Countermovement Jump Assessment (CMJ)

The CMJ test serves as a means to assess the maximal speed, strength, and explosiveness of the lower body. During this evaluation, participants were directed to place their hands on their hips and position their feet on a Kistler force-measuring platform (9260AA, Kistler GmbH, Winterthur, Switzerland). Subsequently, they were instructed to execute a squat until their knees reached a 90-degree bend, followed with an immediate jump with maximal force to attain the highest possible vertical height. Each participant performed three replicates of the CMJ test, and data were acquired at specified points. The instrument was calibrated for each individual's weight to ensure accuracy [26]. The parameters measured included the rate of force development (RFD), relative peak force, and jump height. These metrics collectively provided insights into the dynamic capabilities of the lower body, reflecting the participant's ability to generate force and power during vertical jumping movements.

2.8. Isometric Mid-Thigh Pull (IMTP)

The isometric mid-thigh pull (IMTP) serves as a force–time diagnostic tool [27]. For this purpose, customized IMTP test equipment and two force plates (type 9287BA, Kistler Instruments AG, Winterthur, Switzerland) were employed. All participants were positioned with their feet at a consistent width and a rod was placed between the thighs, ensuring an upright torso, neutral spine, and knee and hip angles set at 140° to acquaint them with the IMTP test method. A careful evaluation was conducted to confirm the symmetry of the force trace in the practice test and the stability of the weighing cycle. To minimize the risk of measurement errors associated with changes in posture, each test phase involved repeating the specific measurement at 2 min intervals. The recorded parameters encompassed the average relative peak force (N/kg), rate of force development (RFD), and peak rate of force development (pRFD). The collection of these parameters aimed to provide a nuanced understanding of the isometric force production capabilities of the participants and their potential alterations in response to the strenuous exercise regimen.

2.9. Wingate Anaerobic Test (WAnT)

Following a standardized warm-up routine, all participants underwent assessment using the classical Wingate Anaerobic Test (WAnT) on a cycloergometer (Monark 894E, Varberg, Sweden) in a 30 s "go all out" ultramax test. The seat height was adjusted to each participant's satisfaction, and toe clips were utilized to prevent their feet from slipping off the pedals. Prior to the initial test, a 5 min warm-up at approximately 50 W was conducted. Following the warm-up, two 3 s preparation exercises, with a load of 3% of their own body weight, were administered to familiarize the participants with the resistance [28]. The test commenced with the resistance set on the friction belt of the dynamometer. External loading was individually estimated at 5% of body weight. The recorded results included the relative mean power (W/kg), relative peak power (W/kg), and fatigue index (%), which provide insights into the anaerobic power of the participants, showcasing their mean and peak power outputs, as well as their resistance to fatigue during intense, short-duration efforts.

2.10. Body Composition

To assess body composition, a multi-frequency approach utilizing a bioelectrical impedance analyzer (BIA) was employed; specifically, an InBody 570 (In-body, Seoul, Republic of Korea) was utilized. This advanced device conducts measurements at frequencies of 1, 5, 50, 260, 500, and 1000 kHz within a 60 s timeframe. During the measurement process, participants stood on the foothold electrodes following clearance of the palms and soles. Holding the sensing handle with both hands, participants maintained an open-arm position at a 30° angle away from their body, refraining from speaking or making movements throughout the measurement period. Additionally, participants observed an 8 h fasting period prior to the tests, ensuring optimal conditions for the assessment of body composition. We conducted measurements every two weeks during the experimental period.

Because BIA is measured in skeletal muscle, it may be affected by the amount of water in the body. Therefore, we additionally used comprehensive and reliable dual-energy X-ray absorptiometry (DXA) measurements before and after the intervention and facilitated the acquisition of additional bone quality data. The Lunar iDXA system from GE Healthcare (Chicago, IL, USA) was used. Subjects were instructed to lie flat on the test bed, aligning their bodies along the centerline with limbs positioned within the detection range to measure the distribution of lean tissue, fat mass, and bone density. The DXA technology involved the use of two X-ray beams with different energies to scan the designated body regions. The scintillation detector received the X-rays that penetrated the inspected parts, facilitating the calculation and analysis of parameters such as muscle mass and body fat.

2.11. Ultrasonic Measurement of Muscle and Fascia Thickness

Muscle and fascia thickness measurements of the biceps brachii and quadriceps femoris muscles were conducted using an ultrasonic scanner (H1300, BENQ, Taipei, Taiwan). Participants were positioned lying flat on their backs with palms facing upward, assuming an anatomical position. For the biceps brachii, the measurement position involved determining the straight-line distance from the acromion to the transverse crease of the elbow. The ultrasonic probe was then placed at one-third of the transverse crease of the elbow. Regarding the quadriceps femoris, the muscle probe was positioned from the anterior superior iliac spine (ASIS) to the patella at a one-half straight-line distance from the edge [29]. Muscle thickness analysis was comprised of calculating the average thickness by dividing the sum of the thickness at both ends of the muscle by two. The method for calculating muscle fascia thickness was based on the muscle fascia calculation method [30], defined as muscle thickness $\times$ sin(θ), where θ represents the angle between the muscle fascia and the horizontal line. Image analysis was performed using ImageJ version 2.3.0/1.53f (National Institutes of Health, Bethesda, MD, USA).

2.12. Clinical Biochemistry and Hematology Analysis

Blood samples were obtained from arm venous catheters both before and after the 6-week intervention as part of the study. The collected samples underwent a comprehensive analysis encompassing liver function, renal function, blood lipids, and glucose levels to assess the metabolic and health status of the subjects. To ensure standardized conditions for blood analysis, all participants were instructed to observe a fasting period of 8 h at minimum prior to blood collection. The blood serum obtained from the samples was then assessed using an autoanalyzer (Hitachi 717, Tokyo, Japan) for various biochemical markers, including aspartate transaminase (AST), aminotransferase (ALT), total cholesterol (TC), triglyceride (TG), high-density lipoprotein cholesterol (HDL-C), low-density lipoprotein cholesterol (LDL-C), blood urea nitrogen (BUN), creatinine (CREA), uric acid (UA), and glucose levels. Complete blood count (CBC) profiles (Sysmex XE-2100, Sysmex Corporation, Kobe, Japan) were also analyzed at the 90 min time point in the recovery phase.

2.13. Statistics

All data are expressed as the mean $\pm$ SD. Statistical analyses were performed using SPSS Statistics 25 (IBM Co., Armonk, NY, USA). To assess differences within the groups before and after the intervention, a paired Student's t-test was applied for parametric analysis. Post hoc comparisons were analyzed using the Bonferroni test. For non-parametric data, the Wilcoxon signed-rank test was employed. Between-group comparisons were conducted using an unpaired Student's t-test for parametric analyses, while non-parametric data were analyzed using the Mann–Whitney U-test. A significance level of $p < 0.05$ was considered indicative of a statistically significant difference in all analyses.

3. Results

3.1. Effects of 6-Week Supplementation with GKEX on Body Composition Change

First, Table 2 shows the results of the Inbody body composition measurements every two weeks. We found that, under the same resistance exercise training intervention, there were no significant differences between the placebo and GKEX groups in terms of body weight, BMI, muscle mass, and body fat percentage before the intervention and at 2, 4, and 6 weeks after the intervention. However, through comparison before and after within the groups, it was found that the weight and body fat percentage in the placebo group increased significantly at 2 and 4 weeks after the intervention compared with before the intervention, and the BMI had increased significantly after 4 weeks of intervention ($p < 0.05$). On the other hand, at the 2nd, 4th, and 6th weeks of intervention, compared with before the intervention, the GKEX group presented significantly increased muscle mass (one and one-hundredths-fold, $p = 0.0005$; one and two-hundredths-fold, $p < 0.0001$; and one and three-hundredths-fold, $p < 0.0001$, respectively) and reduced body fat rate (1.76%, $p = 0.0075$; 4.68%, $p = 0.0001$; and 5.23%, $p = 0.0001$, respectively). In addition, increments were calculated separately for weeks 2, 4, and 6 to gain further insight into the changes. The results showed that the placebo group significantly increased in weight, BMI, and body fat percentage compared with the GKEX group at weeks 2, 4, and 6 of the intervention, respectively ($p < 0.05$); meanwhile, the GKEX group had a significantly increased change in muscle mass compared with the placebo group at weeks 2, 4, and 6 of the intervention ($p < 0.05$).

On the other hand, the measurement results from whole body scans using DXA are shown in Table 3. There was no significant difference between the placebo and GKEX groups in terms of LMB, FBM, and bone weight. However, after 6 weeks of supplementation with GKEX, LBM was significantly increased (by one and one-hundredths-fold, $p = 0.0231$) and FBM was significantly reduced (by 4.88%, $p = 0.0002$) compared to before the intervention. Moreover, regarding the changes in total body fat, the values for the placebo and GKEX groups at 6 weeks were -1.1 ± 1.3 and 0.0 ± 0.9, respectively; notably, the GKEX group presented a significantly reduced total LBM when compared to the placebo group ($p = 0.0011$).

Table 2. Body composition of subjects, measured every two weeks using Inbody, under the 6-week GKEX combined with resistance training intervention.

Body Composition	Week	Placebo	GKEX	Delta	
				Placebo	GKEX
BW (kg)	Before	61.9 ± 9.2^{a}	65.3 ± 12.6^{a}	0.0 ± 0.0^{a}	0.0 ± 0.0^{a}
	Week 2	$62.2 \pm 9.4^{a,*}$	65.0 ± 12.4^{a}	0.4 ± 0.9^{b}	-0.3 ± 0.8^{a}
	Week 4	$62.4 \pm 9.5^{a,*}$	64.9 ± 12.4^{a}	0.5 ± 0.8^{b}	-0.3 ± 1.1^{a}
	After	62.2 ± 9.5^{a}	64.7 ± 12.3^{a}	0.3 ± 0.9^{b}	-0.5 ± 1.4^{a}
BMI (kg/m^2)	Before	22.4 ± 2.7^{a}	22.8 ± 2.6^{a}	0.0 ± 0.0^{a}	0.0 ± 0.0^{a}
	Week 2	22.5 ± 2.7^{a}	22.7 ± 2.5^{a}	0.1 ± 0.3^{b}	-0.1 ± 0.3^{a}
	Week 4	$22.5 \pm 2.7^{a,*}$	22.6 ± 2.5^{a}	0.2 ± 0.3^{b}	-0.1 ± 0.4^{a}
	After	22.5 ± 2.7^{a}	22.6 ± 2.5^{a}	0.1 ± 0.3^{b}	-0.2 ± 0.4^{a}
Muscle Weight (kg)	Before	27.0 ± 4.9^{a}	28.6 ± 6.9^{a}	0.0 ± 0.0^{a}	0.0 ± 0.0^{a}
	Week 2	26.9 ± 4.9^{a}	$28.8 \pm 6.8^{a,*}$	-0.1 ± 0.4^{a}	0.3 ± 0.3^{b}
	Week 4	27.0 ± 5.0^{a}	$29.1 \pm 6.8^{a,*}$	0.0 ± 0.3^{a}	0.5 ± 0.4^{b}
	After	27.0 ± 5.0^{a}	$29.3 \pm 6.9^{a,*}$	0.0 ± 0.5^{a}	0.8 ± 0.5^{b}
Fat (%)	Before	21.2 ± 6.9^{a}	22.4 ± 5.4^{a}	0.0 ± 0.0^{a}	0.0 ± 0.0^{a}
	Week 2	$21.4 \pm 6.7^{a,*}$	$22.1 \pm 5.5^{a,*}$	0.2 ± 0.7^{b}	-0.4 ± 0.7^{a}
	Week 4	$21.4 \pm 6.9^{a,*}$	$21.4 \pm 5.4^{a,*}$	0.2 ± 0.8^{b}	-1.1 ± 1.2^{a}
	After	21.1 ± 6.9^{a}	$21.3 \pm 5.3^{a,*}$	-0.1 ± 1.0^{b}	-1.2 ± 1.3^{a}

Data are presented as mean ± SD. Different superscript letters (a, b) indicate significant differences among groups at the same time point ($p < 0.05$); * indicates a significant effect after the intervention compared to before the intervention ($p < 0.05$). Delta is the value after the intervention minus before the intervention. BW, body weight; BMI, body mass index; LBM, lean body mass; FBM, fat body mass.

Table 3. Body composition of subjects before and after the 6-week GKEX combined with resistance training intervention, measured through DXA.

Body Composition	Before		After		Delta	
	Placebo	GKEX	Placebo	GKEX	Placebo	GKEX
LBM (kg)	46.1 ± 7.6^{a}	47.7 ± 10.2^{a}	46.4 ± 7.8^{a}	$48.3 \pm 10.1^{a,*}$	0.3 ± 0.9^{a}	0.6 ± 1.2^{a}
FBM (kg)	21.1 ± 6.9^{a}	22.4 ± 5.6^{a}	21.1 ± 6.9^{a}	$21.3 \pm 5.3^{a,*}$	-0.2 ± 5.5^{b}	-1.1 ± 1.3^{a}
Mineral Weight (kg)	2.5 ± 0.4^{a}	2.7 ± 0.5^{a}	2.5 ± 0.4^{a}	2.7 ± 0.5^{a}	0.0 ± 0.0^{a}	0.0 ± 0.0^{a}

Data are presented as mean ± SD. Different superscript letters (a, b) indicate significant differences among groups at the same time point ($p < 0.05$). * indicates a significant effect after the intervention compared to before the intervention ($p < 0.05$). Delta is the value after the intervention minus before the intervention. LBM, lean body mass; FBM, fat body mass.

3.2. *Effects of 6-Week Supplementation with GKEX on Muscle Thickness and Fascia*

According to the results shown in Table 4, after 6 weeks of supplementation with GKEX, the right and left quadriceps thicknesses were significantly increased compared to those in the placebo group (by one and nine-hundredths-fold, $p = 0.0124$, and one and nine-hundredths-fold, $p = 0.0231$, respectively). Within-group comparisons revealed that, after 6 weeks of GKEX supplementation, the right biceps were significantly increased (by one and eight-hundredths-fold, $p < 0.0001$), as well as the right quadriceps (by one and five-hundredths-fold, $p < 0.0001$), left biceps (by one and eight-hundredths-fold, $p < 0.0001$), and left quadriceps (by one and three-hundredths-fold, $p = 0.0020$), when compared with those before the intervention. In addition, the change values for the GKEX group were significantly greater than those for the placebo group in the right biceps ($p < 0.0001$), right quadriceps ($p < 0.0001$), left biceps ($p < 0.0001$), and left quadriceps ($p = 0.0037$). There was a significant time effect ($p < 0.05$) and group × time interaction effect ($p < 0.05$) for the right biceps, right quadriceps, left biceps, and left quadriceps, respectively. However, no significant differences between groups were detected with the post hoc test.

Table 4. Effects of 6-week GKEX combined with resistance training intervention on muscle and fascia thickness determined by ultrasound.

Muscle Ultrasound	Before		After		Delta	
	Placebo	GKEX	Placebo	GKEX	Placebo	GKEX
			Muscle thickness (cm)			
Right biceps	2.19 ± 0.40 [a]	2.12 ± 0.41[a]	2.21 ± 0.43 [a,*]	2.29 ± 0.42 [a,*]	0.02 ± 0.05 [a]	0.16 ± 0.07 [b]
Right quadriceps	2.35 ± 0.29 [a]	2.44 ± 0.30 [a]	2.35 ± 0.28 [a]	2.57 ± 0.32 [a,*]	0.00 ± 0.07 [a]	0.12 ± 0.07 [b]
Left biceps	2.13 ± 0.38 [a]	2.13 ± 0.42 [a]	1.95 ± 0.37 [a]	2.29 ± 0.51[a,*]	0.00 ± 0.08 [a]	0.16 ± 0.14 [b]
Left quadriceps	2.31 ± 0.26 [a]	2.44 ± 0.36 [a]	2.31 ± 0.26 [a]	2.52 ± 0.37 [b,*]	0.00 ± 0.06 [a]	0.08 ± 0.12 [b]
			Fascia (cm)			
Right biceps	0.13 ± 0.07 [a]	0.11 ± 0.09 [a]	0.12 ± 0.07 [a]	0.13 ± 0.09 [a,*]	0.00 ± 0.02 [a]	0.02 ± 0.02 [b]
Right quadriceps	0.15 ± 0.06 [a]	0.16 ± 0.06 [a]	0.16 ± 0.06 [a]	0.18 ± 0.07 [a,*]	0.01 ± 0.02 [a]	0.02 ± 0.04 [a]
Left biceps	0.13 ± 0.06 [a]	0.15 ± 0.08 [a]	0.14 ± 0.06 [a,*]	0.17 ± 0.08 [b,*]	0.01 ± 0.02 [a]	0.01 ± 0.02 [a]
Left quadriceps	0.16 ± 0.06 [a]	0.19 ± 0.07 [b]	0.16 ± 0.06 [a]	0.20 ± 0.06 [b]	0.01 ± 0.03 [a]	0.01 ± 0.03 [a]

Data are presented as mean ± SD. Different superscript letters (a, b) indicate significant differences among groups at the same time point ($p < 0.05$). * indicates a significant effect after the intervention compared to before the intervention ($p < 0.05$). Delta is the value after the intervention minus before the intervention.

For the fascia, there was no significant difference between the placebo and GKEX groups regarding the right biceps, right quadriceps, and left biceps, respectively, before and after the intervention. The left quadriceps thickness of the GKEX group was significantly higher than that of the placebo group before the intervention and at 6 weeks after the intervention. In addition, the GKEX group presented significant increases in the right biceps (by 1.20-fold, $p < 0.0001$), right quadriceps (by 1.12-fold, $p = 0.0092$), and left biceps (by 1.09-fold, $p = 0.0067$) compared with before the intervention. The change values in the GKEX group were only significantly greater than those in the placebo group for the right biceps ($p < 0.0001$).

3.3. Effects of 6-Week Supplementation with GKEX on Maximum Handgrip Strength

As shown on Figure 2, at both before the intervention and after 6 weeks of intervention, there were no significant differences between the placebo and GKEX groups in terms of left and right maximum handgrip strength. However, in terms of the within-group comparison, only the GKEX supplementation group presented significantly increased right and left maximum handgrip grip strength (by one and four-hundredths-fold, $p = 0.0374$, and one and six-hundredths-fold, $p = 0.0055$, respectively) compared with before the intervention (Figure 2A,B).

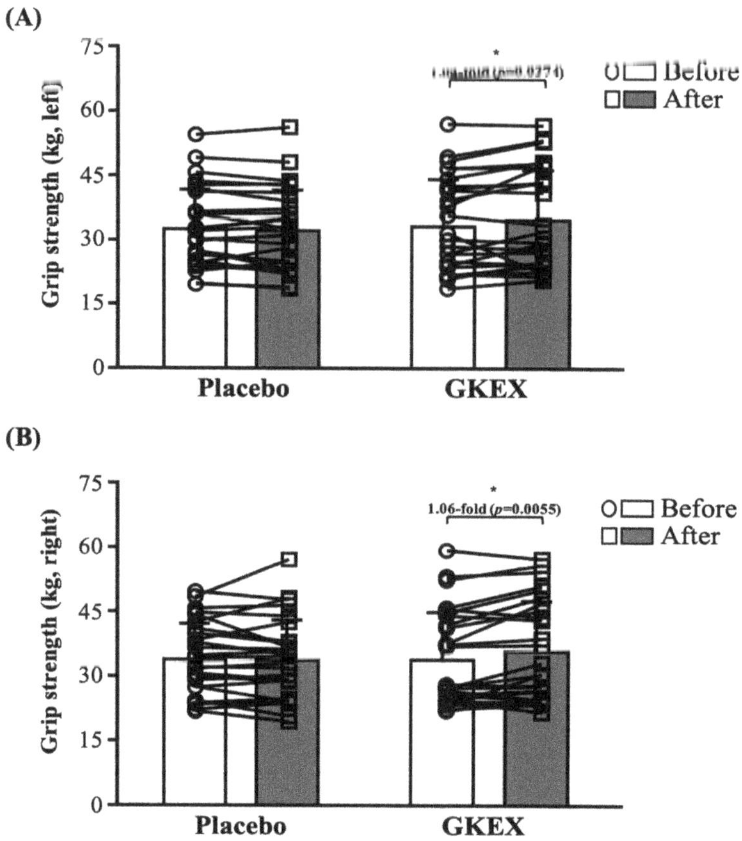

Figure 2. Experimental procedure description. Effects of GKEX combined with resistance training intervention on (**A**) left handgrip strength and (**B**) right handgrip strength. Data are shown as mean ± SD. * indicates significant difference before and after the intervention within a group ($p < 0.05$).

3.4. Effects of 6-Week Supplementation with GKEX on Maximal Oxygen Uptake (VO_{2max})

Before the intervention, there were no significant differences in VO_{2max} between the placebo and GKEX groups. After the 6-week intervention, the VO_{2max} values were 42.9 ± 4.9 and 42.5 ± 4.8 (mL/min/kg) in the placebo and GKEX groups, respectively, and there was no significant difference between the two groups. However, within the groups, that in the GKEX group was significantly increased (by one and two-hundredths-fold, $p = 0.0109$) compared to before the intervention (Figure 3). There was a significant time effect ($p < 0.05$) and group × time interaction effect ($p < 0.05$) for VO_{2max}.

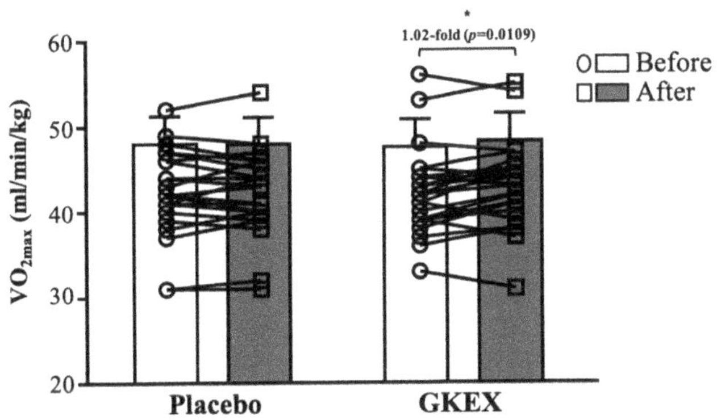

Figure 3. Effects of GKEX combined with resistance training intervention on: VO_{2max}. Data are shown as mean ± SD. * indicates significant difference before and after the intervention within a group ($p < 0.05$).

3.5. Effects of 6-Week Supplementation with GKEX on Maximum Vertical Jump Height of Exercise Performance

Although there were no significant differences in the relative force peak, RFD, and jump height between the placebo and the GKEX groups before and after the intervention, after 6 weeks of supplementation with GKEX, significant increases in the relative force peak (by one and two-hundredths-fold, $p = 0.0005$) and RFD (by one and seven-hundredths-fold, $p < 0.0001$) were observed (Table 5). In addition, through the percentage change, we found that the GKEX group presented a significantly increased relative force peak ($p = 0.0024$; placebo: 2.1 ± 2.8% vs. GKEX: 0.0 ± 1.8%) and RFD ($p = 0.0004$; placebo: 6.5 ± 6.0% vs. GKEX: 0.7 ± 5.1%) (Figure 4A).

Table 5. Effects of 6-week GKEX combined with resistance training intervention on the CMJ, IMTP, and Wingate tests.

Exercise Performance	Before		After	
	Placebo	GKEX	Placebo	GKEX
CMJ				
Relative peak force (N/kg)	14.2 ± 2.1 [a]	14.3 ± 1.9 [a]	14.2 ± 2.2 [a]	14.6 ± 1.9 [a,*]
RFD (N/kg/s)	7.5 ± 1.7 [a]	7.4 ± 1.6 [a]	7.5 ± 1.7 [a]	7.9 ± 1.6 [a,*]
Jump height (cm)	28.4 ± 6.0 [a]	29.7 ± 5.4 [a]	29.1 ± 6.4 [a]	30.3 ± 5.9 [a]
IMTP				
Relative peak force (N/kg)	13.5 ± 2.7 [a]	13.2 ± 3.0 [a]	13.6 ± 2.7 [a]	13.9 ± 3.2 [a,*]
RFD (N/s)	8039 ± 940 [a]	8057 ± 855 [a]	8120 ± 976 [a]	8261 ± 892 [a,*]
Wingate				
Relative mean power (W/kg)	5.8 ± 1.0 [a]	5.9 ± 1.0 [a]	5.8 ± 1.0 [a]	6.1 ± 1.0 [a,*]
Relative peak power (W/kg)	8.5 ± 1.9 [a]	8.9 ± 1.7 [a]	8.6 ± 1.8 [a]	9.0 ± 1.6 [a]
Fatigue index (%)	48.1 ± 7.5 [a]	46.1 ± 11.1 [a]	47.9 ± 6.9 [a]	45.2 ± 11.0 [a]

Data are presented as mean ± SD. Superscript letter (a) indicates significant differences among groups at the same time point ($p < 0.05$); * indicates a significant effect after the intervention compared to before the intervention ($p < 0.05$). Delta is the value after the intervention minus that before the intervention. CMJ, countermovement jump; IMTP, isometric mid-thigh pull; RFD, rate of force development.

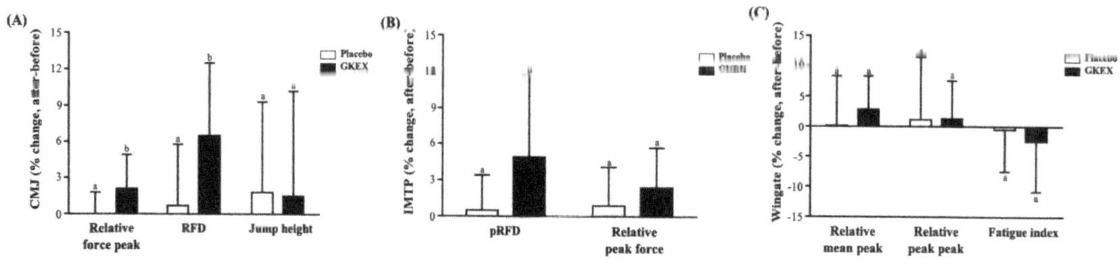

Figure 4. Effects of GKEX combined with resistance training intervention on percentage change in the following: (**A**) CMJ test; (**B**) IMTP test; (**C**) Wingate test. Data are shown as mean ± SD. Different superscript letters (a, b) above the bars indicate significant differences between groups at the same time point ($p < 0.05$). RFD, rate of force development.

3.6. Effects of 6-Week Supplementation with GKEX on Isotonic Muscle Strength of Exercise Performance

Table 5 indicates that there were no significant differences in the relative peak force and RFD between the placebo and GKEX groups before and after the intervention; however, after 6 weeks of supplementation with GKEX, significant increases in the relative peak force (by one and five-hundredths-fold, $p = 0.0032$) and RFD (by one and three-hundredths-fold, $p = 0.0009$) were observed. We found that the GKEX group had a significantly increased percentage change in the relative peak force ($p = 0.0046$; placebo: 4.9 ± 6.8% vs. GKEX: 0.5 ± 2.9%) (Figure 4B).

3.7. Effects of 6-Week Supplementation with GKEX on Anaerobic Exercise Performance

There were no significant differences in the relative mean power, relative peak power, and fatigue index between the placebo and GKEX groups before and after the intervention. However, after 6 weeks of supplementation with GKEX, there was only a significant increase in the relative mean power (by one and three-hundredths-fold, $p = 0.0112$) (Table 5), and no significant difference in the percentage change was observed (Figure 4C).

3.8. Effects of 6-Week GKEX Intervention on Biochemical and Hematological Characteristics of Subjects

All participants underwent blood parameter analysis before and after the intervention to determine their physiological status. This was carried out to ensure that the 6-week GKEX intervention did not result in any adverse side-effects. The study confirmed that all subjects were in good health during the trial, and the intervention did not result in any changes in the various blood indicators measured. No significant changes were observed within each group after the intervention when compared to the pre-intervention period. Moreover, no adverse events related to the intervention were reported during the study (Table 6).

Table 6. Blood biochemical parameters and blood count profiles of subjects before and after the GKEX combined with resistance training intervention.

Characteristics	Before		After	
	Placebo	GKEX	Placebo	GKEX
	Blood Biochemical Parameters			
AST (U/L)	18 ± 4	18 ± 4	18 ± 3	18 ± 4
ALT (U/L)	17 ± 3	16 ± 4	16 ± 2	15 ± 4
TC (mg/dL)	161 ± 13	162 ± 12	160 ± 13	161 ± 13
TG (mg/dL)	89 ± 17	88 ± 22	88 ± 18	87 ± 17
HDL (mg/dL)	57.4 ± 6.9	56.6 ± 8.1	57.7 ± 7.2	56.9 ± 7.5
LDL (mg/dL)	85.2 ± 14.4	87.8 ± 8.7	85.0 ± 14.3	86.0 ± 10.6

Table 6. *Cont.*

Characteristics	Before		After	
	Placebo	GKEX	Placebo	GKEX
BUN (mg/dL)	15.2 ± 2.0	15.5 ± 2.4	15.0 ± 2.1	15.0 ± 2.9
CREA (mg/dL)	0.98 ± 0.12	0.97 ± 0.12	0.98 ± 0.12	0.97 ± 0.10
UA (mg/dL)	5.2 ± 0.7	5.0 ± 0.6	5.3 ± 0.7	5.0 ± 0.6
TP (g/dL)	7.2 ± 0.3	7.3 ± 0.3	7.2 ± 0.4	7.3 ± 0.3
Glucose (mg/dL)	85 ± 7	86 ± 5	84 ± 7	85 ± 5
Insulin (μU/mL)	10.93 ± 2.24	11.00 ± 3.1	10.81 ± 2.46	10.93 ± 3.10
Lactate (mmol/L)	1.57 ± 0.27	1.64 ± 0.19	1.56 ± 0.29	1.63 ± 0.20
NH_3 (μmol/L)	56.8 ± 9.9	53.3 ± 9.3	56.7 ± 10.3	53.1 ± 9.6
Complete Blood Count				
WBC (10^3 cumm)	6.0 ± 0.9	6.5 ± 1.1	6.2 ± 0.9	6.5 ± 1.5
Neutrophils (%)	53.4 ± 7.2	55 ± 6.4	55.4 ± 6.9	53 ± 6.6
Lymphocytes (%)	36.2 ± 6.6	33.8 ± 6.4	34.0 ± 5.9	36.4 ± 6.0
Monocytes (%)	7.5 ± 1.5	7.8 ± 1.7	7.9 ± 1.7	7.3 ± 1.8
Eosinophils (%)	2.3 ± 1.3	2.7 ± 1.5	2.0 ± 1.2	2.5 ± 1.2
Basophils (%)	0.6 ± 0.2	0.6 ± 0.3	0.6 ± 0.3	0.7 ± 0.3
RBC (MIL/cumm)	4.8 ± 0.4	4.7 ± 0.4	4.8 ± 0.4	15.0 ± 2.9
Hemoglobin (gm/dL)	14.1 ± 1.2	14.0 ± 1.2	14.1 ± 1.3	13.9 ± 1.1
Hematocrit (%)	42.1 ± 2.9	41.9 ± 3.7	41.9 ± 3.8	42.2 ± 3.8
M.C.V (fl)	88.0 ± 2.7	89.1 ± 2.5	88.1 ± 2.5	89.1 ± 2.6
M.C.H (pg)	29.7 ± 1.1	29.9 ± 0.9	29.5 ± 1.1	29.8 ± 1.0
M.C.H.C (%)	33.7 ± 0.6	33.4 ± 0.7	33.6 ± 0.7	33.4 ± 0.6
Platelet count (10^3/Cumm)	273 ± 51	278 ± 51	262 ± 62	273 ± 55

Data are presented as mean ± SD. AST, aspartate aminotransferase; ALT, alanine aminotransferase; BUN, blood urea nitrogen; CREA, creatine; UA, uric acid; TP, total protein; TC, total cholesterol; TG, triacylglycerol; HDL, high-density lipoprotein; LDL, low-density lipoprotein; WBC, white blood cell.

4. Discussion

In recent years, a growing body of research has explored the concept of the gut–muscle axis [31], and further evidence has emerged to support the beneficial effects of probiotics in promoting muscle growth and improving exercise performance [32]. However, this study is the first to examine the benefits of supplementation with *L. brevis* to increase muscle mass and exercise performance. In the current study, we found that 6 weeks of supplementation with GKEX and combined with resistance training three times/week could significantly improve muscle strength performance, explosive force, muscle mass, and muscle thickness compared with before the intervention. Supplementation with GKEX combined with resistance exercise training not only effectively increased muscle mass and strength, but it was also found to effectively reduce the body fat.

Resistance training has also been shown to improve the diversity and composition of the gut microbiota. It can increase the richness of SCFA-producing gut microbiota and reduce the relative abundance of pro-inflammatory-inducing species [33]. Previous research has shown that elite rugby players have a higher gut microbial diversity compared to sedentary individuals. Additionally, athletes have a greater proportion of bacteria that are involved in the carbohydrate and amino acid metabolism, resulting in an increased production of SCFAs in the gut, such as acetate, butyrate, and propionate [34]. Among these, butyrate has been shown to prevent cell apoptosis and muscle protein degradation by inhibiting histone deacetylase, and can also increase ATP and muscle fiber metabolic efficiency by activating the UCP2-AMPK-ACC and PGC1-α pathways [18]. In addition, some studies have shown that butyric acid can activate AMPK signaling pathways, stimulate mitochondrial fatty acid oxidation, enhance the glucose metabolism in muscle cells, reduce lipid deposition in muscles, and thereby improve muscle quality and function [18]. It may also promote the proliferation of C2C12 myoblasts by upregulating ERK phosphorylation, in which the expression of myogenic regulatory factors including Myf5 and MyoD are significantly increased, thereby regulating the activation of muscle satellite cells [35]. In one of our previous animal studies, we found that supplementation with *L. plantarum* PL-02 for 4 weeks combined with resistance exercise training significantly increased the muscle mass and strength performance in mice [36]. Moreover, a study in humans without regular exercise habits has also found that supplementation with *L. plantarum* TWK10 for 6 weeks significantly increased muscle mass and endurance performance [4]. Although the *L. brevis* intervention in our study was combined with resistance exercise training, similar results were found in terms of the promotion of muscle mass gain (Tables 2 and 3). However, we need further research to verify the mechanism by which GKEX promotes muscle growth.

Improvements in muscle strength are thought to be positively correlated with increases in muscle mass [37]. Furthermore, there is also a strong positive correlation between the ability of muscles to produce force and their cross-sectional area (CSA) [38]. Previous research has confirmed that muscle strength is typically determined by the product of muscle mass and the relative strength to muscle volume. Therefore, increasing muscle thickness serves as the basis for achieving strength gains [39]. Despite the use of a muscle ultrasound in our study for measurements, we observed a significant increase in muscle thickness in the GKEX supplementation group compared to before the intervention, which was accompanied by an increase in fascial thickness (Table 4). As greater muscle strength and body mass require thicker connective tissue to withstand and transmit greater forces, increases in muscle thickness may also be accompanied by increases in fascia thickness [40]. This seems to explain our results. In a previous study, continuous supplementation of pea peptides combined with resistance exercise training for 8 weeks significantly enhanced the expression of the insulin-like growth factor 1 receptor and AMP-activated protein kinase. Simultaneously, there was a significant reduction in the expression of myostatin—a muscle growth inhibitor—in the skeletal muscles. This led to an increase in the cross-sectional area of muscles and muscle strength in rats [41]. Although one possible way that probiotics can enhance muscle mass or strength is by improving protein digestion and the absorption of amino acids into the bloodstream [3], the *L. brevis* used in the intervention of this study has not been proven to have this effect. Increased muscle mass is thought to have a positive impact on the rate of strength development, thereby improving movements such as sprinting, jumping, and the ability to change direction [42]. In this study, grip strength, CMJ, IMTP, and anaerobic power tests were performed to assess changes in muscle strength and explosive performance [43]. Although there were no significant differences in various measurements between the two groups, we observed a notable increase in performance only in the group supplemented with GKEX when compared to that before the intervention (Figure 1 and Table 5). Moreover, the GKEX supplementation group exhibited more significant improvements in muscle strength and explosive performance compared to the placebo group, as demonstrated by the magnitude of changes (Figure 4). Previous research in patients with sarcopenia has reported that daily supplementation for 16 weeks with a mixture of various strains totaling 1.12 trillion led to significant improvements in grip strength and gait speed [44]. Another study, considering supplementation with inactivated and coagulated spore-forming bacteria and conducted in soldiers participating in a self-defense course, observed an increase in vertical jump power [45]. These findings support the potential of probiotics in promoting the development of muscle strength. However, further exploration is needed to understand the mechanisms through which *L. brevis* can enhance muscle mass and strength performance.

In addition to influencing muscle mass, probiotics are thought to have the potential to improve endurance performance, although several potential mechanisms remain uncertain. Previous research has suggested that supplementation with a mixture of different probiotics can attenuate the exercise-induced decrease in circulating tryptophan concentrations. This is thought to maintain normal serotonin metabolism, thereby slowing fatigue and preserving exercise performance [46]. Studies have suggested that an increase in training volume and compliance might lead to a reduction in the frequency of upper respiratory tract infections (URTIs). In a trial involving female adolescent swimmers, it has been observed that continuous daily supplementation with 400 mL of probiotic yogurt containing

4×10^{10} CFU/mL (including *Lactobacillus acidophilus* SPP, *Lactobacillus delbrueckii Bulgaricus*, *Bifidobacterium bifidum*, and *Streptococcus salivarius subsp. thermophilus*) for 8 weeks led to a significant reduction in the frequency and duration of URTIs, as well as an improvement in maximal oxygen consumption [47]. In our previous study, we found that continuous supplementation with *L. plantarum* TWK10 for six weeks led to a significant improvement in exercise endurance performance in mice. Additionally, there were notable increases in muscle mass and glycogen storage [48]. Glycogen is a polymer of glucose that is related to muscle mass. It is catabolized by glycogen to produce adenosine triphosphate (ATP), which provides the primary fuel source for muscles during exercise [49]. Bacteria in the gut may produce SCFAs, of which butyrate can maintain blood sugar stability and promote liver glucose metabolism through the GPR43-AKT-GSK3 signaling pathway [50]. This can improve exercise endurance performance, delay fatigue, and accelerate recovery [51]. Although it was found that 6 weeks of resistance exercise training combined with GKEX supplementation significantly increased VO_{2max} (Figure 3) in this study, further research is required to gain a deeper understanding of the mechanisms involved.

Although this study confirmed the benefits of *L. brevis* GKEX in terms of improving muscle mass and strength performance, there are limitations and possible mechanisms that require further exploration. To ensure gender equality, no special restrictions were imposed on recruitment, resulting in a slightly different gender ratio between the two groups of subjects. However, the implications of gender differences in the sample and the composition of the sample being comprised of physical education non-majors, specifically, physiological differences between males and females, such as in skeletal muscles, energy metabolism, and hormonal profiles, can lead to larger standard deviations within the group. These differences can potentially affect the statistical results and interpretations of the study [52]. This study primarily investigated the impact of GKEX on enhancing muscle mass and exercise performance, necessitating the inclusion of both male and female subjects in the same group for comparison. Therefore, it may be necessary to include more subjects in future trials to further compare the effects and mechanisms of this sample in both genders. Second, all participants were non-physical education majors from the same university. Although their living conditions were similar and variability was low, this does not necessarily mean that the results apply to the wider ethnic group or population. Third, the original purpose of this study was to explore efficacy and, so, the group design did not compare whether sedentary control or supplementation alone had substantial benefits. Furthermore, apart from this study, no research has been conducted on the correlation between *L. brevis* GKEX and muscle and exercise performance. In summary, the current study demonstrates that *L. brevis* GKEX supplementation combined with resistance exercise training improves muscle mass and strength performance. Although there have been many studies in the past showing the benefits of combining probiotics with exercise training on athletic performance and even competitive athletes [53], this study aimed to investigate its efficacy. To minimize the burden on participants, we collected a limited number of specimens. Once the efficacy test results are confirmed, we plan to delve deeper into the mechanisms using more rigorous experimental designs and analysis projects. Finally, to gain a deeper understanding of the cellular mechanisms underlying our findings, it may be necessary to conduct muscle biopsies or in vitro animal studies. Based on the above, a broader perspective is needed to confirm our findings. We aim to delve deeper into the mechanisms of muscle growth and the role of the gut–muscle axis by analyzing the gut microbiota. We anticipate that, through continued research into these mechanisms and the efficacy results obtained from this study, our findings can be applied to athletes, offering a potential avenue for sports nutrition supplements. Further research in the future needs to develop more stringent experimental design and condition restrictions, and make appropriate adjustments based on the characteristics of the probiotics and the intervention time [54].

5. Conclusions

This study demonstrated the effectiveness of combining *L. brevis* GKEX with resistance exercise training for the improvement of muscle mass and strength performance. After six weeks of supplementation and training, GKEX improved muscle mass, thickness, strength, and power. Additionally, it significantly reduced body fat percentage. Further research is needed in the future to elucidate the key mechanism of *L. brevis* GKEX in the gut–muscle axis through in vivo and in vitro studies, and to analyze the relationship between intestinal microbial analysis and muscle synthesis pathways.

Author Contributions: M.-C.L., Y.-J.H. and C.-C.H. designed the experiments; M.-C.L., C.-S.H. and C.-C.H. performed the laboratory experiments; Y.-S.T. screened strain and sample preparation. M.-C.L., Y.-J.H., C.-S.H., C.-C.C. and C.-C.H. analyzed the data and interpreted the results; M.-C.L. and C.-C.H. prepared the figures and wrote the manuscript. All authors have read and agreed to the published version of the manuscript.

Funding: This study was funded by the University–Industry Cooperation Fund, National Taiwan Sport University, Taoyuan, Taiwan (NTSU No. 1111013).

Institutional Review Board Statement: The study was performed in accordance with the guidelines of the Declaration of Helsinki and was approved by the Institutional Review Board of Landseed International Hospital (Taoyuan, Taiwan; LSHIRB No. 22-022-A2). Approval date: 16 June 2022.

Informed Consent Statement: Before the experiment, the researchers explained the experimental process in detail and obtained informed consent from all subjects participating in the study, and the experiment started after the subjects signed the consent form.

Data Availability Statement: The original contributions presented in the study are included in the article, further inquiries can be directed to the corresponding author.

Acknowledgments: The authors are grateful to the graduate students at the Sport Nutrition Laboratory, National Taiwan Sport University, for their technical assistance in conducting the analysis experiments. And we are very grateful to Hsiao-Yun Chang for assisting with muscle ultrasound detection and analysis.

Conflicts of Interest: You-Shan Tsai and Chin-Chu Chen are employed by the Grape King Bio Ltd. The other authors declare that they have no known competing financial interests or personal relationships that could have appeared to influence the work reported in this paper. The company provided materials and equipment essential for the research, and it did offer funding support.

References

1. Anderson, E.; Durstine, J.L. Physical activity, exercise, and chronic diseases: A brief review. *Sports Med. Health Sci.* **2019**, *1*, 3–10. [CrossRef] [PubMed]
2. Kim, C.B.; Park, J.H.; Park, H.S.; Kim, H.J.; Park, J.J. Effects of whey protein supplement on 4-week resistance exercise-induced improvements in muscle mass and isokinetic muscular function under dietary control. *Nutrients* **2023**, *15*, 1003. [CrossRef] [PubMed]
3. Prokopidis, K.; Giannos, P.; Kirwan, R.; Ispoglou, T.; Galli, F.; Witard, O.C.; Triantafyllidis, K.K.; Kechagias, K.S.; Morwani-Mangnani, J.; Ticinesi, A.; et al. Impact of probiotics on muscle mass, muscle strength and lean mass: A systematic review and meta-analysis of randomized controlled trials. *J. Cachexia Sarcopenia Muscle* **2023**, *14*, 30–44. [CrossRef] [PubMed]
4. Huang, W.C.; Lee, M.C.; Lee, C.C.; Ng, K.S.; Hsu, Y.J.; Tsai, T.Y.; Young, S.L.; Lin, J.S.; Huang, C.C. Effect of *Lactobacillus plantarum* TWK10 on exercise physiological adaptation, performance, and body composition in healthy humans. *Nutrients* **2019**, *11*, 2836. [CrossRef]
5. Hill, C.; Guarner, F.; Reid, G.; Gibson, G.R.; Merenstein, D.J.; Pot, B.; Morelli, L.; Canani, R.B.; Flint, H.J.; Salminen, S.; et al. Expert consensus document. The international scientific association for probiotics and prebiotics consensus statement on the scope and appropriate use of the term probiotic. *Nat. Rev. Gastroenterol. Hepatol.* **2014**, *11*, 506–514. [CrossRef] [PubMed]
6. Maughan, R.J.; Burke, L.M.; Dvorak, J.; Larson-Meyer, D.E.; Peeling, P.; Phillips, S.M.; Rawson, E.S.; Walsh, N.P.; Garthe, I.; Geyer, H.; et al. IOC consensus statement: Dietary supplements and the high-performance athlete. *Br. J. Sports Med.* **2018**, *52*, 439–455. [CrossRef] [PubMed]
7. Dukes, A.; Davis, C.; El Refaey, M.; Upadhyay, S.; Mork, S.; Arounleut, P.; Johnson, M.H.; Hill, W.D.; Isales, C.M.; Hamrick, M.W. The aromatic amino acid tryptophan stimulates skeletal muscle IGF1/p70s6k/mTor signaling in vivo and the expression of myogenic genes in vitro. *Nutrition* **2015**, *31*, 1018–1024. [CrossRef] [PubMed]

8. Jäger, R.; Zaragoza, J.; Purpura, M.; Iametti, S.; Marengo, M.; Tinsley, G.M.; Anzalone, A.J.; Oliver, J.M.; Fiore, W.; Biffi, A.; et al. Probiotic Administration increases amino acid absorption from plant protein: A placebo-controlled, randomized, double-blind, multicenter, crossover study. *Probiotics Antimicrob. Proteins* **2020**, *12*, 1330–1339. [CrossRef] [PubMed]
9. Prokopidis, K.; Chambers, E.; Ni Lochlainn, M.; Witard, O.C. Mechanisms linking the gut-muscle axis with muscle protein metabolism and anabolic resistance: Implications for older adults at risk of sarcopenia. *Front. Physiol.* **2021**, *12*, 770455. [CrossRef]
10. Storelli, G.; Defaye, A.; Erkosar, B.; Hols, P.; Royet, J.; Leulier, F. *Lactobacillus plantarum* promotes drosophila systemic growth by modulating hormonal signals through TOR-dependent nutrient sensing. *Cell Metabolism* **2011**, *14*, 403–414. [CrossRef]
11. Chen, L.H.; Huang, S.Y.; Huang, K.C.; Hsu, C.C.; Yang, K.C.; Li, L.A.; Chan, C.H.; Huang, H.Y. *Lactobacillus paracasei* PS23 decelerated age-related muscle loss by ensuring mitochondrial function in SAMP8 mice. *Aging* **2019**, *11*, 756–770. [CrossRef] [PubMed]
12. Shokri-Mashhadi, N.; Navab, F.; Ansari, S.; Rouhani, M.H.; Hajhashemy, Z.; Saraf-Bank, S. A meta-analysis of the effect of probiotic administration on age-related sarcopenia. *Food Sci. Nutr.* **2023**, *11*, 4975–4987. [CrossRef] [PubMed]
13. Mante, E.S.; Sakyi-Dawson, E.; Amoa-Awua, W.K. Antimicrobial interactions of microbial species involved in the fermentation of cassava dough into agbelima with particular reference to the inhibitory effect of lactic acid bacteria on enteric pathogens. *Int. J. Food Microbiol.* **2003**, *89*, 41–50. [CrossRef] [PubMed]
14. Liu, H.; Ji, H.F.; Zhang, D.Y.; Wang, S.X.; Wang, J.; Shan, D.C.; Wang, Y.M. Effects of *Lactobacillus brevis* preparation on growth performance, fecal microflora and serum profile in weaned pigs. *Livest. Sci.* **2015**, *178*, 251–254. [CrossRef]
15. Zhang, Q.; Fan, X.Y.; Cao, Y.J.; Zheng, T.T.; Cheng, W.J.; Chen, L.J.; Lv, X.C.; Ni, L.; Rao, P.F.; Liang, P. The beneficial effects of Lactobacillus brevis FZU0713-fermented Laminaria japonica on lipid metabolism and intestinal microbiota in hyperlipidemic rats fed with a high-fat diet. *Food Funct.* **2021**, *12*, 7145–7160. [CrossRef] [PubMed]
16. Frampton, J.; Murphy, K.G.; Frost, G.; Chambers, E.S. Short-chain fatty acids as potential regulators of skeletal muscle metabolism and function. *Nat. Metab.* **2020**, *2*, 840–848. [CrossRef] [PubMed]
17. den Besten, G.; Gerding, A.; van Dijk, T.H.; Ciapaite, J.; Bleeker, A.; van Eunen, K.; Havinga, R.; Groen, A.K.; Reijngoud, D.J.; Bakker, B.M. Protection against the metabolic syndrome by guar gum-derived short-chain fatty acids depends on peroxisome proliferator-activated receptor γ and glucagon-like peptide-1. *PLoS ONE* **2015**, *10*, e0136364. [CrossRef] [PubMed]
18. Walsh, M.E.; Bhattacharya, A.; Sataranatarajan, K.; Qaisar, R.; Sloane, L.; Rahman, M.M.; Kinter, M.; Van Remmen, H. The histone deacetylase inhibitor butyrate improves metabolism and reduces muscle atrophy during aging. *Aging Cell* **2015**, *14*, 957–970. [CrossRef]
19. Li, H.; Qiu, T.; Huang, G.; Cao, Y. Production of gamma-aminobutyric acid by *Lactobacillus brevis* NCL912 using fed-batch fermentation. *Microb. Cell Fact.* **2010**, *9*, 85. [CrossRef]
20. Godfrey, R.J.; Madgwick, Z.; Whyte, G.P. The exercise-induced growth hormone response in athletes. *Sports Med.* **2003**, *33*, 599–613. [CrossRef]
21. Sakashita, M.; Nakamura, U.; Horie, N.; Yokoyama, Y.; Kim, M.; Fujita, S. Oral supplementation using gamma-aminobutyric acid and whey protein improves whole body fat-free mass in men after resistance training. *J. Clin. Med. Res.* **2019**, *11*, 428–434. [CrossRef] [PubMed]
22. Arvanitoyannis, I.S. *HACCP and ISO 22000: Application to Foods of Animal Origin*; John Wiley & Sons: Hoboken, NJ, USA, 2009.
23. Brzycki, M. Strength testing—Predicting a one-rep max from reps-to-fatigue. *J. Phys. Educ. Recreat. Dance* **1993**, *64*, 88–90. [CrossRef]
24. Bruce, R.A.; Kusumi, F.; Hosmer, D. Maximal oxygen intake and nomographic assessment of functional aerobic impairment in cardiovascular disease. *Am. Heart J.* **1973**, *85*, 546–562. [CrossRef] [PubMed]
25. Lee, M.C.; Ho, C.S.; Hsu, Y.J.; Wu, M.F.; Huang, C.C. Effect of 8-week frequency-specific electrical muscle stimulation combined with resistance exercise training on muscle mass, strength, and body composition in men and women: A feasibility and safety study. *PeerJ* **2023**, *11*, e16303. [CrossRef] [PubMed]
26. Lee, M.C.; Ho, C.S.; Hsu, Y.J.; Huang, C.C. Live and Heat-Killed Probiotic *Lactobacillus paracasei* PS23 accelerated the improvement and recovery of strength and damage biomarkers after exercise-induced muscle damage. *Nutrients* **2022**, *14*, 4563. [CrossRef] [PubMed]
27. Lee, M.C.; Hsu, Y.J.; Ho, C.S.; Chang, C.H.; Liu, C.W.; Huang, C.C.; Chiang, W.D. Evaluation of the efficacy of supplementation with planox® lemon verbena extract in improving oxidative stress and muscle damage: A randomized double-blind controlled trial. *Int. J. Med. Sci.* **2021**, *18*, 2641–2652. [CrossRef] [PubMed]
28. Huang, C.C.; Lee, M.C.; Ho, C.S.; Hsu, Y.J.; Ho, C.C.; Kan, N.W. Protective and recovery effects of resveratrol supplementation on exercise performance and muscle damage following acute plyometric exercise. *Nutrients* **2021**, *13*, 3217. [CrossRef] [PubMed]
29. Nakamura, M.; Ikezoe, T.; Takeno, Y.; Ichihashi, N. Acute and prolonged effect of static stretching on the passive stiffness of the human gastrocnemius muscle tendon unit in vivo. *J. Orthop. Res.* **2011**, *29*, 1759–1763. [CrossRef] [PubMed]
30. Zhu, H.; Qu, F.; Zhu, L.H. Isolation of genomic DNAs from plants, fungi and bacteria using benzyl chloride. *Nucleic Acids Res.* **1993**, *21*, 5279–5280. [CrossRef]
31. Chew, W.; Lim, Y.P.; Lim, W.S.; Chambers, E.S.; Frost, G.; Wong, S.H.; Ali, Y. Gut-muscle crosstalk. A perspective on influence of microbes on muscle function. *Front. Med.* **2023**, *9*, 1065365. [CrossRef]
32. Przewłócka, K.; Folwarski, M.; Kaźmierczak-Siedlecka, K.; Skonieczna-Żydecka, K.; Kaczor, J.J. Gut-muscle axisexists and may affect skeletal muscle adaptation to training. *Nutrients* **2020**, *12*, 1451. [CrossRef] [PubMed]

33. Bycura, D.; Santos, A.C.; Shiffer, A.; Kyman, S.; Winfree, K.; Sutliffe, J.; Pearson, T.; Sonderegger, D.; Cope, E.; Caporaso, J.G. Impact of different exercise modalities on the human gut microbiome. *Sports* **2021**, *9*, 14. [CrossRef]
34. Clarke, S.F.; Murphy, E.F.; O'Sullivan, O.; Lucey, A.J.; Humphreys, M.; Hogan, A.; Hayes, P.; O'Reilly, M.; Jeffery, I.B.; Wood-Martin, R.; et al. Exercise and associated dietary extremes impact on gut microbial diversity. *Gut* **2014**, *63*, 1913–1920. [CrossRef]
35. Guan, L.; Cao, Z.; Pan, Z.; Zhao, C.; Xue, M.; Yang, F.; Chen, J. Butyrate promotes C2C12 myoblast proliferation by activating ERK/MAPK pathway. *Mol. Omics* **2023**, *19*, 552–559. [CrossRef] [PubMed]
36. Lee, M.C.; Hsu, Y.J.; Ho, H.H.; Kuo, Y.W.; Lin, W.Y.; Tsai, S.Y.; Chen, W.L.; Lin, C.L.; Huang, C.C. Effectiveness of human-origin *Lactobacillus plantarum* PL-02 in improving muscle mass, exercise performance and anti-fatigue. *Sci Rep.* **2021**, *11*, 19469. [CrossRef]
37. Chen, L.; Nelson, D.R.; Zhao, Y.; Cui, Z.; Johnston, J.A. Relationship between muscle mass and muscle strength, and the impact of comorbidities: A population-based, cross-sectional study of older adults in the United States. *BMC Geriatr.* **2013**, *13*, 74. [CrossRef] [PubMed]
38. Maughan, R.J.; Watson, J.S.; Weir, J. Strength and cross-sectional area of human skeletal muscle. *J. Physiol.* **1983**, *338*, 37–49. [CrossRef]
39. Ikai, M.; Fukunaga, T. A study on training effect on strength per unit cross-sectional area of muscle by means of ultrasonic measurement. *Int. Z. Angew. Physiol.* **1970**, *28*, 173–180. [CrossRef] [PubMed]
40. Wilke, J.; Macchi, V.; De Caro, R.; Stecco, C. Fascia thickness, aging and flexibility: Is there an association? *J. Anat.* **2019**, *234*, 43–49. [CrossRef]
41. Jia, S.; Wu, Q.; Wang, S.; Kan, J.; Zhang, Z.; Zhang, X.; Zhang, X.; Li, J.; Xu, W.; Du, J.; et al. Pea Peptide Supplementation in Conjunction with Resistance Exercise Promotes Gains in Muscle Mass and Strength. *Front. Nutr.* **2022**, *9*, 878229. [CrossRef]
42. Suchomel, T.J.; Nimphius, S.; Bellon, C.R.; Stone, M.H. The importance of muscular strength: Training considerations. *Sports Med.* **2018**, *48*, 765–785. [CrossRef] [PubMed]
43. Marián, V.; Katarína, L.; Dávid, O.; Matúš, K.; Simon, W. Improved maximum strength, vertical jump and sprint performance after 8 weeks of jump squat training with individualized loads. *J. Sports Sci. Med.* **2016**, *15*, 492–500. [PubMed]
44. Karim, A.; Muhammad, T.; Shahid Iqbal, M.; Qaisar, R. A multistrain probiotic improves handgrip strength and functional capacity in patients with COPD: A randomized controlled trial. *Arch. Gerontol. Geriatr.* **2022**, *102*, 104721. [CrossRef] [PubMed]
45. Hoffman, J.R.; Hoffman, M.W.; Zelicha, H.; Gepner, Y.; Willoughby, D.S.; Feinstein, U.; Ostfeld, I. the effect of 2 weeks of inactivated probiotic *Bacillus coagulans* on endocrine, inflammatory, and performance responses during self-defense training in soldiers. *J. Strength Cond. Res.* **2019**, *33*, 2330–2337. [CrossRef] [PubMed]
46. Vaisberg, M.; Paixão, V.; Almeida, E.B.; Santos, J.M.B.; Foster, R.; Rossi, M.; Pithon-Curi, T.C.; Gorjão, R.; Momesso, C.M.; Andrade, M.S.; et al. Daily intake of fermented milk containing *Lactobacillus casei* shirota (lcs) modulates systemic and upper airways immune/inflammatory responses in marathon runners. *Nutrients* **2019**, *11*, 1678. [CrossRef] [PubMed]
47. Salarkia, N.; Ghadamli, L.; Zaeri, F.; Sabaghian Rad, L. Effects of probiotic yogurt on performance, respiratory and digestive systems of young adult female endurance swimmers: A randomized controlled trial. *Med. J. Islam. Repub. Iran* **2013**, *27*, 141–146. [PubMed]
48. Chen, Y.M.; Wei, L.; Chiu, Y.S.; Hsu, Y.J.; Tsai, T.Y.; Wang, M.F.; Huang, C.C. *Lactobacillus plantarum* TWK10 supplementation improves exercise performance and increases muscle mass in mice. *Nutrients* **2016**, *8*, 205. [CrossRef]
49. Murray, B.; Rosenbloom, C. Fundamentals of glycogen metabolism for coaches and athletes. *Nutr. Rev.* **2018**, *76*, 243–259. [CrossRef]
50. Zhang, W.Q.; Zhao, T.T.; Gui, D.K.; Gao, C.L.; Gu, J.L.; Gan, W.J.; Huang, W.; Xu, Y.; Zhou, H.; Chen, W.N.; et al. Sodium butyrate improves liver glycogen metabolism in type 2 diabetes mellitus. *J. Agric. Food Chem.* **2019**, *67*, 7694–7705. [CrossRef]
51. Lawler, T.P.; Cialdella-Kam, L. Non-carbohydrate Dietary Factors and Their Influence on Post-Exercise Glycogen Storage: A Review. *Curr. Nutr. Rep.* **2020**, *9*, 394–404. [CrossRef]
52. Ansdell, P.; Thomas, K.; Hicks, K.M.; Hunter, S.K.; Howatson, G.; Goodall, S. Physiological sex differences affect the integrative response to exercise: Acute and chronic implications. *Exp. Physiol.* **2020**, *105*, 2007–2021. [CrossRef] [PubMed]
53. de Paiva, A.K.F.; de Oliveira, E.P.; Mancini, L.; Paoli, A.; Mota, J.F. Effects of probiotic supplementation on performance of resistance and aerobic exercises: A systematic review. *Nutr. Rev.* **2023**, *81*, 153–167. [CrossRef] [PubMed]
54. Mohr, A.E.; Pugh, J.; O'Sullivan, O.; Black, K.; Townsend, J.R.; Pyne, D.B.; Wardenaar, F.C.; West, N.P.; Whisner, C.M.; McFarland, L.V. Best practices for probiotic research in athletic and physically active populations: Guidance for future randomized controlled trials. *Front. Nutr.* **2022**, *9*, 809983. [CrossRef] [PubMed]

foods

MDPI

Article

Evaluation of a Standard Dietary Regimen Combined with Heat-Inactivated *Lactobacillus gasseri* HM1, Lactoferrin-Producing HM1, and Their Sonication-Inactivated Variants in the Management of Metabolic Disorders in an Obesity Mouse Model

Wei-Chen Shiu [1,†], Zhen-Shu Liu [2,3,4,†], Bo-Yuan Chen [1], Yu-We Ku [1,5] and Po-Wen Chen [1,*]

1 Department of Veterinary Medicine, College of Veterinary Medicine, National Chung Hsing University, Taichung 402202, Taiwan; sponge77625@gmail.com (W.-C.S.); jackynike061548@gmail.com (B.-Y.C.); kyw168@mail.e-land.gov.tw (Y.-W.K.)
2 Department of Safety, Health and Environmental Engineering, Ming Chi University of Technology, New Taipei City 24301, Taiwan; zsliu@mail.mcut.edu.tw
3 Chronic Diseases and Health Promotion Research Center, Chang Gung University of Science and Technology, Chiayi 61363, Taiwan
4 Center for Sustainability and Energy Technologies, Chang Gung University, Taoyuan 33302, Taiwan
5 Animal and Plant Disease Control Center Yilan County, Wujie Township, Yilan County 268015, Taiwan
* Correspondence: powenchen@nchu.edu.tw; Tel.: +886-4-22840368-36
† These authors contributed equally to this work.

Abstract: This study investigated the impact of incorporating various inactivated probiotic formulations, with or without recombinant lactoferrin (LF) expression, into a standard chow diet on metabolic-related disorders in obese mice. After inducing obesity through a 13-week high-fat diet followed by a standard chow diet, mice received daily oral administrations of different probiotics for 6 weeks using the oral gavage approach. These probiotic formulations consisted of a placebo (MRS), heat-inactivated *Lactobacillus gasseri* HM1 (HK-HM1), heat-killed LF-expression HM1 (HK-HM1/LF), sonication-killed HM1 (SK-HM1), and sonication-killed LF-expression HM1 (SK-HM1/LF). The study successfully induced obesity, resulting in worsened glucose tolerance and insulin sensitivity. Interestingly, the regular diet alone improved glucose tolerance, and the addition of inactivated probiotics further enhanced this effect, with SK-HM1/LF demonstrating the most noticeable improvement. However, while regular dietary intervention alone improved insulin sensitivity, probiotic supplementation did not provide additional benefits in this aspect. Inflammation in perirenal and epididymal fat tissues was partially alleviated by the regular diet and further improved by probiotics, particularly by SK-HM1, which showed the most significant reduction. Additionally, HK-HM1 and HK-HM1/LF supplements could contribute to the improvement of serum total triglycerides or total cholesterol, respectively. Overall, incorporating inactivated probiotics into a regular diet may enhance metabolic indices, and recombinant LF may offer potential benefits for improving glucose tolerance.

Keywords: anti-inflammation; obesity; metabolic disorder; probiotic; postbiotics; *Lactobacillus gasseri*; high-fat diet; lactoferrin

1. Introduction

Obesity denotes a modified health state stemming from irregularities in energy intake, energy balance, gut microbiota, and an inappropriate diet, influenced by genetic and environmental factors. It is operationally characterized as the accumulation of excessive fat, adversely affecting health status [1]. Nowadays, obesity has been characterized as a chronic disease, reaching pandemic proportions and being associated with various metabolic disorders, including type 2 diabetes (T2DM), cardiovascular ailments, and cancer [2,3]. Hence, there lies imperative significance in the treatment or prevention of obesity and metabolic disorder. To date, various interventions, including caloric restriction, aerobic exercise, pharmacological treatments, and bariatric surgery, have shown efficacy in promoting weight loss and mitigating the associated risks of disease in individuals grappling with obesity [3–6]. However, at both individual and community levels, the prevention and treatment strategies for obesity have not achieved success in the long run [7]. Nonetheless, lifestyle modifications remain the primary therapy for obesity and related metabolic disorders. For instance, calorie restriction (CR) diets have gained recognition for their role in promoting health, extending lifespan, and averting the onset of metabolic and age-related ailments [8]. However, maintaining long-term weight loss has proven to be a formidable challenge, primarily owing to intricate interplays between hormones and behavior [9–11]. These interactions frequently culminate in gradual or rapid weight regain following a CR intervention [12,13].

One potentially ideal strategy for treating obesity involves the manipulation of gut microbiota. For example, accumulating scientific evidence supports the role of probiotics, probiotic components, and their metabolites in regulating host lipid, sugar, and cholesterol metabolism, as well as in suppressing oxidative stress and mitigating endoplasmic reticulum stress. Certain probiotics also induce improvements in gut microbiota composition and inhibit chronic low-grade inflammation, thereby preventing and ameliorating obesity and related chronic diseases [14–17]. The advantages of manipulating gut microbiota or implementing probiotic-related interventions have been thoroughly reviewed and succinctly summarized as follows. First, this therapeutic approach is deemed safe, with no reported adverse effects, and it is well tolerated, making it suitable for long-term use. Second, the modulation of gut microbiota through probiotic treatment or dietary intervention, owing to its beneficial effects, can impact body weight, influence glucose and fat metabolism, enhance insulin sensitivity, and mitigate chronic systemic inflammation [18]. However, a comprehensive literature review reveals that the favorable effects of probiotic supplementation in individuals with obesity are linked to various factors. These include the nature of the probiotic strain, the composition of the probiotic formula (such as single or multiple strains, with or without prebiotics), the duration of the intervention, dosage, and other supplementary activities, such as CR and use of weight loss medications [19].

Probiotics, as defined by the International Scientific Association of Probiotics and Prebiotics (ISAPP), are active micro-organisms, which, when ingested in sufficient quantities, confer health benefits to the host [20]. In recent years, the scientific community has introduced various terms, such as non-viable probiotics, paraprobiotics, ghostbiotics, and notably, postbiotics, to describe inanimate micro-organisms and/or their constituents conferring health benefits [21,22]. Postbiotics are compounds originating from the metabolites or cell wall fragments produced by probiotics. This term encompasses a diverse array of bioactive molecules, such as non-viable or inactivated microbial cells, short-chain fatty acids, vitamins, enzymes, teichoic acid, peptides, and exopolysaccharides, all derived from beneficial micro-organisms [23,24]. Notably, postbiotics have been recognized as safe for use, and documented data highlight the potential of postbiotics and derivatives from probiotics in managing metabolic disorders and associated changes [22,25,26]. Remarkably, the utilization of specific prebiotics and postbiotics has demonstrated efficacy in long-term weight management for the control of obesity [27].

Lactoferrin (LF) is an 80 kDa iron-binding glycoprotein predominantly found in the milk and various exocrine fluids of mammals. It manifests diverse pleiotropic activities, encompassing anti-inflammatory, antimicrobial, antioxidant, immune-regulating, and prebiotic properties [28–30]. Additionally, in both low- and high-dose combinations, LF and inulin have demonstrated an additive or supra-additive effect in reducing energy intake, body weight, and adiposity in a diet-induced obese rat model [31]. Furthermore, dietary lactalbumin and LF have also proven effective in enhancing the energy balance, metabolism, and reducing adiposity. Furthermore, the impact of LF appears to be partly independent of caloric intake in a diet-induced obese rat model [32]. To facilitate the concurrent display of functional attributes inherent to both LF and probiotics, we recently developed engineered probiotics proficient in expressing bovine, human, or porcine LF [33]. Subsequently, we further formulated three distinct probiotic supplements, namely lactic acid bacteria (LAB), LAB/LF, and inactivated LAB/LF. The LAB supplement comprised 10 viable LAB without recombinant LF expression; the LAB/LF supplement involved 10 viable LF-expressing probiotics; and the inactivated LAB/LF supplement was derived from 10 inactivated LF-expressing probiotics. Utilizing the NAFLD mice model, the administration of the aforementioned live or inactivated probiotic mixtures derived from host probiotics or LF-expressing probiotics demonstrated a significant and distinct amelioration of hepatic steatosis and inflammation [34].

In light of the observed improvements in metabolic-related disorders following the administration of inactivated probiotics (postbiotics) or LF, we hypothesized that the inactivated singular probiotic strain, specifically *Lactobacillus gasseri* HM1 (HM1), which acts as the host strain for LF-expressing probiotic, may also demonstrate anti-obesity capabilities. Hence, in the present study, we aim to investigate the anti-obesity effects of the inactivated HM1 strain. This comparison extends to evaluating heat-killed and sonication-killed HM1, with a particular emphasis on those expressing LF, representing two distinct formulations of inactivated HM1/LF. Secondly, given the well-known association between dietary control and improved metabolic indices, we also aim to explore the impact of a combination of regular diet (standard chow diet) and supplementation with one of four inactivated probiotic formulations in addressing metabolic disorders.

2. Materials and Methods

2.1. Probiotic Strain and Postbiotics Preparation

The probiotic strain (*L. gasseri* HM1, our laboratory stock), sourced from human milk [35,36], was cultivated under anaerobic conditions at 37 °C in De Man, Rogosa, and Sharpe (MRS) broth (BD Biosciences, San Jose, CA, USA) for a duration of 24 h. Another *L. gasseri*/BLF strain, the LF-expressing HM1 (HM1/LF), was activated and cultured anaerobically in MRS broth at 37 °C without agitation [33]. The expression of recombinant LF was induced by supplementing cultures of recombinant LF-expressing probiotics with nisin at a concentration of 1 ng/mL (supplemented in fresh medium) for 16 h at 30 °C, as previously described in our report. Moreover, the confirmation of LF expression in probiotics was achieved through SDS page and Western blotting analysis, utilizing an anti-LF antibody [33]. To investigate the varied effects of two probiotic inactivation methods on improving the indicators of metabolic disease, two distinct approaches were employed: high-temperature (heat-killing) and non-high-temperature (sonication-killing) methods were utilized to deactivate the probiotics. Initially, probiotic strains were harvested via centrifugation (Hermle Z 366 K, Wehingen, Germany) at 12,000× *g* for 10 min. The pellets underwent two washes with 10 mL of PBS buffer. Following a cell count, the bacterial concentration was adjusted to 5×10^{10} cells/mL using MRS broth. Then, 1 mL of HM1 or HM1/LF probiotics (5×10^{10} cells/mL) underwent heat treatment in an autoclave (TM-321, Tomin Autoclave, Taipei, Taiwan) under the following conditions: 121 °C and 1.5 atmospheres of pressure for 15 min. Additionally, another 1 mL of HM1 or HM1/LF probiotics (5×10^{10} cells/mL) was subjected to sonication for 5 min at 4 °C using a sonicator (Sonic Material VCX600, Newtown, CT, USA) with a power output of 20 W. Subsequently, the supernatant from both treatments was collected through centrifugation at 15,000× *g* for 5 min at 4 °C to eliminate any precipitates. This procedure yielded four inactivated probiotic formulations, including heat-killed HM1 (HK-HM1), heat-killed LF-expressing HM1 (HK-HM1/LF), sonication-killed HM1 (SK-HM1), and sonication-killed HM1/LF (SK-HM1/LF).

2.2. Animal, Experimental Design, and Sample Collection

The animal experiments were conducted with the approval of the Institutional Animal Care and Use Committee of the National Chung Hsing University, under approval number IACUC No. 110-080R. A total of 36 six-week-old male C57BL/6J mice, obtained from the National Laboratory Animal Center in Taiwan, were utilized for this experiment. The mice were housed in a controlled animal facility with an environmental temperature of 26 °C, a relative humidity of 63%, and a 12 h light–dark cycle, providing them with unrestricted access to food and drinking water. After a two-week acclimatization period in this controlled environment, the animals were introduced to the experimental procedures.

To induce an obesity animal model, 30 mice were exposed to a 13-week high-fat diet (HFD), comprising 60% kcal from fat, 20% kcal from protein, and 20% kcal from carbohydrates (D12492; Research Diets, New Brunswick, NJ, USA). Following this 13-week period, the mice were transitioned to a standard diet, specifically the standard chow diet (SCD), for a 6-week duration. The SCD contained 13% kcal from fat, 28% kcal from protein, and 59% kcal from carbohydrates, adhering to the formulation of the 5001 Laboratory Rodent Diet from Lab Diet in the USA. Over the course of these six weeks, five groups of mice underwent daily oral treatments administered via oral gavage, with a dosage of 0.2 mL per mouse: placebo (MRS; Placebo; n = 6), HK-HM1 (n = 6), SK-HM1 (n = 6), HK-HM1/LF (n = 6), and SK-HM1/LF (n = 6). Additionally, another six mice were designated as the normal control group (Normal control), and they were provided with a regular diet for a total of 19 weeks, with the administration of a placebo (MRS) through oral gavage oral ingestion for 6 weeks, starting from the 13th week.

Throughout the experimental period, weekly assessments were conducted to monitor changes in body weight and food consumption. The final weight recorded at the conclusion of the experiment served as the basis for evaluating the effectiveness of weight loss or gain in each group. Following this assessment, the mice were anesthetized, and euthanasia was performed by collecting blood from the heart. The obtained blood samples underwent additional analysis.

Subsequently, liver, epididymal fat, and perirenal fat were harvested from the mice. These tissue samples were preserved in 10% formalin for 24 h, followed by embedding in paraffin and sectioning for histological examination. The tissue sections underwent staining with hematoxylin and eosin (H&E) to observe histopathological changes. The histological analysis employed semi-quantitative criteria in histopathological examination, assessed blindly by a certified veterinary pathologist in Taiwan (CSVP Vet Path No. 0019). Moreover, the degree of lesions was graded from zero to four based on severity: 0 = normal; 1 = slight (<10%); 2 = moderate (10–33%); 3 = moderate/severe (33–66%); 4 = severe/high (66–100%) [37].

2.3. Biochemical Analyses

Biochemical analyses were performed on blood specimens collected in BD Vacutainer® SST II Advance blood collection tubes, equipped with coagulation enhancers and separating gels. Following gentle mixing, the samples stood at room temperature for 30 min. Subsequently, they underwent centrifugation at 2000× *g* for 10 min to facilitate serum separation. The resulting serum samples were then forwarded to the National Laboratory Animal Center (NLAC Taiwan) for the quantification of total cholesterol (TC), total triglycerides (TG), low-density lipoprotein cholesterol (LDL-C), high-density lipoprotein cholesterol (HDL-C), and glucose on a 7080 biochemical analyzer (Hitachi, Tokyo, Japan), according to the manufacturer's instructions.

2.4. Testing for Glucose Tolerance and Insulin Sensitivity

We adhered to established experimental protocols, as outlined in prior publications, for assessing glucose and insulin tolerance to evaluate insulin sensitivity in mice before and after a 6-week dietary and probiotic intervention [38]. Glucose tolerance was assessed at weeks 12 and 17 of the experiment, while insulin tolerance was evaluated biweekly during weeks 13 and 18. For the glucose tolerance test (GTT), mice underwent a 6 h fasting period followed by intraperitoneal injections of a 2 g/kg glucose solution. Blood glucose levels were monitored at 0, 15, 30, 60, and 120 min post-injection using a blood glucose monitor (OneTouch, Select Plus; LifeScan, Milpitas, Santa Clara, CA, USA). Regarding the insulin tolerance test (ITT), mice also experienced a 6 h fasting period, followed by intraperitoneal injections of 1 IU/kg insulin. Blood glucose levels were then monitored at 0, 15, 30, 60, and 120 min post-injection using the same blood glucose monitor (OneTouch, Select Plus; LifeScan, Milpitas, Santa Clara, CA, USA).

2.5. Statistical Analysis

The results are presented as mean values ± standard deviation (SD). Analysis of variance (ANOVA) was utilized to assess differences between the groups, and Tukey's post hoc test was employed to evaluate the significance of differences between specific data points. The analysis was performed using SPSS (Version 20.0). Statistical significance was established with a p-value of less than 0.05, indicating the presence of meaningful distinctions.

3. Results

3.1. Weight Fluctuations in Mice throughout the Entire Experimental Period

In this study, our objective was to evaluate the impact of combining a regulated diet with inactivated natural or recombinant LF expression on metabolic indices. Initially, we measured weight variations among mouse groups at the beginning, after a 13-week high-fat diet induction period, and following a 6-week regular diet with four inactivated probiotic supplementations. The body weight values of mice at significant time points during the experiment are presented in Table 1. At week 0, all mouse groups exhibited comparable weights ($p > 0.05$). Subsequently, following a 13-week high-fat diet, including placebo, HK-HM1, SK-HM1, HK-HM1/LF, and SK-HM1/LF, all groups showed significant weight gains, with weights statistically higher than those of Normal control groups ($p < 0.05$). These weights increased by over 42% compared to the normal group, indicating successful induction of obesity in all mouse groups. After an additional 6-week regulated diet with any of the four probiotic formulations (week 19), the weights of the five groups declined considerably, yet remaining higher than those of the normal mouse group. Although the HK-HM1/LF group had a significantly lower weight than the placebo group, the weight gain index did not differ between the placebo and HK-HM1/LF groups.

Table 1. Body weight values of mice at important time points during the experiment. Results are presented as means ± SDs.

Weight (g)	Normal Control	HFD Groups [a]				
		Placebo	HK-HM1	SK-HM1	HK-HM1/LF	SK-HM1/LF
0 week	23.24 ± 1.2	24.25 ± 1.49	23.44 ± 1.31	22.65 ± 0.61	22.94 ± 1.18	22.96 ± 0.36
13 weeks	26.98 ± 1.39	41.99 ± 4.24 #	40.02 ± 2.74 #	39.12 ± 3.48 #	39.74 ± 3.17 #	38.56 ± 1.86 #
19 weeks—I	27.09 ± 1.18	32.3 ± 1.29 #	30.92 ± 1.23 #	31.49 ± 1.56 #	30.12 ± 1.34 #*	31.8 ± 1.09 #
19 weeks—II (weight gain)	0.1 ± 0.46	−9.69 ± 3.16 #	−9.1 ± 2.95 #	−7.63 ± 2.16 #	−9.62 ± 2.74 #	−6.76 ± 1.45 #

[a] Various groups of mice were subjected to different diets and probiotic formulations to assess the potential effects of probiotics under diverse conditions. The healthy control group mice (n = 6) were maintained on a regular diet throughout the 19-week experiment. Additionally, four groups of mice (n = 6 for each group) initially received a high-fat diet for 13 weeks, followed by a switch to a standard chow diet (SCD), along with daily oral administration of different probiotic formulations for 6 weeks (from week 13 to 19). Probiotic strain: HK-HM1: heat-killed *Lactobacillus gasseri* HM1; SK-HM1: sonication-killed HM1; HK-HM1/LF: heat-killed lactoferrin-expressing HM1; SK-HM1/LF: sonication-killed lactoferrin-expressing HM1. # Statistically significant difference compared to Normal control ($p < 0.05$); * Statistically significant difference compared to placebo + SCD ($p < 0.05$).

In Figure 1, we further compared the trends in the body weight change curve of mice during the experiment. The results indicate that during the induction of obesity with a high-fat feed, the weight gain trends in the five groups of mice were similar. Additionally, after the regular diet and probiotic intervention, the weight loss trends in the five groups of mice were also similar. Overall, these findings suggest that while a standard chow diet can contribute to weight loss, the administration of any of the four probiotic formulations does not provide additional benefits in reducing body weight. Furthermore, based on the weight reduction curve after 13 weeks, the optimal weight loss effect of dietary control can be maintained for approximately 3 weeks (from week 13 to 16), with the downward trend becoming more moderate after 16 weeks. Nevertheless, we also assessed additional metabolic indices, which could be influenced by the administration of a regular diet or probiotics, as outlined below.

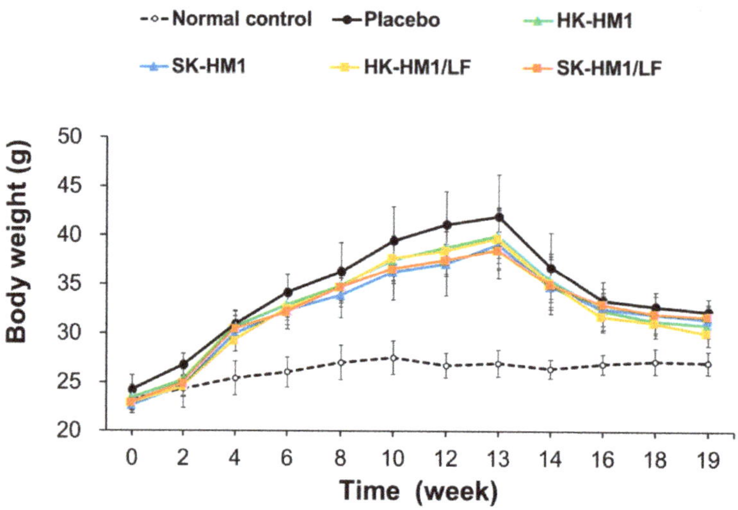

Figure 1. Trends in body weight change curve of mice during the experiment. See Table 1 for the key.

3.2. Glucose Tolerance among Treatments

The glucose tolerance test (GTT) is a diagnostic tool used to assess glucose metabolism and pancreatic function in the body. It is commonly employed to identify signs of diabetes or abnormalities in pancreatic function. In Figure 2, we conducted GTT evaluations among different groups of mice both before and after a 6-week period of regular diet and probiotic intervention. As described in Table 1 and Figure 1, following a 13-week high-fat diet induction, all five groups of mice (n = 30) were successfully induced into obesity. Given the comparable backgrounds of these obese mice, we specifically analyzed GTT in four mouse groups: placebo, HK-HM1, SK-HM1, and HK-HM1/LF. As expected, mice in the placebo, HK-HM1, SK-HM1, and HK-HM1/LF groups demonstrated consistent patterns in the elevation and decline of GTT variation (Figure 2A). Moreover, the GTT curves for these four groups at the designated time points of testing (15, 30, 60, and 120 min) were all higher compared to those of the normal mouse group. This outcome indicates that the 13-week high-fat diet intervention indeed worsened the GTT in these mice. To validate these observations, we computed the total area under the GTT curve among the mouse groups based on the data presented in Figure 2A. As depicted in Figure 2B, the placebo, HK-HM1, SK-HM1, and HK-HM1/LF groups indeed exhibited a significantly higher area under the GTT curve compared to the Normal control ($p < 0.05$). Additionally, when considering the surface area under the curve in healthy mice as 1, the respective surface areas under the curves for the placebo, HK-LM1, SK-LM1, and HK-LM1/LF groups (all obese mice) were 1.48, 1.52, 1.40, and 1.41. These results further support the notion that obesity indeed exacerbates glucose intolerance in mice by up to 40–50%.

The glucose tolerance of mice treated with a regular diet combined with probiotic formulations for six weeks is depicted in Figure 2C,D. In Figure 2C, HK-HM1, HK-HM1/LF, and SK-HM1/LF exhibited lower blood glucose levels than the Normal control at most tested time points. To validate this, we also calculated the total area under the glucose curve for each mouse group. As shown in Figure 2D, the placebo group demonstrated glucose tolerance similar to that of the normal mice, suggesting that dietary control alone (regular diet) contributes to improved glucose tolerance. Remarkably, the HK-HM1/LF and SK-HM1/LF mouse groups displayed a significantly lower area under the GTT curve compared to the normal mice ($p < 0.05$). Additionally, the SK-HM1/LF mouse groups exhibited a statistically lower area under the GTT curve compared to the placebo mice group ($p < 0.05$). Finally, assigning the area under the GTT curve of normal mice as 1, our analysis indicates that the probiotics HK-HM1, SK-HM1, HK-HM1/LF, and SK-HM1/LF could improve glucose tolerance by approximately 11%, 10.2%, 12.5%, and 17.4%, respectively.

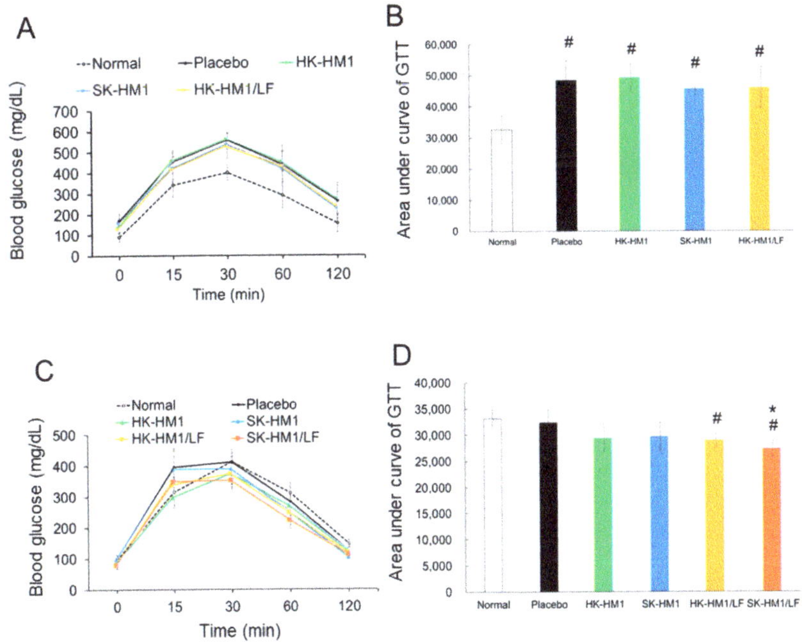

Figure 2. The results of the glucose tolerance test (GTT) in obese mice before and after a 6-week intervention with a combination of probiotics and a conventional diet. (**A**): Variations in blood glucose levels within the obese mice group prior to the intervention of probiotics and a standard diet; (**B**): Total area under the curve computed from A; (**C**): Changes in blood glucose levels within the obese mice group after the intervention of probiotics and a standard diet; (**D**): Total area under the curve calculated from C. # Statistically significant difference compared to Normal control ($p < 0.05$); * Statistically significant difference compared to placebo + SCD ($p < 0.05$).

In summary, these findings suggest that supplementation with SK-HM1/LF probiotic formulations leads to additional improvements in glucose tolerance.

3.3. Insulin Sensitivity across Treatments

We also assessed insulin sensitivity before and after a 6-week period of regular dietary intake combined with one of four probiotic formulations (Figure 3). Figure 3A,B depict the ITT results in mice prior to the initiation of dietary and probiotic interventions. Considering that all 30 mice were induced into an obese state (Table 1 and Figure 1), ITTs were conducted on only 18 mice to reveal the baseline insulin sensitivity in these obese mice (Figure 3A,B). In Figure 3A, the glucose response curves for the three obese mice groups (placebo, HK-LM1, and SK-LM1) showed a consistent and significant elevation in blood glucose levels at 15, 30, 60, and 120 min post-glucose administration compared to the healthy control group. Additionally, the glucose concentration area under the curve (AUC) from 0 to 120 min (Figure 3B) demonstrated a significantly higher AUC in all three obese groups compared to the healthy group ($p < 0.05$), with no statistical differences among the obese groups. These results indicate a consistent reduction in insulin sensitivity among obese mice, confirming that obesity indeed worsens insulin sensitivity. Furthermore, normalizing the AUC of healthy mice to 1, the AUC values for the placebo, HK-LM1, and SK-LM1 groups were 1.42, 1.49, and 1.46, respectively. These findings reinforce the notion that obesity exacerbates insulin sensitivity in mice. Comparing these results with Figure 2B, it is evident that obesity similarly deteriorates both insulin sensitivity and glucose tolerance, with both parameters being affected by approximately 40–50%.

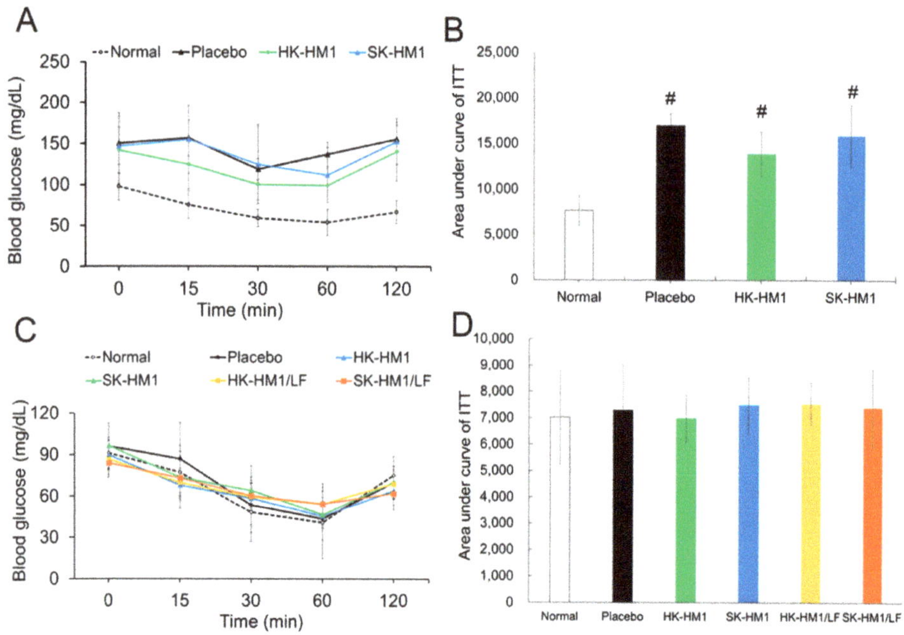

Figure 3. The outcomes of the insulin sensitivity test (ITT) conducted on obese mice both before and after a 6-week intervention involving a combination of probiotics and a conventional diet. (**A**) depicts the fluctuations in blood glucose levels within the obese mice group before the initiation of the probiotics and standard diet intervention. (**B**) represents the total area under the curve computed from the data in A. (**C**) shows the changes in blood glucose levels within the obese mice group following the intervention of probiotics and a standard diet. (**D**) displays the total area under the curve calculated from the data in C. # Statistically significant difference compared to Normal control ($p < 0.05$).

Figure 3C illustrates the insulin sensitivity results in obese mice after 6 weeks of dietary control and supplementation with probiotics. The glucose response curves among the six groups exhibit strikingly similar trends, indicating comparable insulin sensitivity across these groups. Further analysis of the overall insulin sensitivity among different groups, measured by the area under the curves (Figure 3D), reveals no statistically significant differences among the six groups of mice. These findings suggest that dietary control alone effectively improves insulin sensitivity in obese mice. However, regardless of the probiotic formulation administered, the mice did not experience additional benefits in insulin sensitivity.

3.4. Improvement Benefits of Dietary Control and Oral Probiotics in Tissue Pathological Changes in Obese Mice

To understand the beneficial effects of dietary control combined with oral probiotic supplementation on tissue pathological changes, we analyzed the extent of tissue lesions in the liver, perirenal fat, and epididymal fat among different experimental groups. Figure 4 illustrates representative results of the aforementioned tissue lesions. As shown in Figure 4A, the livers of mice in both the five experimental groups and the control group exhibited no lesions or inflammatory cell infiltration, and the morphology of their liver cells resembled that of the healthy control group. Since both the placebo group and the probiotic intervention group showed no liver cell lesions, these results support the substantial improvement in liver cell lesions achieved solely through dietary intervention. Additionally, Figure 4B,C, respectively, demonstrate inflammatory pathological changes in epididymal fat and perirenal fat among different groups. Notably, we did not observe significant differences in the size and quantity of adipose tissue among different groups (Figure 4B,C); additionally, there were no notable variations in adipose tissue weight between the groups. Conversely, inflammatory pathological changes in adipose tissue were more pronounced (Figure 4B,C). For example, compared to the healthy group, the placebo group still exhibited tissue or inflammatory lesions, while the lesions in the probiotic intervention group were relatively mild. We further calculated the tissue lesion scores and the incidence rates of relevant lesions for all groups (Table 2). First, regarding liver lesions, after 6 weeks of dietary control and any probiotic intervention, all mice in the six groups showed no liver lesions. For epididymal fat lesions, the epididymal fat of healthy mice showed no observed lesions, while in the placebo group, four mice exhibited Grade 1 lesions (66%). However, after additional oral administration of HK-HM1, SK-HM1, HK-HM1/LF, and SK-HM1/LF to obese mice, the probability of developing lesions decreased to 17%, 0%, 50%, and 50%, respectively, with lesion grading reduced to between 0 and 0.5. Regarding perirenal fat lesions, the perirenal fat of healthy mice also showed no lesions, while in the placebo group, four mice exhibited lesions of approximately Grade 0.85 (66%). However, after additional oral administration of HK-HM1, SK-HM1, HK-HM1/LF, and SK-HM1/LF to obese mice, the probability of developing lesions decreased to 17%, 0%, 33.3%, and 50%, respectively, with lesion grading reduced to between 0 and 0.5.

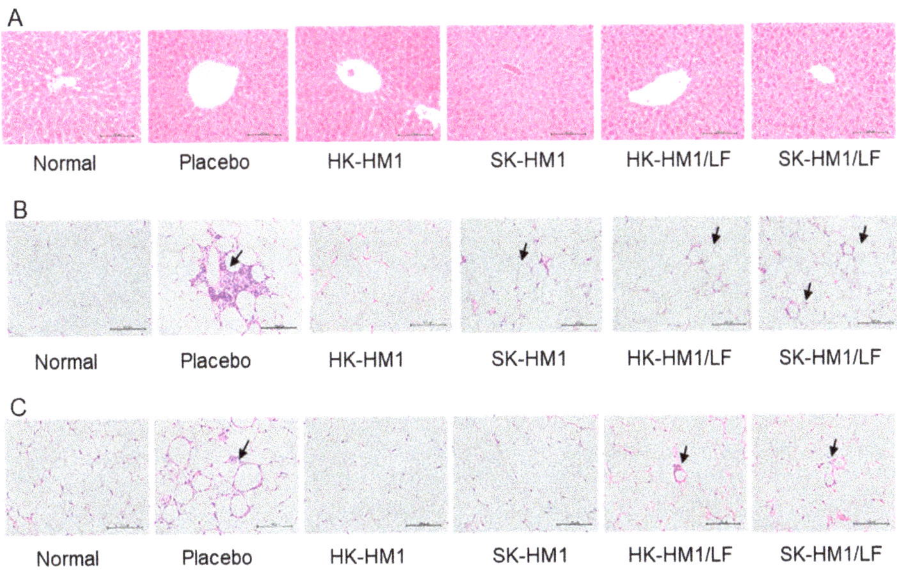

Figure 4. Pathological analysis of liver, epididymal, and perirenal adipose tissues in obese mice following 6 weeks of standard diet and oral administration of inactivated probiotics. (**A**) Liver histopathological sections; (**B**) Epididymal fat histopathological sections; (**C**) Perirenal fat histopathological sections. Pathological lesions are indicated by black arrows in the figure, where black arrows in the fat tissue histopathological sections represent pathological changes related to fat inflammation.

Table 2. The incidence rate and mean score of histopathological lesions or inflammation in the liver, perirenal, and epididymal fat of diet-induced obese mice supplemented with a standard chow diet, along with oral supplementation of various inactivated probiotic strains for 6 weeks.

		Normal	Standard Chow Diet				
			Placebo	HK-HM1	SK-HM1	HK-HM1/LF	SK-HM1/LF
Liver fatty change or inflammation	Incidence rate [1] Mean score [2]	0/6 0 ± 0 [2,3]	0/6 0 ± 0	0/6 0 ± 0	0/6 0 ± 0	0/6 0 ± 0	0/6 0 ± 0
Epididymis fat inflammation, Multifocal	Incidence rate Mean score	0/6 [1] 0 ± 0 [2,3]	4/6 1.00 ± 0.89 #	1/6 0.17 ± 0.41	0/6 0 ± 0 *	3/6 0.5 ± 0.55	3/6 0.5 ± 0.55
Perirenal fat inflammation, Multifocal	Incidence rate Mean score	0/6 [1] 0 ± 0 [2,3]	4/6 0.83 ± 0.75 #	1/6 0.17 ± 0.41	0/6 0 ± 0 *	2/6 0.5 ± 0.84	3/6 0.5 ± 0.55

[1] Incidence rate: this was calculated as the number of affected mice divided by the total number of mice examined. [2] Lesion severity was graded on a scale from zero to four, based on the following criteria: 0 = normal; 1 = slight (<10%); 2 = moderate (10–33%); 3 = moderate/severe (33–66%); 4 = severe/high (66–100%); # Statistically significant difference compared to Normal control ($p < 0.05$); [3] The final numerical score was calculated by dividing the sum of the number per grade of affected mice by the total number of examined mice. * Statistically significant difference compared to placebo control ($p < 0.05$).

In summary, our data reveal that a mere dietary change to a regular diet for 6 weeks is sufficient to restore liver tissue slices to a state similar to that of healthy mice. However, in the group that underwent only a dietary change to a regular diet for 6 weeks (placebo group), up to 66% of mice still exhibited mild lesions in epididymal and perirenal fat (approximately between Grade 0.8 and 1). Nevertheless, oral administration of probiotics in mice could significantly reduce the severity and incidence of these lesions. It is noteworthy that the probiotic SK-HM1 demonstrated the most effective improvement in epididymal and perirenal fat lesions, as mice in this group did not exhibit any lesions.

3.5. Impact of Diverse Probiotic Formulations, Coupled with a Standard Chow Diet, on Serum Biochemical Values in Diet-Induced Obese Mice

After 6 weeks of administering a regular diet and probiotics to obese mice, we evaluated their serum biochemical values (Table 3). First, the placebo group exhibited significantly higher serum total triglyceride and total cholesterol levels compared to normal mice ($p < 0.05$). Consequently, dietary control alone did not lead to a reduction in serum total triglyceride and total cholesterol in these mice. However, mice that received HK-HM1 demonstrated a similar level of serum total triglycerides to that of the normal group ($p > 0.05$). Furthermore, mice that received HK-HM1/LF demonstrated a similar level of serum total cholesterol to that of the normal group ($p > 0.05$). These data support the notion that HK-HM1 and HK-HM1/LF supplements could contribute to the improvement in serum total triglycerides or total cholesterol, respectively. On the other hand, a higher level of HDL-c in the serum is generally considered desirable. The results revealed that the placebo group's serum HDL-c value was significantly higher than that of normal mice ($p < 0.05$). Additionally, the HDL-c values of HK-HM1 and SK-HM1 were also significantly higher than those of normal mice ($p < 0.05$). However, no statistical difference in HDL-C was observed among the placebo, HK-HM1, SK-HM1, and SK-HM1/LF groups. These findings suggest that the higher serum HDL-c levels may be attributed to the effect of the regular diet. Furthermore, concerning LDL-C, there was no statistical difference in LDL-C levels between the placebo group, the four probiotic groups, and normal mice. Finally, in terms of blood glucose indicators, despite higher glucose levels in both the placebo and the four probiotic intervention groups compared to the normal mouse group, only the blood glucose level in HK-HM1/LF significantly exceeded the values observed in normal mice ($p < 0.05$). Moreover, there was no statistical difference in serum glucose among the placebo, HK-HM1, SK-HM1, HK-LM1/LF, and SK-HM1/LF groups ($p > 0.05$), and all were significantly higher than normal mice values ($p < 0.05$). These data suggest that the administration of any of the four inactivated probiotic formulations did not contribute to further control of the blood glucose level with a regular diet. However, as shown by the results of

GTT (Figure 2), the glucose tolerance of the placebo group was already similar to that of normal mice. Additionally, after oral administration of HK-HM1/LF and SK-HM1/LF, the glucose tolerance of obese mice was even better than that of normal mice ($p < 0.05$).

Table 3. Effects of various probiotic formulations in combination with regular diet (standard chow diet) on serum biochemical values in diet-induced obese mice.

	Normal	Standard Chow Diet				
		Placebo	**HK-HM1**	**SK-HM1**	**HK-HM1/LF**	**SK-HM1/LF**
Serum total triglyceride (mg/dL)	11.4 ± 4.49	29.57 ± 9.62 [#,a]	24.17 ± 4.03 [a]	28.08 ± 8.12 [#,a]	26.1 ± 11.01 [#,a]	33.65 ± 4.65 [#,a]
Serum total cholesterol (mg/dL)	61.68 ± 9.45	86.32 ± 12.79 [#,a]	79.47 ± 5.98 [#,a]	82.52 ± 4.24 [#,a]	78.43 ± 14.33 [a]	79.03 ± 6.77 [#,a]
Serum HDL-C (mg/dL)	50.27 ± 8.27	66.9 ± 6.39 [#,a]	66.78 ± 5.46 [#,a]	68.85 ± 4.37 [#,a]	64.72 ± 13.59	63.8 ± 5.25 [a]
Serum LDL-C (mg/dL)	9.68 ± 2.74	13.07 ± 4.64 [a]	9.83 ± 0.98 [a]	9.68 ± 2.02 [a]	10.58 ± 2.57 [a]	10.85 ± 3.37 [a]
Serum glucose (mg/dL)	106.47 ± 12.98	157.77 ± 54.87 [a]	146.33 ± 39.56 [a]	180.97 ± 45.38 [a]	201.03 ± 54.31 [#,a]	155.88 ± 38.66 [a]

[#] Statistically significant difference compared to Normal group ($p < 0.05$). [a] There was no statistically significant difference among placebo, HK-HM1, SK-HM1, HK-HM1/LF, and SK-HM1/LF mouse groups.

In summary, the combined strategy of a regular diet and HK-HM1 showed a tendency toward reducing serum total triglyceride. Furthermore, the combination of a regular diet and HK-HM1/LF also demonstrated a tendency toward decreasing serum total cholesterol. Additionally, dietary control alone increased serum HDL-c and helped maintain LDL-c levels similar to those of normal mice.

4. Discussion

The effectiveness of dietary strategies, such as calorie restriction, in controlling weight gain and metabolic disorders is well established [39–41]. However, adherence to such regimens can be challenging, necessitating exploration of alternative approaches. The present study explores the potential metabolic improvements associated with a conventional diet, characterized by standard caloric intake, when supplemented with inactivated probiotics. Specifically, we sought to understand whether this approach could yield beneficial effects on various metabolic indicators without strict calorie reduction. Our results indicated that the current inactivated probiotic formulation provided limited assistance in weight reduction. Notably, the majority of probiotics currently utilized, particularly those employed in sectors such as aquaculture, livestock farming, or economic animal industries, generally exhibit growth-promoting effects, enhancing body weight, improving feed efficiency, and bolstering host immunity against intestinal infections while increasing defense capabilities against specific diseases [42–44]. In fact, the probiotic effect on body weight was found to be species- and strain-specific, as highlighted in a recent systematic review. For instance, particular strains may contribute to weight reduction, while other strains might exhibit anti-obesity effects [45]. The limited impact of our probiotics on the body weight of obese mice might also possibly be attributed to the use of inactivated probiotics here. These inactivated probiotics have lost their ability to colonize the gut with live bacteria, potentially resulting in a weaker effect on improving gut microbiota. Supporting this notion, our recent publication revealed that mice receiving live LAB/LF probiotics while being induced into obesity with a high-fat diet consistently exhibited significantly higher body weights than placebo mice throughout most of the induction period. Conversely, mice administered inactivated LAB/LF probiotics showed a significant reduction in body weight gains compared to mice receiving the placebo and live LAB/LF probiotics [34]. While these experiments utilized a mixture of multiple probiotic strains, the results indicated that the oral administration of live and inactivated probiotic formulations plays distinct roles in influencing weight gain. Collectively, concerning the probiotic strains we employed, we believe that the impact of diet on body weight outweighs the simultaneous use of inactivated probiotics. Nevertheless, the current inactivated probiotics we used do not contribute to additional weight gain in obese mice.

In this study, we confirmed that mice subjected to a high-fat diet for 13 weeks not only experienced a weight gain exceeding 40% but also exhibited a decline in glucose tolerance and insulin sensitivity compared to normal mice, aligning with observations reported by previous studies [46,47]. These outcomes substantiate the successful establishment of an obese mouse or non-alcoholic fatty liver disease (NAFLD) mouse model. Within the established obese mouse model, transitioning to a regular diet (standard caloric intake) for 6 weeks not only sustained weight reduction but also improved glucose tolerance and insulin sensitivity, restoring them to levels comparable to normal mice. Importantly, it is noteworthy that mice orally administered with HK-HM1/LF and SK-HM1/LF exhibited glucose tolerance even superior to the normal mice group. Our data also support the idea that orally administering SK-HM1/LF, but not HK-HM1/LF, to mice can further enhance glucose tolerance compared to the placebo control. Notably, the distinction between HK-HM1/LF and SK-HM1/LF lies in the inactivation process of HM1/LF, with HK-HM1/LF undergoing autoclaving. This heating process could disrupt the structure and denature LF partially. We posit that LF in SK-HM1/LF remains active during sonication (in an ice bath) in the inactivation procedure, contributing to the observed improvement in glucose tolerance. Overall, these findings support the potential roles of LF in improving glucose tolerance. To support the advantageous effects of LF in improving glucose tolerance, some previous studies also suggested that LF potentially facilitates glucose regulation [48–50]. Moreover, a previous report discovered the anti-diabetic effectiveness of LF through significant improvement in the baseline hemoglobin A1C, body mass index, and lipid profile of an obese pediatric cohort [51]. Additionally, LF has been shown to improve hepatic insulin resistance and pancreatic dysfunction in high-fat diet and streptozotocin-induced diabetic mice [52]. Furthermore, recombinant human LF has been found to attenuate the progression of hepatic steatosis and hepatocellular death by regulating iron and lipid homeostasis in obese mice [53]. In contrast, our data reveal that adding either of the probiotic formulations did not increase insulin sensitivity. It is known that improved glucose tolerance is mostly attributed to two factors: increased glucose-stimulated insulin secretion or increased insulin sensitivity. Since insulin sensitivity was not improved, as observed in the results of the ITT here, it is crucial to determine whether the administration of SK-HM1/LF could contribute to insulin secretion in our next study. Furthermore, investigating the effects of SK-HM1/LF on iron burden or deposition would also aid in dissecting the potential mechanisms of inactivated probiotics on ITT or GTT.

We intend to investigate this aspect further in our upcoming study, as our current findings suggest that while HK-HM1/LF and HK-HM1/LF supplementation contributed to improved glucose tolerance, they did not enhance insulin sensitivity. On the other hand, our results confirm that improving insulin sensitivity in obese mice can be achieved solely through dietary control (regular diet; normal calorie intake). While the use of inactivated probiotics did not further enhance insulin sensitivity, the fact that improvement was achieved with a regular diet, and not a calorie-restricted one, is noteworthy. This suggests that transitioning from a high-calorie to a normal-calorie diet can be beneficial for health.

As indicated by previous studies, obese mice may encounter challenges in reverting to their pre-obesity state, even after weight loss, owing to imbalances in gut microbiota and inflammatory responses in fat tissues, which occurred during the period of obesity [54–56]. In alignment with prior findings, while our dietary control facilitated weight loss and improved insulin sensitivity and glucose tolerance, it proved insufficient for fully restoring the incidence rate and pathological scores related to inflammation in perirenal and epididymal fat tissues. Significantly, our results suggest that inflammation in these fat tissues—a consequence of obesity—was partially alleviated by the regular diet alone and further improved by the administration of all inactivated probiotic formulations. Notably, SK-HM1 demonstrated the most substantial reduction in inflammation, restoring it to levels comparable to the normal mice group. Nevertheless, our present study emphasizes the advantages of incorporating regular diets with inactivated probiotic formulations in ameliorating inflammation in fat tissues.

Additionally, we observed that dietary control alone, with a regular diet (normal calorie intake), did not reduce the serum total triglyceride and serum total cholesterol levels in mice. However, it led to an increase in serum HDL-c and helped maintain LDL-c levels similar to those of normal mice. Notably, the utilization of HK-HM1 or HK-HM1/LF showed comparable levels of serum total triglyceride or cholesterol values, respectively, compared to those of normal mice groups. Therefore, the use of HK-HM1 or HK-HM1/LF could provide additional benefits in improving the two aforementioned serum biochemical indices.

The current study has certain limitations. It is acknowledged as a pilot study, and the primary objective was to assess whether combining a regular diet with inactivated probiotics could yield improvements in several metabolic indicators. Consequently, the mechanisms underlying the effects of this strategy on improving glucose tolerance and reducing inflammation in perirenal and epididymal fat tissues have not been thoroughly analyzed yet.

In conclusion, this pilot study demonstrates that the combination of a regular diet with non-viable probiotics, whether expressing LF or not, can provide certain benefits in improving various metabolic disorders. For instance, SK-HM1, as opposed to SK-HM1/LF, significantly mitigates the inflammatory status in the examined tissues. Conversely, SK-HM1/LF, rather than SK-HM1, notably improves glucose tolerance, thereby indicating the potential advantages of LF in this context. Our findings also suggest that the deactivation methods applied to probiotics could contribute to varying efficacies in improving metabolic-related indices. However, further efforts are needed to confirm these findings. Nevertheless, the use of inactivated probiotics expressing LF offers several advantages, including addressing concerns related to the use of genetically modified foods. Additionally, these inactivated probiotics exhibit enhanced tolerance to processing methods, increasing their practical applicability. Further research is warranted to delve deeper into the mechanisms and potential long-term effects of this dietary strategy. Finally, the potential roles of LF must be further examined.

Author Contributions: Conceptualization: P.-W.C.; Data curation: W.-C.S., B.-Y.C. and Y.-W.K.; Funding acquisition: P.-W.C. and Z.-S.L.; Resources: P.-W.C. and Z.-S.L.; Writing—original draft preparation: P.-W.C., W.-C.S. and Z.-S.L.; Writing—review and editing: P.-W.C., W.-C.S. and Z.-S.L. All authors have read and agreed to the published version of the manuscript.

Funding: This work was supported by Taiwan National Science and Technology Council: NSTC 112-2313-B-005-037.

Institutional Review Board Statement: The animal experiments were conducted with the approval of the Institutional Animal Care and Use Committee of National Chung Hsing University, under approval number IACUC No. 110-080R.

Informed Consent Statement: Not applicable.

Data Availability Statement: The original contributions presented in the study are included in the article, further inquiries can be directed to the corresponding author.

Conflicts of Interest: There are no conflicts of interest in relation to this study.

References

1. Hou, Y.-P.; He, Q.-Q.; Ouyang, H.-M.; Peng, H.-S.; Wang, Q.; Li, J.; Lv, X.-F.; Zheng, Y.-N.; Li, S.-C.; Liu, H.-L. Human gut microbiota associated with obesity in Chinese children and adolescents. *BioMed Res. Int.* **2017**, *2017*, 7585989. [CrossRef] [PubMed]
2. Chooi, Y.C.; Ding, C.; Magkos, F. The epidemiology of obesity. *Metabolism* **2019**, *92*, 6–10. [CrossRef]
3. Lin, X.; Li, H. Obesity: Epidemiology, pathophysiology, and therapeutics. *Front. Endocrinol.* **2021**, *12*, 706978. [CrossRef] [PubMed]
4. Duggan, C.; Tapsoba, J.D.; Shivappa, N.; Harris, H.R.; Hebert, J.R.; Wang, C.Y.; McTiernan, A. Changes in Dietary Inflammatory Index Patterns with Weight Loss in Women: A Randomized Controlled Trial. *Cancer Prev. Res.* **2021**, *14*, 85–94. [CrossRef]
5. de Souza Kraemer, M.; Moehlecke, M.; Rheinheimer, J.; Canani, L.H.; Leitao, C.B.; Nicoletto, B.B. Plasma progranulin levels in obese patients before and after Roux-en-Y gastric bariatric surgery: A longitudinal study. *Surg. Obes. Relat. Dis.* **2020**, *16*, 1655–1660. [CrossRef] [PubMed]

6. Scheen, A.J. Sodium–glucose cotransporter type 2 inhibitors for the treatment of type 2 diabetes mellitus. *Nat. Rev. Endocrinol.* **2020**, *16*, 556–577. [CrossRef]
7. Bluher, M. Obesity: Global epidemiology and pathogenesis. *Nat. Rev. Endocrinol.* **2019**, *15*, 288–298. [CrossRef] [PubMed]
8. Colman, R.J.; Anderson, R.M.; Johnson, S.C.; Kastman, E.K.; Kosmatka, K.J.; Beasley, T.M.; Allison, D.B.; Cruzen, C.; Simmons, H.A.; Kemnitz, J.W. Caloric restriction delays disease onset and mortality in rhesus monkeys. *Science* **2009**, *325*, 201–204. [CrossRef]
9. Weiss, E.C.; Galuska, D.A.; Khan, L.K.; Gillespie, C.; Serdula, M.K. Weight regain in US adults who experienced substantial weight loss, 1999–2002. *Am. J. Prev. Med.* **2007**, *33*, 34–40. [CrossRef]
10. Sarlio-Lähteenkorva, S.; Rissanen, A.; Kaprio, J. A descriptive study of weight loss maintenance: 6 and 15 year follow-up of initially overweight adults. *Int. J. Obes.* **2000**, *24*, 116–125. [CrossRef]
11. Kraschnewski, J.L.; Boan, J.; Esposito, J.; Sherwood, N.E.; Lehman, E.B.; Kephart, D.K.; Sciamanna, C.N. Long-term weight loss maintenance in the United States. *Int. J. Obes.* **2010**, *34*, 1644–1654. [CrossRef] [PubMed]
12. Anderson, J.W.; Konz, E.C.; Frederich, R.C.; Wood, C.L. Long-term weight-loss maintenance: A meta-analysis of US studies. *Am. J. Clin. Nutr.* **2001**, *74*, 579–584. [CrossRef] [PubMed]
13. Marlatt, K.L.; Redman, L.M.; Burton, J.H.; Martin, C.K.; Ravussin, E. Persistence of weight loss and acquired behaviors 2 y after stopping a 2-y calorie restriction intervention. *Am. J. Clin. Nutr.* **2017**, *105*, 928–935. [CrossRef] [PubMed]
14. Tang, C.; Kong, L.; Shan, M.; Lu, Z.; Lu, Y. Protective and ameliorating effects of probiotics against diet-induced obesity: A review. *Food Res. Int.* **2021**, *147*, 110490. [CrossRef] [PubMed]
15. Fabersani, E.; Abeijon-Mukdsi, M.C.; Ross, R.; Medina, R.; Gonzalez, S.; Gauffin-Cano, P. Specific Strains of Lactic Acid Bacteria Differentially Modulate the Profile of Adipokines In Vitro. *Front. Immunol.* **2017**, *8*, 266. [CrossRef] [PubMed]
16. Fabersani, E.; Russo, M.; Marquez, A.; Abeijon-Mukdsi, C.; Medina, R.; Gauffin-Cano, P. Modulation of intestinal microbiota and immunometabolic parameters by caloric restriction and lactic acid bacteria. *Food Res. Int.* **2019**, *124*, 188–199. [CrossRef]
17. Dahiya, D.K.; Renuka; Puniya, M.; Shandilya, U.K.; Dhewa, T.; Kumar, N.; Kumar, S.; Puniya, A.K.; Shukla, P. Gut Microbiota Modulation and Its Relationship with Obesity Using Prebiotic Fibers and Probiotics: A Review. *Front. Microbiol.* **2017**, *8*, 563. [CrossRef] [PubMed]
18. Kobyliak, N.; Conte, C.; Cammarota, G.; Haley, A.P.; Styriak, I.; Gaspar, L.; Fusek, J.; Rodrigo, L.; Kruzliak, P. Probiotics in prevention and treatment of obesity: A critical view. *Nutr. Metab.* **2016**, *13*, 14. [CrossRef] [PubMed]
19. Sivamaruthi, B.S.; Kesika, P.; Suganthy, N.; Chaiyasut, C. A Review on Role of Microbiome in Obesity and Antiobesity Properties of Probiotic Supplements. *BioMed Res. Int.* **2019**, *2019*, 3291367. [CrossRef]
20. Hill, C.; Guarner, F.; Reid, G.; Gibson, G.R.; Merenstein, D.J.; Pot, B.; Morelli, L.; Canani, R.B.; Flint, H.J.; Salminen, S. Expert consensus document: The International Scientific Association for Probiotics and Prebiotics consensus statement on the scope and appropriate use of the term probiotic. *Nat. Rev. Gastroenterol. Hepatol.* **2014**, *11*, 506–514. [CrossRef]
21. Vinderola, G.; Sanders, M.E.; Salminen, S. The Concept of Postbiotics. *Foods* **2022**, *11*, 1077. [CrossRef] [PubMed]
22. Cuevas-Gonzalez, P.F.; Liceaga, A.M.; Aguilar-Toala, J.E. Postbiotics and paraprobiotics: From concepts to applications. *Food Res. Int.* **2020**, *136*, 109502. [CrossRef] [PubMed]
23. Aguilar-Toalá, J.; Garcia-Varela, R.; Garcia, H.; Mata-Haro, V.; González-Córdova, A.; Vallejo-Cordoba, B.; Hernández-Mendoza, A. Postbiotics: An evolving term within the functional foods field. *Trends Food Sci. Technol.* **2018**, *75*, 105–114. [CrossRef]
24. Nataraj, B.H.; Ali, S.A.; Behare, P.V.; Yadav, H. Postbiotics-parabiotics: The new horizons in microbial biotherapy and functional foods. *Microb. Cell Fact.* **2020**, *19*, 168. [CrossRef]
25. Bourebaba, Y.; Marycz, K.; Mularczyk, M.; Bourebaba, L. Postbiotics as potential new therapeutic agents for metabolic disorders management. *Biomed. Pharmacother.* **2022**, *153*, 113138. [CrossRef] [PubMed]
26. Lee, M.; Yun, Y.R.; Choi, E.J.; Song, J.H.; Kang, J.Y.; Kim, D.; Lee, K.W.; Chang, J.Y. Anti-obesity effect of vegetable juice fermented with lactic acid bacteria isolated from kimchi in C57BL/6J mice and human mesenchymal stem cells. *Food Funct.* **2023**, *14*, 1349–1356. [CrossRef] [PubMed]
27. Aoun, A.; Darwish, F.; Hamod, N. The influence of the gut microbiome on obesity in adults and the role of probiotics, prebiotics, and synbiotics for weight loss. *Prev. Nutr. Food Sci.* **2020**, *25*, 113. [CrossRef] [PubMed]
28. Sienkiewicz, M.; Jaskiewicz, A.; Tarasiuk, A.; Fichna, J. Lactoferrin: An overview of its main functions, immunomodulatory and antimicrobial role, and clinical significance. *Crit. Rev. Food Sci. Nutr.* **2022**, *62*, 6016–6033. [CrossRef] [PubMed]
29. Kowalczyk, P.; Kaczynska, K.; Kleczkowska, P.; Bukowska-Osko, I.; Kramkowski, K.; Sulejczak, D. The Lactoferrin Phenomenon-A Miracle Molecule. *Molecules* **2022**, *27*, 2941. [CrossRef]
30. Liu, Z.S.; Chen, P.W. Featured Prebiotic Agent: The Roles and Mechanisms of Direct and Indirect Prebiotic Activities of Lactoferrin and Its Application in Disease Control. *Nutrients* **2023**, *15*, 2759. [CrossRef]
31. Singh, A.; Zapata, R.C.; Pezeshki, A.; Chelikani, P.K. Dietary lactalbumin and lactoferrin interact with inulin to modulate energy balance in obese rats. *Obesity* **2017**, *25*, 1050–1060. [CrossRef] [PubMed]
32. Zapata, R.C.; Singh, A.; Pezeshki, A.; Nibber, T.; Chelikani, P.K. Whey Protein Components—Lactalbumin and Lactoferrin—Improve Energy Balance and Metabolism. *Sci. Rep.* **2017**, *7*, 9917. [CrossRef] [PubMed]
33. Liu, Z.-S.; Lin, C.-F.; Lee, C.-P.; Hsieh, M.-C.; Lu, H.-F.; Chen, Y.-F.; Ku, Y.-W.; Chen, P.-W. A single plasmid of nisin-controlled bovine and human lactoferrin expressing elevated antibacterial activity of lactoferrin-resistant probiotic strains. *Antibiotics* **2021**, *10*, 120. [CrossRef]

34. Liu, Z.-S.; Li, P.-L.; Ku, Y.-W.; Chen, P.-W. Oral Administration of Recombinant Lactoferrin-Expressing Probiotics Ameliorates Diet-Induced Lipid Accumulation and Inflammation in Non-Alcoholic Fatty Liver Disease in Mice. *Microorganisms* **2022**, *10*, 2215. [CrossRef] [PubMed]
35. Chen, P.W.; Lin, Y.L.; Huang, M.S. Profiles of commensal and opportunistic bacteria in human milk from healthy donors in Taiwan. *J. Food Drug Anal.* **2018**, *26*, 1235–1244. [CrossRef] [PubMed]
36. Van, V.T.H.; Liu, Z.-S.; Hsieh, Y.J.; Shiu, W.-C.; Chen, B.-Y.; Ku, Y.-W.; Chen, P.-W. Therapeutic effects of orally administration of viable and inactivated probiotic strains against murine urinary tract infection. *J. Food Drug Anal.* **2023**, *31*, 2. [CrossRef] [PubMed]
37. Brunt, E.M.; Janney, C.G.; Di Bisceglie, A.M.; Neuschwander-Tetri, B.A.; Bacon, B.R. Nonalcoholic steatohepatitis: A proposal for grading and staging the histological lesions. *Am. J. Gastroenterol.* **1999**, *94*, 2467–2474. [CrossRef] [PubMed]
38. Benedé-Ubieto, R.; Estévez-Vázquez, O.; Ramadori, P.; Cubero, F.J.; Nevzorova, Y.A. Guidelines and considerations for metabolic tolerance tests in mice. *Diabetes Metab. Syndr. Obes.* **2020**, *13*, 439–450. [CrossRef] [PubMed]
39. Brehm, B.J.; Seeley, R.J.; Daniels, S.R.; D'Alessio, D.A. A randomized trial comparing a very low carbohydrate diet and a calorie-restricted low fat diet on body weight and cardiovascular risk factors in healthy women. *J. Clin. Endocrinol. Metab.* **2003**, *88*, 1617–1623. [CrossRef]
40. Davoodi, S.H.; Ajami, M.; Ayatollahi, S.A.; Dowlatshahi, K.; Javedan, G.; Pazoki-Toroudi, H.R. Calorie shifting diet versus calorie restriction diet: A comparative clinical trial study. *Int. J. Prev. Med.* **2014**, *5*, 447.
41. Masuo, K.; Rakugi, H.; Ogihara, T.; Lambert, G.W. Different mechanisms in weight loss-induced blood pressure reduction between a calorie-restricted diet and exercise. *Hypertens. Res.* **2012**, *35*, 41–47. [CrossRef] [PubMed]
42. Million, M.; Angelakis, E.; Paul, M.; Armougom, F.; Leibovici, L.; Raoult, D. Comparative meta-analysis of the effect of Lactobacillus species on weight gain in humans and animals. *Microb. Pathog.* **2012**, *53*, 100–108. [CrossRef] [PubMed]
43. Ezema, C. Probiotics in animal production: A review. *J. Vet. Med. Anim. Health* **2013**, *5*, 308–316.
44. Angelakis, E. Weight gain by gut microbiota manipulation in productive animals. *Microb. Pathog.* **2017**, *106*, 162–170. [CrossRef] [PubMed]
45. Rouxinol-Dias, A.L.; Pinto, A.R.; Janeiro, C.; Rodrigues, D.; Moreira, M.; Dias, J.; Pereira, P. Probiotics for the control of obesity—Its effect on weight change. *Porto Biomed. J.* **2016**, *1*, 12–24. [CrossRef] [PubMed]
46. Pereira, R.M.; da Cruz Rodrigues, K.C.; Sant'Ana, M.R.; Peruca, G.F.; Morelli, A.P.; Simabuco, F.M.; da Silva, A.S.; Cintra, D.E.; Ropelle, E.R.; Pauli, J.R. Strength exercise reduces hepatic pyruvate carboxylase and gluconeogenesis in DIO mice. *J. Endocrinol.* **2020**, *247*, 127–138. [CrossRef] [PubMed]
47. Cao, M.; Pan, Q.; Dong, H.; Yuan, X.; Li, Y.; Sun, Z.; Dong, X.; Wang, H. Adipose-derived mesenchymal stem cells improve glucose homeostasis in high-fat diet-induced obese mice. *Stem Cell Res. Ther.* **2015**, *6*, 208. [CrossRef] [PubMed]
48. Moreno-Navarrete, J.; Ortega, F.; Bassols, J.; Ricart, W.; Fernández-Real, J. Decreased circulating lactoferrin in insulin resistance and altered glucose tolerance as a possible marker of neutrophil dysfunction in type 2 diabetes. *J. Clin. Endocrinol. Metab.* **2009**, *94*, 4036–4044. [CrossRef]
49. Maekawa, Y.; Sugiyama, A.; Takeuchi, T. Lactoferrin potentially facilitates glucose regulation and enhances the incretin effect. *Biochem. Cell Biol.* **2017**, *95*, 155–161. [CrossRef]
50. Moreno-Navarrete, J.M.; Ortega, F.J.; Bassols, J.; Castro, A.; Ricart, W.; Fernández-Real, J.M. Association of circulating lactoferrin concentration and 2 nonsynonymous LTF gene polymorphisms with dyslipidemia in men depends on glucose-tolerance status. *Clin. Chem.* **2008**, *54*, 301–309. [CrossRef]
51. Mohamed, W.A.; Schaalan, M.F. Antidiabetic efficacy of lactoferrin in type 2 diabetic pediatrics; controlling impact on PPAR-gamma, SIRT-1, and TLR4 downstream signaling pathway. *Diabetol. Metab. Syndr.* **2018**, *10*, 89. [CrossRef] [PubMed]
52. Du, Y.; Li, D.; Chen, J.; Li, Y.H.; Zhang, Z.; Hidayat, K.; Wan, Z.; Xu, J.Y.; Qin, L.Q. Lactoferrin improves hepatic insulin resistance and pancreatic dysfunction in high-fat diet and streptozotocin-induced diabetic mice. *Nutr. Res.* **2022**, *103*, 47–58. [CrossRef] [PubMed]
53. Guo, C.; Xue, H.; Guo, T.; Zhang, W.; Xuan, W.-Q.; Ren, Y.-T.; Wang, D.; Chen, Y.-H.; Meng, Y.-H.; Gao, H.-L.J.F.; et al. Recombinant human lactoferrin attenuates the progression of hepatosteatosis and hepatocellular death by regulating iron and lipid homeostasis in ob/ob mice. *Food Funct.* **2020**, *11*, 7183–7196. [CrossRef] [PubMed]
54. Thaiss, C.A.; Itav, S.; Rothschild, D.; Meijer, M.T.; Levy, M.; Moresi, C.; Dohnalová, L.; Braverman, S.; Rozin, S.; Malitsky, S. Persistent microbiome alterations modulate the rate of post-dieting weight regain. *Nature* **2016**, *540*, 544–551. [CrossRef] [PubMed]
55. Fischer, I.; Irmler, M.; Meyer, C.; Sachs, S.; Neff, F.; Hrabě de Angelis, M.; Beckers, J.; Tschöp, M.; Hofmann, S.; Ussar, S. A history of obesity leaves an inflammatory fingerprint in liver and adipose tissue. *Int. J. Obes.* **2018**, *42*, 507–517. [CrossRef]
56. Li, M.; Wang, S.; Li, Y.; Zhao, M.; Kuang, J.; Liang, D.; Wang, J.; Wei, M.; Rajani, C.; Ma, X. Gut microbiota-bile acid crosstalk contributes to the rebound weight gain after calorie restriction in mice. *Nat. Commun.* **2022**, *13*, 2060. [CrossRef]

foods

MDPI

Article

Effect of *Lactiplantibacillus plantatum* HFY11 on Colitis in Mice

Fang Tan [1], Xianrong Zhou [1], Lixuan Ren [1] and Chang-Suk Kong [1,2,3,*]

[1] Department of Bioscience, Silla University, Busan 46958, Republic of Korea; g-202311663@sillain.ac.kr (F.T.); zxr960032412@sillain.ac.kr (X.Z.); g-202311664@sillain.ac.kr (L.R.)

[2] Department of Food and Nutrition, Silla University, Busan 46958, Republic of Korea

[3] Marine Biotechnology Center for Pharmaceuticals and Foods, Silla University, Busan 46958, Republic of Korea

* Correspondence: cskong@silla.ac.kr; Tel.: +82-51-999-5429

Abstract: This study aimed to examine the potential impact of the intervention of *Lactiplantibacillus plantatum* HFY11 (LP-HFY11) on colitis using in vivo animal trials. The impact of LP-HFY11 intervention on colitis was determined by measuring the levels of relevant indicators in the intestine, colon, and blood after oxazolone-induced colitis in BALB/c mice. The results of the trial show that LP-HFY11 improved the colon weight-to-length ratio, reduced the colitis-induced colon length shortening, and reduced colonic abstinence. Furthermore, it decreased the levels of myeloperoxidase, nitric oxide, and malondialdehyde activities while increasing the glutathione content in the colon tissue of colitis-affected animals. LP-HFY11 lowered the interleukin-10 (IL-10) level and increased the IL-2 level in the serum of colitis mice. LP-HFY11 also upregulated the expression of neuronal nitric oxide synthase, endothelial nitric oxide synthase, c-Kit, and stem cell factor (SCF), and downregulated the expression of IL-8, C-X-C chemokine receptor type 2 (CXCR2), and inducible nitric oxide synthase (iNOS) in the colon tissue of mice with colitis. LP-HFY11 decreased the expression of *Firmicutes* in the gut while increasing the expression of *Bacteroidetes*, *Bifidobacteria*, and *Lactobacillus*. This indicates that LP-HFY11 could control physiological alterations in the serum and colon tissue, as well as the expression of gut microorganism.

Keywords: *Lactiplantibacillus plantatum*; colitis; intestinal microecology; expression; cytokine

1. Introduction

Rich in nutrients, yak yogurt is a naturally fermented dish often found in the minority areas of the Qinghai–Tibet Plateau [1]. Yak yogurt offers several physiological benefits, including decreasing cholesterol, boosting the immune system, and acting as an antioxidant. The taste and quality of yak yogurt differ greatly from regular fermented milk due to the distinct natural fermentation environment of the Qinghai–Tibet Plateau, the use of yak milk and special fermentation vessels, and the unique fermentation microorganisms of the Tibetan people [2]. The *Lactiplantibacillus plantatum* HFY11 used in this study was a strain of lactic acid bacteria that can be used for food after the isolation and identification of yak yogurt microorganisms in the Hongyuan area of the Qinghai–Tibet Plateau in Sichuan Province, China.

Colitis is an inflammatory intestinal disease, mainly involving inflammation and damage to the colon (large intestine) mucosa. The inflammation and damage to the mucosa interfere with the normal digestion and absorption of food, resulting in discomfort and malnutrition. At the same time, patients with colitis often have problems such as loss of appetite and indigestion, leading to weight loss [3]. If colitis is not effectively treated or controlled, the inflammation may further expand and lead to serious complications, such as intestinal obstruction, bleeding, and perforation. This increases the health risks and the need for medical intervention [4]. In addition, patients with colitis may often experience abdominal pain, abdominal discomfort, and colic. Colitis has a multifaceted impact on the human body, which is not only limited to digestive system problems, but may also involve malnutrition, weight changes, pain, and discomfort. Therefore, timely diagnosis, active treatment, and disease management are essential for patients with colitis [5]. Oxazolone can induce a T-cell-mediated immune response. The cell-mediated type 2 (Th2) immune response and interleukin-4 (IL-4) and IL-5 production increase significantly, accompanied by weight loss, diarrhea, ulcers, and a decreased number of colorectal epithelial cells [6]. The oxazolone-induced ulcerative colitis model is frequently used to identify the physiological action of medications and functional foods in colitis in humans because both show similar symptoms.

Probiotics have a certain impact and potential benefits on colitis; they can help restore the balance of intestinal microbiota, enhance the integrity of the intestinal mucosal barrier, and regulate the immune system [7]. Probiotics can reduce the inflammatory response and symptoms in patients with colitis by enhancing the composition and function of the intestinal microbiota, inhibiting the growth of pathogenic bacteria, and increasing the number of beneficial bacteria [8]. They may enhance the quality of life of patients and aid in the relief of digestive system issues such as constipation, diarrhea, and stomach discomfort [9–11]. In addition, probiotics can also enhance the integrity of the intestinal mucosal barrier and reduce the penetration of harmful substances and bacteria, and thus reduce the degree of inflammation. They can also regulate the immune response, reduce the release of inflammatory factors, and promote the resolution of inflammation, thus helping to control the progression of colitis [12].

Lactobacillus with good physiological activity can be developed and used as a probiotic. Our group investigated the intestinal physiological activity of lactic acid bacteria found in yak yogurt from the Qinghai–Tibet Plateau. The findings demonstrate that lactic acid bacteria separated from yak yogurt had an antioxidant and constipation-relieving effect. Thus, this study was performed to investigate the impact of an *L. plantarum* HFY11 (LP-HFY11) strain isolated from yak yogurt on oxazolone-induced colitis. The findings of this study might give us a theoretical foundation for the development and use of LP-HFY11 in the future.

2. Materials and Methods

2.1. Experimental Microbial Species

In Hongyuan County, Aba Tibetan and Qiang Autonomous Prefecture, Sichuan Province, China, naturally fermented yak yogurt yielded *L. plantarum* HFY11, which was isolated and stored at the China General Microbiological Culture Collection Center (CGMCC, collection number: CGMCC No. 16644). The frozen P2-generation *L. plantarum* HFY11 strains were inoculated into MRS medium at a 2% inoculation rate and incubated at 36 °C for 16 h, and then freeze-dried. Then, the strain's dry powder was mixed with other feed ingredients in proportion to qualitatively produce mouse feed, and the resulting feed was stored in a refrigerator at −20 °C for use.

2.2. Animal Experiment

Hunan Slake Jingda Experimental Animal Co., Ltd. provided 50 specific pathogen-free (SPF) male BALB/c mice (aged 7 weeks, weighing 25–30 g), bearing animal license number SCXK (Xiang) 2019-0004. Ten BALB/c mice were randomly assigned to each of five groups: the LP-HFY11 low-concentration treatment (LP-HFY11L) group, LP-HFY11 high-concentration treatment (LP-HFY11H) group, gatifloxacin group (positive control group), model group, and normal group. Following a week of adaptive feeding, a patch measuring 2×2 cm^2 was shaved from the abdomen of each mouse. Then, 0.2 mL of analytical pure ethanol was used to smear the mice in the normal and model group, whereas 0.2 mL of a 3% oxazolone solution (w/w) was used to smear the animals in the other groups. The mice were permitted to consume water freely on the fifth day, despite fasting beforehand. Following a 24 h fast, the mice were given 0.1 mL/10 g of chloral hydrate to induce unconsciousness. Following this, a silicone tube with a blunt end was inserted 3.5 cm deep into the anus of the animals. Then, 0.15 mL of a 50% ethanol solution was injected into the normal group of mice, whereas 0.15 mL of a 1% oxazolone solution (mass ratio, 50% ethanol as the solvent) was injected into the other groups of mice. The catheter was removed after 20 s, and the mice were inverted for 30 s [13]. The mice in the gatifloxacin treatment group were allowed to eat and drink 0.1% (*w*/*w*) gatifloxacin. The mice in the LP-HFY11L and LP-HFY11H groups were allowed to eat and drink 0.01% (*w*/*w*, 10^7 CFU LP-HFY11/g feed) and 0.1% (*w*/*w*, 10^8 CFU LP-HFY11/g feed) lyophilized LP-HFY11 powder, and fed with the same diets for 7 days. Subsequently, all of the mice were killed by dissecting their necks, and the plasma was extracted. The colon's weight and length were then measured using the extracted colon tissue. The disease activity index (DAI) was established concurrently using the following formula: DAI = (weight loss score + stool traits score + fecal blood score)/3 (Table 1). The operations conformed to the experimental ethics requirements of the animal ethics committee of the Chongqing Collaborative Innovation Center for Functional Food (license number: 202311003B).

Table 1. DAI score benchmark.

Percentage of Body Weight Loss (%)	Stool Consistency	Presence of Blood in Stool	Score
No decrease	Normal	Occult blood (−)	0
1–5			1
5–10	Semi-loose stool	Occult blood (+)	2
10–15			3
>15	Loose stools	Naked-eye bloody stool	4

2.3. Determination of MPO, NO, GSH, MDA, and SOD Levels in Mouse Colon Tissue

The cleaned mouse colon was filled nine times with normal saline, and ultrasonic crushing was used to homogenize the colon tissue. Next, the myeloperoxidase (MPO), nitric oxide (NO), glutathione (GSH), malonaldehyde (MDA), and superoxide dismutase (SOD) levels in the mouse colon tissue were determined using corresponding kits following the manufacturer's protocols (Thermo Fisher Scientific, Waltham, MA, USA).

2.4. Determination of Inflammation-Related Cytokines in Mouse Serum

The blood was extracted from the orbit and centrifuged at 4 °C for 10 min at 1500 rpm. After separating the upper layer of the serum, the levels of cytokines IL-2 and IL-10 in the mouse serum were measured using enzyme-linked immunosorbent assay following the kit's instructions (Thermo Fisher Scientific).

2.5. Colon Tissue Slice Detection

Once dissected, the colon tissue from the mice was fixed in 10% formalin. The tissue samples were dehydrated for 48 h before being embedded in paraffin, cut, and stained with hematoxylin and eosin (H&E) (BX53, Olympus, Tokyo, Japan) [14].

2.6. Determination of Related Expression in Colon Tissue Using qPCR

Furthermore, 100 mg of tissue from the middle segment of the mouse colon was taken. The tissue was washed with normal saline and then with clean normal saline at a ratio of 1:9. Following the homogenization of the tissue, 1.0 mL of the RNAzol reagent was applied to extract mouse tissue RNA, and the extracted RNA concentration was adjusted to 1 µg/µL. After reverse transcription, cDNA was obtained, and the reaction system was set up. SYBR Green Polymerase Chain Reaction (PCR) Master Mix (10 µL), sterile deionized water (7 µL), cDNA (1 µL), and PCR primers (1 µL each of the forward and reverse primers, each with a concentration of 10 µmol/L, Thermo Fisher Scientific) were all included in the system solution. The amplification conditions for the prepared reaction solution were as follows: 95 °C for 60 s; 95 °C for 15 s, 40 cycles; 55 °C for 30 s; 72 °C for 35 s; 95 °C for 30 s; and 55 °C for 35 s (SteponePlus, Thermo Fisher Scientific). The internal reference gene employed in this study was glyceraldehyde-3-phosphate dehydrogenase (GAPDH)

(Table 2), and the $2^{-\Delta\Delta Ct}$ technique was used to determine the relative expression of each gene [15].

Table 2. Primer sequence for the reverse transcription–polymerase chain reaction.

Gene	Primer Sequence
nNOS	F: 5′-TGCTCTGTGAATCCACCACC-3′ R: 5′-GAGGAAGCGAAAGCACTCCAG-3′
eNOS	F: 5′-TCAGCCATCACAGTGTTCCC-3′ R: 5′-ATAGCCCGCATAGCGTATCAG-3′
iNOS	F: 5′-GGAGTGACGGCAAACATGACT-3′ R: 5′-TCGATGCACAACTGGGTGAAC-3′
c-Kit	F: 5′-AGACCGAACGCAACTT-3′ R: 5′-GGTGCCATCCACTTCA-3′
SCF	F: 5′-AAACTGGTGGCGAATC-3′ R: 5′-CACGGGTAGCAAGAAC-3′
IL-8	F: 5′-GGCCTTGTAGACACCTTGGT-3′ R: 5′-GAACAA AGGCAAGGCTAA-3′
CXCR2	F: 5′-ATGCCCTCTATTCTGCCAGAT-3′ R: 5′-GTGCTCCGGTTGTATAAGATGAC-3′
GAPDH	F: 5′-AATGGATTTGGACGCATTGGT-3′ R: 5′-TTTGCACTGGTACGTGTTGAT-3′
Firmicutes	F: 5′-GCGTGAGTGAAGAAGT-3′ R: 5′-CTACGCTCCCTTTACAC-3′
Bacteroidetes	F: 5′-ACGCTAGCTACAGGCTTAACA-3′ R: 5′-ACGCTACTTGGCTGGTTCA-3′
Lactobacillus	F: 5′-CACCGCTACACATGGAG-3′ R: 5′-AGCAGTAGGGAATCTTCCA-3′
Bifidobacterium	F: 5′-TCGCGTCYGGTGTGAAAG-3′ R: 5′-CCACATCCAGCRTCCAC-3′
Total bacteria	F: 5′-ACTCCTACGGGAGGCAGCAGT-3′ R: 5′-ATTACCGCGGCTGCTGGC-3′

2.7. Determination of Microbial mRNA Expression in the Intestinal Contents of Mice

Furthermore, 1.0 g of mouse intestinal contents was weighed, and the microbial mRNA expression in the mouse intestinal contents was determined.

2.8. Statistical Analysis

Following three iterations of the experiment, the findings of three parallel experiments were averaged. SAS9.1 statistical software (SAS Institute Inc., Cary, NC, USA) was used to determine whether the data from each group differed substantially at the $p < 0.05$ level using one-way analysis of variance.

3. Results

3.1. Effects of LP-HFY11 on Colitis Symptoms

The experimental results showed that the DAI value of the normal-group mice remained at 0.00 ± 0.00 throughout the experimental cycle (Table 3). On the contrary, the DAI value of the model-group mice increased steadily following oxazolone enema administration and remained the highest of all the groups. However, after administering gatifloxacin and LP-HFY11 via gavage, the DAI value of the oxazolone-induced colitis mice showed a downward trend ($p < 0.05$). Specifically, the DAI value declined more noticeably in the gatifloxacin and LP-HFY11H groups, but the LP-HFY11L group had a lower DAI value than the two treatment groups mentioned earlier.

Table 3. DAI value for each group of mice ($n = 10$).

Group	On the 1st Day after Enema	On the 3rd Day after Enema	On the 7th Day after Enema
Normal	0.00 ± 0.00^{c}	0.00 ± 0.00^{c}	0.00 ± 0.00^{c}
Model	1.83 ± 0.18^{a}	2.71 ± 0.21^{a}	3.96 ± 0.12^{a}
Gatifloxacin	0.58 ± 0.30^{ab}	1.50 ± 0.25^{ab}	1.88 ± 0.25^{ab}
LP-HFY11L	1.04 ± 0.28^{b}	1.95 ± 0.28^{b}	2.42 ± 0.30^{b}
LP-HFY11H	0.67 ± 0.25^{ab}	1.63 ± 0.21^{ab}	1.96 ± 0.21^{ab}

$^{a–c}$ The different letters denote statistically significant differences ($p < 0.05$) in the data's mean values between the experimental groups.

Based on the results, the model group exhibited the lowest colon length and colon weight-to-length ratio, whereas the normal group showed the highest values for these measurements. The colon length and colon weight-to-length ratio in the LP-HFY11H group were nearly identical to those in the normal group (Table 4), and the outcome was comparable to that of gatifloxacin.

Table 4. Colon length and weight in every group ($n = 10$).

Group	Colon Length (cm)	Colon Weight/Length (mg/cm)
Normal	9.85 ± 0.33^{a}	39.15 ± 3.5^{a}
Model	5.70 ± 0.27^{d}	23.53 ± 5.10^{c}
Gatifloxacin	8.24 ± 0.25^{b}	35.07 ± 3.43^{ab}
LP-HFY11L	6.65 ± 0.33^{c}	30.46 ± 6.48^{b}
LP-HFY11H	8.15 ± 0.44^{b}	32.95 ± 2.53^{ab}

$^{a–d}$ The different letters denote statistically significant differences ($p < 0.05$) in the data's mean values between the experimental groups.

3.2. *Effects of LP-HFY11 on the Levels of MPO, NO, GSH, and MDA in Mouse Colon Tissue*

The colon tissue in the model group (colitis) mice had the lowest content of GSH and the highest levels of MPO, NO, and MDA, as shown in Table 5. However, the colon tissue in the normal group had the highest concentration of GSH and the lowest levels of MPO, NO, and MDA. In the colon tissue of mice with colitis, gatifloxacin, LP-HFY11L, and LP-HFY11H all decreased the levels of MPO, NO, and MDA, while the content of GSH increased compared with those in the model group. Specifically, the LP-HFY11H had an impact equivalent to that of gatifloxacin, bringing the MPO, NO, GSH, and MDA levels nearly to normal mouse levels.

Table 5. MPO, NO, GSH, and MDA levels in the colon tissue of each mouse group ($n = 10$).

Group	MPO (mU/mg)	NO (µmol/gprot)	GSH (µmol/mg)	MDA (nmol/mg)
Normal	28.33 ± 4.13^{d}	0.58 ± 0.07^{c}	21.79 ± 2.96^{a}	0.52 ± 0.03^{d}
Model	238.39 ± 14.95^{a}	5.28 ± 1.00^{a}	3.49 ± 1.02^{d}	1.79 ± 0.13^{a}
Gatifloxacin	121.99 ± 5.58^{c}	2.62 ± 0.52^{b}	14.58 ± 2.98^{b}	0.94 ± 0.09^{c}
LP-HFY11L	179.51 ± 7.00^{b}	4.36 ± 0.96^{a}	9.19 ± 1.94^{c}	1.47 ± 0.11^{b}
LP-HFY11H	125.43 ± 8.26^{c}	2.67 ± 0.41^{b}	13.73 ± 1.67^{b}	0.97 ± 0.06^{c}

$^{a–d}$ The different letters denote statistically significant differences ($p < 0.05$) in the data's mean values between the experimental groups.

3.3. *Effects of LP-HFY11 on the Levels of Serum Cytokines IL-2 and IL-10 in Mice*

The findings presented in Table 6 indicate that the level of IL-2 in the blood was substantially higher, whereas that of IL-10 was significantly lower in the normal group ($p < 0.05$) compared with the other groups. The serum levels of IL-2 and IL-10 in the gatifloxacin and LP-HFY11H groups were closest to those in the normal group. The IL-2 levels in the LP-HFY11H groups were higher compared to the model group, whereas the IL-10 levels were lower compared with those in the model group and the LP-HFY11L group. The aforementioned conclusions were drawn by comparing the data of each group.

Table 6. IL-2 and IL-10 serum cytokine levels in each group ($n = 10$).

Group	IL-2 (pg/mL)	IL-10 (pg/mL)
Normal	192.99 ± 4.71 [a]	251.91 ± 10.67 [d]
Model	16.03 ± 7.10 [d]	619.34 ± 17.89 [a]
Gatifloxacin	141.75 ± 8.34 [b]	366.29 ± 14.94 [c]
LP-HFY11L	52.63 ± 7.15 [c]	548.01 ± 17.52 [b]
LP-HFY11H	138.49 ± 6.79 [b]	375.07 ± 17.78 [c]

[a–d] The different letters denote statistically significant differences ($p < 0.05$) in the data's mean values between the experimental groups.

3.4. Pathological Observation of Mouse Colon Tissue

The H&E staining results showed that the mucosal epithelial cells of the colon tissue of the normal-group mice were intact, with normal crypts, cleanly distributed glands, and no ulcers (Figure 1). Numerous necrotic lesions and crypt abscesses were the result of a high number of inflammatory cells infiltrating into the colon of the model-group mice. In the gatifloxacin group, less inflammatory cell infiltration and less structural damage to the crypts were observed. The crypt structure was mostly intact, although the LP-HFY11 group had modest inflammatory infiltration. High concentrations of LP-KSFY01 considerably reduced pathological damage to the colon tissue compared with low concentrations of LP-HFY11.

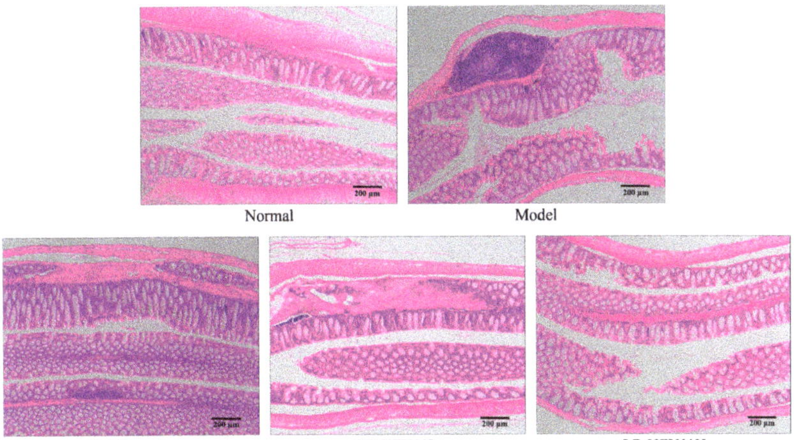

Figure 1. Effects of LR-AFY06 on the histopathology of mouse colon tissues (40×).

3.5. Effects of LP-HFY11 on the mRNA Expression of nNOS, eNOS, and iNOS in Mouse Colon Tissue

Figure 2 shows that the nNOS and eNOS mRNA expression levels were the most significant, whereas the iNOS expression level was the lowest, in the colon tissue of the normal-group mice. On the contrary, the iNOS expression level was the highest, whereas the nNOS and eNOS expression levels were the lowest in the model-group mice. The findings of the quantitative analysis demonstrate that gatifloxacin and LP-HFY11H only marginally reduced the expression levels of nNOS and eNOS in the colon of mice compared with those in the normal group; however, the iNOS expression level was only marginally higher.

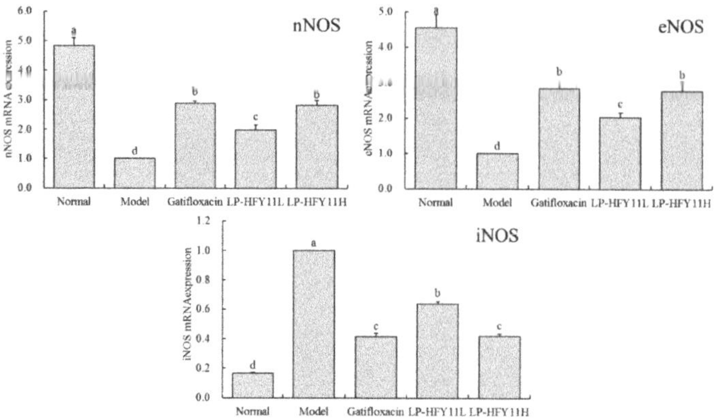

Figure 2. mRNA expression of nNOS, eNOS, and iNOS in mouse colon tissue. [a–d] The different letters denote statistically significant differences ($p < 0.05$) in the data's mean values between the experimental groups ($n = 10$).

3.6. Effects of LP-HFY11 on the mRNA Expression of c-Kit and SCF in Mouse Colon Tissue

Based on the data shown in Figure 3, it was inferred that the c-Kit and SCF mRNA expression levels were considerably higher ($p < 0.05$) in the group treated with gatifloxacin and LP-HFY11 compared with the model group. Furthermore, the colons of mice treated with gatifloxacin and LP-HFY11H had considerably higher expression levels of c-Kit and SCF compared with the colons of mice in the LP-HFY11L group.

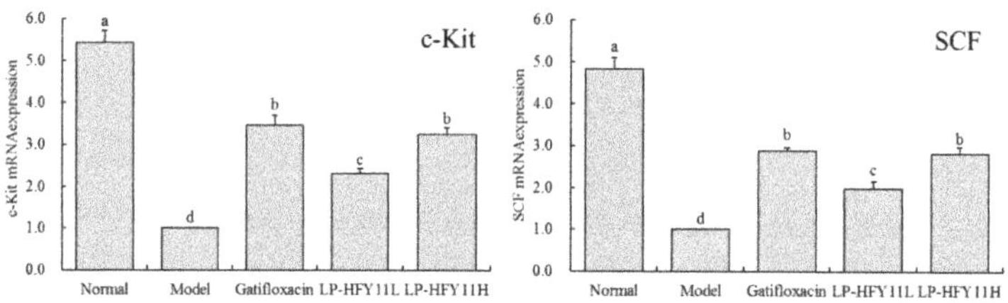

Figure 3. mRNA expression of c-Kit and SCF in mouse colon tissue. [a–d] The different letters denote statistically significant differences ($p < 0.05$) in the data's mean values between the experimental groups ($n = 10$).

3.7. Effects of LP-HFY11 on the mRNA Expression of IL-8 and CXCR2 in Mouse Colon Tissue

The results in Figure 4 show that LP-HFY11H significantly reduced the expression levels of IL-8 and CXCR2 mRNA in the colon of mice ($p < 0.05$), but the levels were marginally higher than those in the gatifloxacin group. This was in contrast to the levels in the model and LP-HFY11L groups.

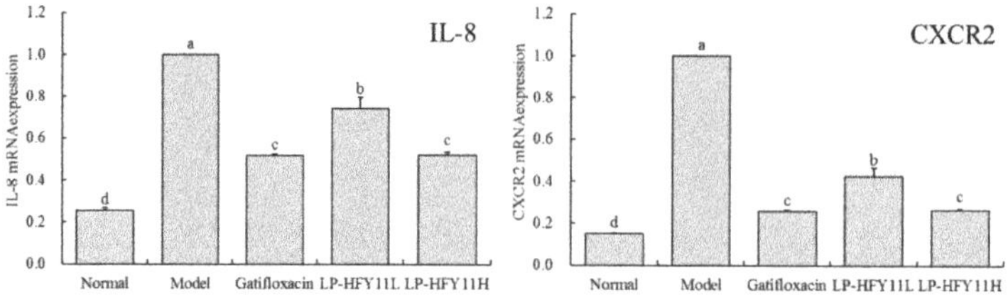

Figure 4. mRNA expression of IL-8 and CXCR2 in mouse colon tissue. [a–d] The different letters denote statistically significant differences ($p < 0.05$) in the data's mean values between the experimental groups ($n = 10$).

3.8. Effects of LP-HFY11 on the mRNA Expression of Microorganisms in Mouse Intestinal Contents

Figure 5 shows that *Firmicutes* mRNA expression was much lower in the intestinal contents of normal mice compared with mice in other groups ($p < 0.05$); however, *Bacteroidetes* and *Bifidobacterium* mRNA expression levels were significantly greater ($p < 0.05$). The *Firmicutes* mRNA expression level was the highest in the intestinal contents of the model-group mice, whereas the *Bacteroidetes, Lactobacillus,* and *Bifidobacterium* expression levels were the lowest. LP-HFY11 and gatifloxacin reduced the mRNA expression of *Firmicutes* in the intestinal contents of mice with thrombosis, while increasing the mRNA expression level of *Bacteroidetes* and *Bifidobacterium*. The mRNA expression level of *Lactobacillus* in the LP-HFY11H group was significantly higher than that in the normal group, and much higher than that in other groups ($p < 0.05$). The mRNA expression levels of *Bifidobacterium, Bacteroidetes,* and *Lactobacillus* in the intestinal contents of the LP-HFY11H group were significantly higher than those in the gatifloxacin group ($p < 0.05$). The expression ratio (F/B) of *Firmicutes/Bacteroidetes* was based on the normal state (ratio = 1). The model group reached 17.48, while the Gatifloxacin, LP-HFY11L, and LP-HFY11H groups were 5.28, 5.13, and 1.73, respectively.

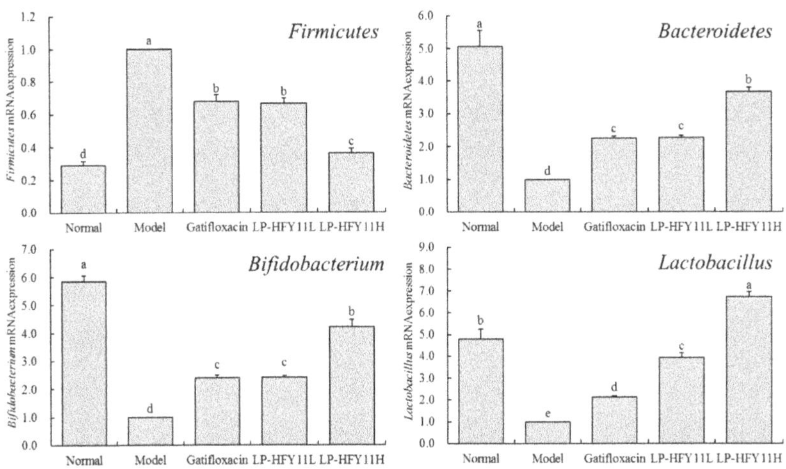

Figure 5. mRNA expression level of intestinal contents in mice. [a–d] The different letters denote statistically significant differences ($p < 0.05$) in the data's mean values between the experimental groups ($n = 10$).

4. Discussion

Colitis can cause weight loss, diarrhea, and other symptoms. DAI is a grading system for determining the severity of colitis based on body weight, stool parameters, and fecal blood [16]. According to the DAI index, we discovered that LP-HFY11 might alleviate the symptoms of oxazolone-induced colitis, with the impact becoming more significant when increasing the concentration of LP-HFY11. In addition, the colon length and colon weight-to-length ratio were used to determine the severity of colitis. Animals with colitis exhibited shorter colon lengths and a lower colon weight-to-length ratio than normal mice [17]. LP-HFY11 might potentially reduce the changes in colon length and weight induced by colitis.

MPO is an enzyme widely present in white blood cells, mainly in the granular body of neutrophils. In colitis, inflammatory responses lead to the aggregation and activation of neutrophils, resulting in the release of a large amount of MPO. MPO plays an essential role in colitis and is closely related to the development of inflammation and tissue damage [18]. It produces reactive oxygen species through oxidative reactions, such as hypochlorite ions and free radicals, which have a strong oxidative capacity and can cause the lipid peroxidation of cell membranes, leading to cell damage and further aggravating the inflammatory process [19]. In addition, MPO also produces nitric oxide (NO) from nitrous oxide (NO) and is involved in the regulation of inflammatory signaling pathways [20]. In colitis, excessive MPO production can damage the intestinal mucosal barrier, leading to cell damage and the further development of inflammation. Studies have shown that the level of MPO in the tissues of patients with colitis is closely related to the severity of inflammation, and high levels of MPO activity are associated with the activity and deterioration of the disease [21]. GSH is an important antioxidant. In colitis, inflammatory reactions and oxidative stress can lead to a decrease in the GSH level. Many free radicals and oxidizing substances are produced in the inflammatory process of colitis, increasing the consumption of GSH within the cell and thereby reducing its antioxidant capacity [22]. MDA is a product of lipid peroxidation and can cause lipid peroxidation in the cell membrane. In colitis, the production of MDA increases due to the increase in the inflammatory reaction and oxidative stress, aggravating the oxidative damage of the cell membrane and thus resulting in an impaired cell function. MDA, as a product of oxidative stress, can cause a further aggravation of inflammatory reactions [23]. It can activate inflammatory cells, such as neutrophils and macrophages, and release inflammatory mediators and cytokines, thus causing the expansion and continuity of inflammatory reactions. In addition, MDA is also associated with tissue damage caused by inflammation, leading to damage to epithelial cells and the destruction of the mucosal barrier function [24]. The regulatory effect of LP-HFY11 on the levels of MPO, NO, GSH, and MDA in mouse colon tissue may also be an essential mechanism underlying its alleviating effect on mouse colitis.

Colitis is intestinal tissue inflammation, and IL-2 is a cytokine promoting the inflammatory response. In colitis, immune cells produce a considerable amount of IL-2 in response to inflammation, exacerbating the inflammatory response. IL-2 is a key T-cell proliferation factor that promotes the proliferation and differentiation of CD^{4+} helper T cells and regulates the activity of various T-cell subsets. Colitis is characterized by an imbalance and aberrant function of T cells, as well as a compromised mucosal barrier function, allowing bacteria and toxins to infiltrate and induce an inflammatory response [25]. IL-2 regulates the immunological response of T cells and is vital for maintaining the immune balance and controlling intestinal inflammation. It is directly associated with the integrity of the intestinal mucosal barrier, emphasizing its importance in preventing the conditions that lead to colitis. IL-2 helps regulate the integrity of the mucosal barrier and maintain a proper intestinal function [26]. In particular, IL-10 is essential for regulating intestinal inflammation and other immune responses. As macrophages are the primary target cells for the anti-inflammatory activity of IL-10, this impact is most noticeable in these cells. Second, IL-10 regulates the metabolic activities of macrophages to alleviate inflammation somewhat. It causes macrophages to switch from glycolysis to oxidative phosphorylation metabolism, which is linked to an anti-inflammatory phenotype. Third, mTOR inhibition is significantly aided by IL-10 [27]. The maintenance of macrophage mitochondrial health and the promotion of phagocytosis depend on this suppression of mTOR. Finally, impaired IL-10 signaling causes the accumulation of damaged mitochondria in macrophages, which adds to the dysregulated inflammatory response seen in patients with inflammatory bowel disease [28]. Hence, it is presumed that LP-HFY11 can regulate the validation cytokines in serum, thereby intervening in the development of colitis.

nNOS mainly exists in nerve tissue, and its function is to synthesize nitric oxide (NO). It regulates neural signal transmission by releasing NO. In colitis, nNOS is involved in regulating intestinal motility, blood flow, and the protection of intestinal mucosa [29]. eNOS is mainly distributed in endothelial cells. It is responsible for synthesizing nitric oxide and plays a role in regulating vasodilation, as well as anti-inflammatory and anticoagulant effects. In colitis, eNOS is involved in regulating the intestinal blood flow, protecting the mucosal barrier, and inhibiting the inflammatory response [30]. iNOS is mainly induced and expressed during the inflammatory process, and its synthesis ability is high, producing a large amount of NO. In colitis, the expression of iNOS is usually associated with an inflammatory response. It can participate in the bactericidal effect of inflammatory cells, regulate the immune response, and regulate inflammatory signaling pathways by releasing NO [31]. nNOS, eNOS, and iNOS play important roles in colitis. They participate in regulating physiological and pathological processes, such as intestinal motility, blood flow, the mucosal barrier, and the inflammatory response, through NO synthesis [32].

c-Kit is mainly expressed in intestinal interstitial cells (Cajal cells) and mucosal epithelial cells. It participates in regulating intestinal motility, maintaining mucosal barrier integrity, and regulating immune responses by binding to its ligand SCF [33]. SCF is mainly produced by intestinal epithelial and stromal cells. It can bind to c-Kit receptors, promote the proliferation and differentiation of Cajal cells, regulate intestinal smooth muscle contraction and movement, and participate in intestinal repair and mucosal barrier protection [34]. In colitis, the functions of c-Kit and SCF may be disrupted. The inflammatory response leads to the abnormal activation of the c-Kit signaling pathway and the abnormal expression of SCF. This may affect the normal motility of the intestine, the integrity of the mucosal barrier, and the regulation of the immune response, further exacerbating the inflammatory process [35].

IL-8 is a chemotactic factor mainly produced by various cells, such as inflammatory, epithelial, and endothelial cells. Its primary purpose is to attract and activate white blood cells, particularly neutrophils, thus promoting their migration to the site of inflammation. Excessive IL-8 production can cause inflammation to persist, thus aggravating colitis [36]. CXCR2 is an IL-8 receptor mainly expressed on the surface of white blood cells. When IL-8 binds to CXCR2, it activates the CXCR2 signaling pathway, triggering various inflammatory responses in cells [37]. In colitis, the activation of CXCR2 can enhance neutrophil chemotaxis and activation, exacerbating the inflammatory response and causing tissue damage. The aberrant expression of IL-8 and CXCR2 may result in the increased aggregation and activation of inflammatory cells, prolonging and exacerbating the inflammatory response. As a result, blocking the IL-8 and CXCR2 signaling pathways may be a useful approach for treating colitis because it can reduce inflammation and tissue damage [38]. The mRNA regulatory effect of LP-HFY11 on colon tissue can effectively intervene in the development of colitis in mice, thereby promoting functional recovery.

Dysfunction of gut microbiota is an essential factor leading to inflammation of the colon. A vast majority of bacteria in the human gut are composed of *Firmicutes* and *Bacteroidetes*, and a higher proportion of *Firmicutes* than *Bacteroidetes* can cause an increase in the levels of inflammatory factors in the blood, increasing the possibility of intestinal tissue damage [39]. The gut *Firmicutes/Bacteroidetes* ratio (F/B) is widely believed to have an important impact on maintaining normal intestinal stability. Experimental colitis has shown that the F/B ratio in the gut increases after the induction of colitis, which can be used as an indicator to evaluate the stability of gut microbiota [40]. *Lactobacillus* and *Bifidobacteria* exert various bioactive effects, especially in protecting the normal metabolic state of the body, avoiding tissue damage caused by inflammation and other abnormalities and thus protecting the intestine [41]. In addition, a healthy gut microbiota can improve the body's resistance and tissue-repair ability by regulating various mechanisms, such as the immune function, and intervene in colitis [42]. This study also confirmed that LP-HFY11 promoted the production of *Bacteroidetes* and *Bifidobacteria* in the mouse intestine, while supplementing with *Lactobacillus* and reducing the expression of *Firmicutes* and the F/B expression ratio, thereby inhibiting inflammation by regulating the intestinal microbiota and exerting an inhibitory effect on colitis formation. In particular, the side effects of gatifloxacin as a drug include symptoms such as nausea, diarrhea, abdominal pain, constipation, and indigestion. As a lactic acid bacteria isolated from food, LP-HFY11 has no side effects and has better application prospects.

5. Conclusions

This preliminary study investigated the inhibitory effect of LP-HFY11 on oxazolone-induced colitis. Animal experiments have shown that LP-HFY11 effectively reduces symptoms in mice with colitis and regulates factors related to inflammation and immunity in colon tissue and serum. Meanwhile, this study further demonstrated the ability of LP-HFY11 to control the gut microbiota composition of mice, improving it in the process. This preserves bodily health, lowers inflammation, and effectively stops the progression of colitis. This indicates that LP-HFY11 has the potential to be developed into a probiotic and can successfully intervene in oxazolone-induced colitis.

Author Contributions: Data curation, writing—original draft preparation, F.T.; writing—review and editing, X.Z.; methodology and software, L.R.; funding acquisition, C.-S.K. All authors have read and agreed to the published version of the manuscript.

Funding: This research was supported by the BB21plus funded by Busan Metropolitan City and Busan Techno Park.

Institutional Review Board Statement: The animal experiment conformed to the experimental ethics requirements of the animal ethics committee of Chongqing Collaborative Innovation Center for Functional Food (license number: 202311003B) on 8 November 2023.

Informed Consent Statement: Not Applicable.

Data Availability Statement: The original contributions presented in the study are included in the article, further inquiries can be directed to the corresponding author.

Conflicts of Interest: The authors declare no conflicts of interest.

References

1. Zhao, X.; Qian, Y.; Li, G.; Yi, R.; Park, K.Y.; Song, J.L. Lactobacillus plantarum YS2 (yak yogurt Lactobacillus) exhibited an activity to attenuate activated carbon-induced constipation in male Kunming mice. *J. Dairy Sci.* **2019**, *102*, 26–36. [CrossRef] [PubMed]
2. Luo, F.; Feng, S.; Sun, Q.; Xiang, W.; Zhao, J.; Zhang, J.; Yang, Z. Screening for bacteriocin-producing lactic acid bacteria from kurut, a traditional naturally-fermented yak milk from Qinghai–Tibet plateau. *Food Cont.* **2011**, *22*, 50–53. [CrossRef]
3. van der Logt, E.M.J.; Blokzijl, T.; van der Meer, R.; Faber, K.N.; Dijkstra, G. Westernized high-fat diet accelerates weight loss in dextran sulfate sodium-induced colitis in mice, which is further aggravated by supplementation of heme. *J. Nutr. Biochem.* **2013**, *24*, 1159–1165. [CrossRef] [PubMed]
4. Xu, P.; Luo, S.; Song, J.; Dai, Z.; Li, D.; Wu, C. Effect of sodium alginate-based hydrogel loaded with lutein on gut microbiota and inflammatory response in DSS-induced colitis mice. *Food Sci. Hum. Well.* **2022**, *12*, 2428–2439. [CrossRef]
5. Floudas, A.; Neto, N.; Orr, C.; Canavan, M.; Gallagher, P.; Hurson, C.; Monaghan, M.G.; Nagpar, S.; Mullan, R.H.; Veale, D.J.; et al. Loss of balance between protective and pro-inflammatory synovial tissue T-cell polyfunctionality predates clinical onset of rheumatoid arthritis. *Ann. Rheum. Dis.* **2022**, *81*, 193–205. [CrossRef] [PubMed]
6. Hunter, M.M.; Wang, A.; McKay, D.M. Helminth infection enhances disease in a murine TH2 model of colitis. *Gastroenterology* **2007**, *132*, 1320–1330. [CrossRef]
7. Rong, L.; Liu, W.; Ding, W.; Jiang, Y.; Zhong, L.; Sun, D. Modulation of dendritic cell phenotype and its function by probiotics in experimental colitis in rats. *Chin. J. Gastroenterol.* **2007**, *12*, 667–670.
8. Dhillon, P.; Singh, K. Therapeutic applications of probiotics in ulcerative colitis: An updated review. *Pharma Nutr.* **2020**, *13*, 100194. [CrossRef]
9. Adolfsson, O.; Meydani, S.N.; Russell, R.M. Yogurt and gut function. *Am. J. Clin. Nutr.* **2004**, *80*, 245–256. [CrossRef]
10. Wildt, S.; Munck, L.K.; Vinter-Jensen, L.; Hanse, B.F.; Nordgaard-Lassen, I.; Christensen, S.; Avnstroem, S.; Rasmussen, S.N.; Rumessen, J.J. Probiotic treatment of collagenous colitis: A randomized, double-blind, placebo-controlled trial with *Lactobacillus acidophilus* and *Bifidobacterium animalis* subsp. Lactis. *Inflamm. Bowel. Dis.* **2006**, *12*, 395–401. [CrossRef]
11. Daniel, C.; Poiret, S.; Goudercourt, D.; Dennin, V.; Leyer, G.; Pot, B. Selecting lactic acid bacteria for their safety and functionality by use of a mouse colitis model. *Appl. Environ. Microbiol.* **2006**, *72*, 5799–5805. [CrossRef] [PubMed]
12. Fitzpatrick, L.R. Probiotics for the treatment of *Clostridium difficile* associated disease. *World J. Gastrointest. Pathophysiol.* **2013**, *4*, 47–52. [CrossRef] [PubMed]
13. Watanabe, T.; Yamamoto, T.; Yoshida, M.; Fujiwara, K.; Kageyama-Yahara, N.; Kuramoto, H.; Shimada, Y.; Kadowaki, M. The traditional herbal medicine saireito exerts its inhibitory effect on murine oxazolone-induced colitis via the induction of Th1-polarized immune responses in the mucosal immune system of the colon. *Int. Arch. Allergy. Immunol.* **2010**, *151*, 98–106. [CrossRef] [PubMed]

14. Zhao, X.; Yi, R.; Zhou, X.; Mu, J.; Long, X.; Pan, Y.; Song, J.L.; Park, K.Y. Preventive effect of Lactobacillus plantarum KSFY02 isolated from naturally fermented yogurt from Xinjiang, China, on d-galactose–induced oxidative aging in mice. *J. Dairy Sci.* **2019**, *102*, 5899–5912. [CrossRef] [PubMed]
15. Hu, T.; Chen, R.; Qian, Y.; Ye, K.; Long, X.; Park, K.Y.; Zhao, X. Antioxidant effect of Lactobacillus fermentum HFY02-fermented soy milk on D-galactose-induced aging mouse model. *Food Sci. Hum. Well.* **2022**, *11*, 1362–1372. [CrossRef]
16. Du, C.; Li, Z.; Zhang, J.; Yin, N.; Tang, L.; Li, J.; Sun, J.; Yu, X.; Chen, W.; Xiao, H.; et al. The protective effect of carnosic acid on dextran sulfate sodium-induced colitis based on metabolomics and gut microbiota analysis. *Food Sci. Hum. Well.* **2023**, *12*, 1212–1223. [CrossRef]
17. Azuma, K.; Osaki, T.; Tsuka, T.; Imagawa, T.; Okamoto, Y.; Minami, S. Effects of fish scale collagen peptide on an experimental ulcerative colitis mouse model. *Pharma Nutr.* **2014**, *2*, 161–168. [CrossRef]
18. Chen, M.; Yao, H.; Tan, G.; Huang, W.; Wu, Q.; Nie, S. Impact of Bifidobacterium longum NSP001 on DSS-induced colitis in conventional and humanised mice. *Food Sci. Hum. Well.* **2022**, *11*, 1109–1118. [CrossRef]
19. Chen, L.; You, Q.; Hu, L.; Gao, J.; Meng, Q.; Liu, W.; Wu, X.; Xu, Q. The antioxidant procyanidin reduces reactive oxygen species signaling in macrophages and ameliorates experimental colitis in mice. *Front. Immunol.* **2018**, *8*, 1910. [CrossRef]
20. Abd Elmaksoud, H.A.; Motawea, M.H.; Desoky, A.A.; Elharrif, M.G.; Ibrahimi, A. Hydroxytyrosol alleviate intestinal inflammation, oxidative stress and apoptosis resulted in ulcerative colitis. *Biomed. Pharmacother.* **2021**, *142*, 112073. [CrossRef]
21. Wang, T.; Wang, P.; Yin, L.; Wang, X.; Shan, Y.; Yi, Y.; Zhou, Y.; Liu, B.; Wang, X.; Lü, X. Dietary Lactiplantibacillus plantarum KX041 attenuates colitis-associated tumorigenesis and modulates gut microbiota. *Food Sci. Hum. Well.* **2023**, *12*, 1626–1636. [CrossRef]
22. Özbeyli, D.; Berberoglu, A.C.; Özen, A.; Erkan, O.; Başar, Y.; Şen, T.; Akakın, D.; Yüksel, M.; Çakır, Ö.K. Protective effect of alpha-lipoic acid, aerobic or resistance exercise from colitis in second hand smoke exposed young rats. *Clin. Exp. Pharmacol. Physiol.* **2017**, *44*, 62–70. [CrossRef] [PubMed]
23. Xie, J.; Liu, L.; Li, H.; Che, H.; Xie, W. Ink melanin from *Sepiapharaonis ameliorates* colitis in mice via reducing oxidative stress, andprotecting the intestinal mucosal barrier. *Food. Res. Int.* **2022**, *151*, 110888. [CrossRef] [PubMed]
24. Wen, Y.; Xiao, H.; Liu, Y.; Yang, Y.; Wang, Y.; Xu, S.; Huang, S.; Hou, S.; Liang, J. Polysaccharides from Dendrobium officinale ameliorate colitis-induced lung injury via inhibiting inflammation and oxidative stress. *Chem. Biol. Interact.* **2021**, *347*, 109615. [CrossRef] [PubMed]
25. Yang, W.; Yao, Y.; Yang, Y.Q.; Lu, F.T.; Li, L.; Wang, Y.H.; Nakajima, T.; Tsuneyama, K.; Ridgway, W.M.; Gershwin, M.E.; et al. Differential modulation by IL-17A of Cholangitis versus Colitis in IL-2Rα deleted mice. *PLoS ONE* **2014**, *9*, e105351. [CrossRef] [PubMed]
26. Yamanaka, K.; Saito, J.; Nakata, W.; Sato, M.; Abe, T.; Mori, N.; Sekii, K.; Yoshioka, T.; Itatani, H. Case of drug-induced colitis like ulcerative colitis during IL-2 therapy for multiple bone metastasis after operation of kidney cancer. *Jpn. J. Urol.* **2009**, *100*, 586–589. [CrossRef] [PubMed]
27. Wu, Q.; Xie, S.; Zhu, Y.; Chen, J.; Tian, J.; Xiong, S.; Wu, C.; Ye, Y.; Peng, Y. Wogonin strengthens the therapeutic effects of mesenchymal stem cells in DSS-induced colitis via promoting IL-10 production. *Oxid. Med. Cell. Longev.* **2021**, *2021*, 5527935. [CrossRef] [PubMed]
28. Hu, S.; Chen, M.; Wang, Y.; Wang, Z.; Pei, Y.; Fan, R.; Liu, X.; Wang, L.; Zhou, J.; Zheng, S.; et al. mTOR inhibition attenuates dextran sulfate sodium-induced colitis by suppressing T cell proliferation and balancing TH1/TH17/Treg profile. *PLoS ONE* **2016**, *11*, e0154564. [CrossRef] [PubMed]
29. Kono, T.; Chisato, N.; Ebisawa, Y.; Asama, T.; Sugawara, M.; Ayabe, T.; Kohgo, Y.; Kasai, S.; Yoneda, M.; Takahashi, T. Impaired nitric oxide production of the myenteric plexus in colitis detected by a new bioimaging system. *J. Surg. Res.* **2004**, *117*, 329–338. [CrossRef]
30. Okaniwa, N.; Sasaki, M.; Mizushima, T.; Ogasawara, N.; Funaki, Y.; Joh, T.; Kasugai, K. eNOS plays an important role in the regulation of colonic inflammation: A novel therapeutic target and a predictive marker for the prognosis of ulcerative colitis. *Free Radic. Res.* **2015**, *49*, 35–44. [CrossRef]
31. Mazzon, E.; Cuzzocrea, S. Role of iNOS in hepatocyte tight junction alteration in mouse model of experimental colitis. *Cell Mol. Biol.* **2003**, *49*, 45–57. [PubMed]
32. Qian, Y.; Yi, R.; Sun, P.; Li, G.; Zhao, X. Lactobacillus plantarum YS2 reduces oxazolone-induced colitis in BALB/c mice. *Biomed. Res.* **2017**, *28*, 9242–9247.
33. Kim, H.J.; Kim, N.; Kim, Y.S.; Nam, R.H.; Lee, S.M.; Park, J.H.; Choi, D.; Hwang, Y.J.; Lee, J.; Lee, H.S.; et al. Changes in the interstitial cells of Cajal and neuronal nitric oxide synthase positive neuronal cells with aging in the esophagus of F344 rats. *PLoS ONE* **2017**, *12*, e0186322. [CrossRef] [PubMed]
34. Chen, Y.; Xu, J.; Liu, S.; Hou, X. Electroacupuncture at ST36 increases contraction of the gastric antrum and improves the SCF/c-kit pathway in diabetic rats. *Am. J. Chin. Med.* **2013**, *41*, 1233–1249. [CrossRef] [PubMed]
35. Long, X.; Pan, Y.; Zhao, X. Prophylactic effect of Kudingcha polyphenols on oxazolone induced colitis through its antioxidant capacities. *Food Sci. Hum. Well.* **2018**, *7*, 209–214. [CrossRef]
36. Li, J.; Moran, T.; Swanson, E.; Julian, C.; Harris, J.; Bonen, D.K.; Hedl, M.; Nicolae, D.L.; Abraham, C.; Cho, J.H. Regulation of IL-8 and IL-1beta expression in Crohn's disease associated NOD2/CARD15 mutations. *Hum. Mol. Genet.* **2004**, *13*, 1715–1725. [CrossRef] [PubMed]

37. Katoh, H.; Wang, D.; Daikoku, T.; Sun, H.; Dey, S.K.; Dubois, R.N. CXCR2-expressing myeloid-derived suppressor cells are essential to promote colitis-associated tumorigenesis. *Cancer Cell* **2013**, *24*, 631–644. [CrossRef] [PubMed]
38. Marchelletta, R.R.; Gareau, M.G.; Okamoto, S.; Guiney, D.G.; Barrett, K.E.; Fierer, J. Salmonella-induced diarrhea occurs in the absence of IL-8 receptor (CXCR2)-dependent neutrophilic inflammation. *J. Infect. Dis.* **2014**, *212*, 128–136. [CrossRef] [PubMed]
39. Parnell, J.A.; Reimer, R.A. Prebiotic fibres dose-dependently increase satiety hormones and alter Bacteroidetes and Firmicutes in lean and obese JCR:LA-cp rats. *Br. J. Nutr.* **2012**, *107*, 601–613. [CrossRef]
40. Yu, Z.T.; Li, D.G.; Sun, H.X. Herba Origani alleviated DSS-induced ulcerative colitis in mice through remolding gut microbiota to regulate bile acid and short-chain fatty acid metabolisms. *Biomed. Pharmacother.* **2023**, *161*, 114409. [CrossRef]
41. Zhou, B.; Jin, G.; Pang, X.; Mo, Q.; Bao, J.; Liu, T.; Wu, J.; Xie, R.; Liu, X.; Liu, J.; et al. *Lactobacillus rhamnosus* GG colonization in early life regulates gut-brain axis and relieves anxiety-like behavior in adulthood. *Pharmacol. Res.* **2022**, *177*, 106090. [CrossRef] [PubMed]
42. Garrett, W.S.; Gallini, C.A.; Yatsunenko, T.; Michaud, M.; DuBois, A.; Delaney, M.L.; Punit, S.; Karlsson, M.; Bry, L.; Glickman, J.N.; et al. Enterobacteriaceae act in concert with the gut microbiota to induce spontaneous and maternally transmitted colitis. *Cell Host Microbe* **2010**, *8*, 292–300. [CrossRef] [PubMed]

foods

MDPI

Article

The Effect of *Lactiplantibacillus plantarum* SHY130 and Konjac Glucomannan on the Physicochemical, Antioxidant, and Sensory Properties of Stirred Yogurt

Penglin He [1,2], Yufei Wang [1,2], Jing Yang [3], Huayi Suo [1,2] and Jiajia Song [1,2,*]

[1] College of Food Science, Southwest University, Chongqing 400715, China
[2] Chongqing Key Laboratory of Speciality Food Co-Built by Sichuan and Chongqing, Chongqing 400715, China
[3] College of Food Science and Engineering, Chongqing Technology and Business University, Chongqing 400067, China
* Correspondence: jiajias@swu.edu.cn

Abstract: The impact of konjac glucomannan (KGM)-based synbiotics on yogurt quality is not well understood. This study investigated the effects of a synbiotic mixture of KGM and the selected probiotic *Lactiplantibacillus plantarum* SHY130 on the physicochemical, antioxidant, and sensory properties of yogurt. The results showed that KGM significantly promoted the growth of *Lactiplantibacillus plantarum* SHY130. The synbiotics dramatically enhanced the count of lactic acid bacteria in yogurt during the 14 days of storage. Texture analysis indicated that the synbiotic supplement had no impact on springiness and cohesiveness but resulted in notable reductions in hardness, gumminess, and chewiness. The synbiotics did not significantly affect the water-holding capacity and syneresis. While the synbiotics initially decreased yogurt viscosity, it increased with storage time. Furthermore, the synbiotics significantly improved the yogurt's antioxidant capacity. Additionally, the supplementation of the synbiotics did not adversely affect sensory properties, although the synbiotics containing 0.02% KGM negatively impacted overall acceptability. Overall, these findings elucidate the effects of KGM-based synbiotics on yogurt quality, providing a foundation for developing novel synbiotic yogurt products.

Keywords: yogurt; synbiotics; konjac glucomannan; lactic acid bacteria

1. Introduction

Yogurt is a dairy product produced through the fermentation of milk by lactic acid bacteria (LAB). It is highly valued for its unique taste and nutritional benefits. In recent years, yogurt has gained significant attention as a carrier of probiotics. Probiotic yogurt is supplemented with strains such as *Bifidobacterium* and *Lactobacillus acidophilus*, which offer numerous health benefits when maintained in adequate numbers [1]. Yogurt supplemented with probiotics has been found to exhibit a better texture compared to yogurt without probiotics [2]. However, probiotics are sensitive to various factors, including pH, storage temperature, and processing conditions, which can significantly impact their viability in yogurt [3].

Prebiotics are defined as indigestible substrates that reach the large intestine without being digested by the host and exert a beneficial effect by selectively stimulating the growth or activity of beneficial bacteria in the colon [4]. Among the many ways to protect probiotics, prebiotics can effectively enhance their viability during storage and while passing through the gastrointestinal tract [5]. Konjac glucomannan (KGM) is a water-soluble, non-ionic macromolecular polysaccharide with unique properties [6]. It is extracted from the roots and tubers of Amorphophallus konjac and consists of β-1,4-linked D-mannosyl and D-glucosyl groups in a molar ratio of 1.6:1.0. Certain branch points are found at the C-3 position of the glucosyl and D-mannosyl residues, and some are found at the C-6 position of the glucosyl unit [7,8]. KGM can promote the multiplication of certain microorganisms and prevent antibiotic-induced gut microbiome perturbation in mice [9]. Additionally, KGM cannot be hydrolyzed by digestive enzymes in the gastrointestinal tract, and KGM-based delivery systems protect probiotics from gastric acid, low pH, and intestinal bile salt damage [10]. Therefore, KGM has great potential as a new prebiotic.

Although studies have demonstrated the potential of KGM as a prebiotic, there are currently few investigations on combining it with probiotics to form a synbiotic and incorporating it into yogurt. The purpose of this study is to investigate the effect of KGM-based synbiotics on the physicochemical, antioxidant, and sensory properties of stirred yogurt. First, probiotics with the ability to utilize KGM were screened. Subsequently, the physicochemical and antioxidant properties of KGM-based synbiotic yogurt were evaluated by determining viable bacterial count, pH, titration acidity, water-holding capacity, syneresis, texture, viscosity, color, and antioxidant activity. Finally, this study also explored the effect of KGM-based synbiotics on the sensory properties of yogurt. This study, for the first time, reveals the impact of KGM-based synbiotics on yogurt quality characteristics and provides a scientific basis for developing new synbiotic yogurts.

2. Materials and Methods

2.1. Materials

The strains (*Lactiplantibacillus plantarum* 17001, *Lactiplantibacillus plantarum* SHY130, *Limosilactobacillus fermentum* H131, *Lacticaseibacillus rhamnosus* 127, *Limosilactobacillus fermentum* 02001, *Lacticaseibacillus casei* 01001, *Lactiplantibacillus plantarum* S58, and *Lacticaseibacillus rhamnosus* T53) used in this experiment were screened from traditional fermented foods (yogurt, cheese, and pickles) in our laboratory, as described previously [11]. The pasteurized milk was provided by Chongqing Guangda Dairy Co., Ltd., Chongqing, China. The starter cultures (*Streptococcus salivarius* subsp. *thermophilus* and *Lactobacillus delbrueckii* subsp. *bulgaricus*) were purchased from DuPont (Wilmington, DE, USA).

2.2. Screening of Strains That Efficiently Metabolize KGM

The activated strains with an inoculum volume of 2% (v/v) were inoculated into an MRS liquid medium containing 0.5% (w/v) KGM as the sole carbon source. The culture was mixed thoroughly and incubated at a constant temperature of 37 °C. The optical density of the bacterial suspension was measured at 600 nm at 0, 3, 6, 9, 12, 24, and 48 h post-inoculation. Subsequently, the bacterial suspension was centrifuged at 1700× g for 10 min, and the supernatant was collected to determine the pH. A medium devoid of a carbon source served as the negative control, while a medium with glucose as the carbon source was used as the positive control.

2.3. Preparation of Stirred Yogurt

The yogurt was prepared as previously described [12]. First, 8% (w/v) sucrose was added to the pasteurized milk and mixed. After it was fully dissolved, it was placed in a 95 °C water bath for 5 min. Once the milk had cooled to 42 °C, 0.03% (w/v) of a starter culture was added for the control group and experimental group. For the experimental group, 2% (v/v) of *Lactiplantibacillus plantarum* SHY130 suspension was supplemented, followed by the addition of 0%, 0.01%, and 0.02% KGM. The inoculated milk was then placed in an incubator set at 42 °C for cultivation. After 8 h, the yogurt was removed from the incubator and gently stirred until no visible particles were observed. It was then placed in a refrigerator at 4 °C for a post-ripening period of 12 h. The properties of the stirred yogurt samples were assessed at the storage intervals of 1, 7, and 14 days [13].

2.4. Determination of Viable Bacterial Count, pH, and Titration Acidity of Yogurt

To quantify the LAB count in each yogurt sample after 14 days of storage at 4 °C, the method described previously was used [12]. A stock solution was prepared by mixing 1 mL of the sample with 9 mL of 0.09% (w/v) sterile NaCl solution. After the gradient dilution of the yogurt, 10 μL of each dilution concentration was applied to the MRS agar medium, cultured at 37 °C for 48 h, and then counted. The pH and titration acidity of the yogurt were analyzed using a previously described method [14] with minor modifications. The pH was measured using a pH meter (PHS-320, Chengdu Century Ark Technology Co., Ltd., Chengdu, Sichuan Province, China). For the determination of titration acidity, 10 g of the yogurt was diluted in 20 g of distilled water, and the mixture was titrated with 0.1 M NaOH until a permanent pink color was observed upon the addition of phenolphthalein as an indicator. The results were calculated using the following formula:

$$X = \frac{c \times v \times 100}{m \times 0.1}$$

where X is the sample's acidity; c is the molar concentration of the standard solution of sodium hydroxide (mol/L); v is the volume of the standard solution of sodium hydroxide consumed by titration; 100 refers to a 100 g sample; m is the mass of the sample.

2.5. Determination of Water-Holding Capacity (WHC) and Syneresis of Yogurt

The WHC and syneresis of each sample were determined using a reported method [12]. Briefly, 20 g of the yogurt was transferred into a centrifuge tube and then centrifuged at 4 °C and 3000 rpm for 10 min. Following centrifugation, the supernatant was carefully decanted, and the weight of the separated whey was recorded. Simultaneously, the weight of the sediment, which represents the yogurt matrix, was also measured. These measurements allowed for the calculation of the WHC, which indicates the yogurt's ability to retain water, as well as the degree of syneresis, which is the loss of whey from the yogurt matrix.

$$\text{Water holding capacity}(\%) = \frac{m_2}{m_1} \times 100$$

$$Syneresis\ (\%) = \frac{m_3}{m_1} \times 100$$

where m_1 is the initial mass of the yogurt (g); m_2 is the mass of the sedimentation product (g); m_3 is the mass of the supernatant (g).

2.6. Determination of Texture, Viscosity, and Color of Yogurt

The textural attributes of the yogurt were quantified using a Texture Analyser (CT325K230, Brookfield Engineering, Middleborough, MA, USA) according to the method of Du et al. with some modifications [15]. The yogurt samples (50 mL) were used for testing at a temperature of 4 °C. The probe model is P/36R, with a diameter of 36 mm. The measurement parameters are as follows: the pre-test speed, test speed, and post-test return speed are all 1.0 mm/s; the test distance is 50%; the trigger force is 5 g; and the relaxation time is 5 s.

For the measurement of yogurt viscosity, a Rapid Visco Analyser (RVA-TecMaster, Perten, Stockholm, Sweden) was employed. The yogurt samples (50 mL) were used for testing at a temperature of 4 °C. A No. 5 rotor was used with a speed of 100 rpm and a time of 30 s.

Additionally, the color characteristics of the yogurt, namely lightness (L*), redness–greenness (a*), and yellowness–blueness (b*), were assessed using a colorimeter (ULtraScan PRO, HunterLab, Reston, VA, USA). A standard white and color board were used to standardize the colorimeter separately.

2.7. Determination of Antioxidant Activity of Yogurt

The yogurt was centrifuged at 4 °C and 1700× g for 10 min. The supernatant was filtered through a 0.45 μm filter to obtain the sample.

The determination of 2,2-diphenyl-1-picrylhydrazyl (DPPH) radical scavenging activity was performed as previously described [16]. Two hundred μL of the sample was combined with 1 mL of 100 μM DPPH and incubated in the dark at room temperature for 30 min. The absorbance was then measured at 515 nm using a spectrophotometer. Each sample was analyzed in triplicate, with distilled water serving as the blank. The DPPH radical scavenging activity (%) was calculated using the following formula:

$$\text{DPPH radical scavenging activity}(\%) = \frac{(Ac - As)}{Ac} \times 100$$

The 2,2′-azino-bis-3-ethylbenzotiazolin-6-sulfonic acid (ABTS) test was carried out according to earlier instructions [16]. To prepare the ABTS radical solution, mix 14.8 mM ABTS and 5 mM potassium persulfate in a 1:1 (v/v) ratio. Allow this mixture to stand at room temperature in the dark for 16 h to ensure the completion of the radical generation process. Before the assay, dilute the ABTS solution with distilled water to achieve an absorbance of approximately 0.70 ± 0.05 at 734 nm. Twenty microliter of the sample was added to 980 μL of diluted ABTS solution and incubated at 37 °C for 15 min. The absorbance was measured at 734 nm. Each sample was tested in triplicate. Distilled water was used instead of the sample as the blank. The ABTS radical scavenging activity (%) was calculated using the following formula:

$$\text{ABTS radical scavenging activity}(\%) = \frac{(Ac - As)}{Ac} \times 100$$

where Ac is the absorbance of the control (distilled water) and As is the absorbance of the sample.

2.8. Determination of Sensory Evaluation of Yogurt

The sensory attributes of the yogurt were assessed by a semi-trained panel comprising 23 individuals. The assessment utilized a nine-point hedonic scale (ranging from 9, indicating 'like very much', to 1, indicating 'dislike very much') to rate each sample based on color, flavor, texture, and odor, as well as overall acceptability. Each sample was labeled with a random three-digit code and served in a random order to each tester.

2.9. Statistical Analysis

The results were presented as the average of three duplicate experiments. The one-way ANOVA and Tukey's post hoc test were applied to the statistical analysis, and significant differences were determined at $p < 0.05$.

3. Results and Discussion

3.1. Strains That Efficiently Metabolize KGM

In this study, the growth of different LAB strains under various carbon source conditions was monitored by measuring the absorbance at a wavelength of 600 nm, and the corresponding pH changes were recorded (Figure 1). In a medium devoid of a carbon source, all eight LAB strains exhibited impaired growth and an inability to produce lactic acid, leading to no significant decline in the pH of the medium. Conversely, when glucose was provided as the sole carbon source, all the strains demonstrated robust growth, reaching a stationary phase after 12 h of incubation, and concurrently produced substantial amounts of lactic acid, which rapidly decreased the pH of the culture medium. In a medium where KGM was the sole carbon source, only four strains—*Lactiplantibacillus plantarum* 17001, *Lactiplantibacillus plantarum* SHY130, *Lactiplantibacillus plantarum* S58, and *Lacticaseibacillus rhamnosus* T53—were capable of utilizing KGM and grew normally. However, the activities of these strains were not as high as those in the glucose medium, and they entered the stationary phase later, producing less lactic acid, which resulted in a small decrease in pH.

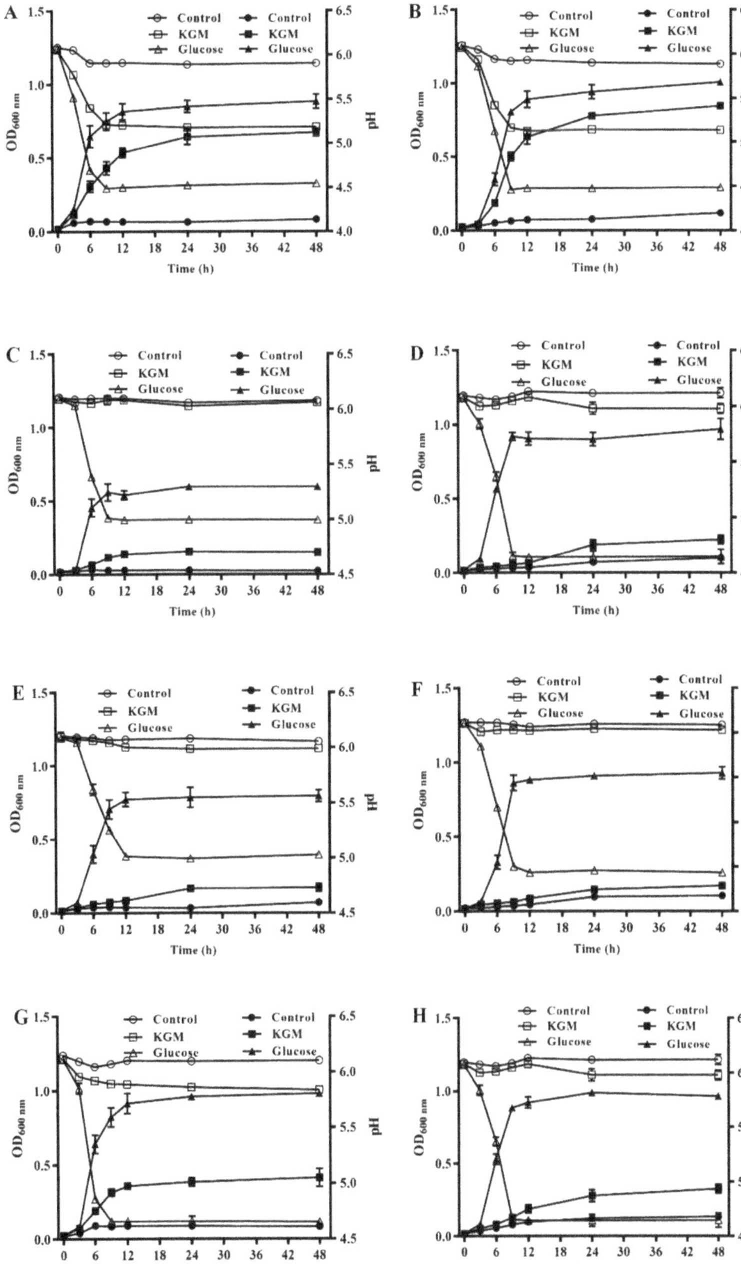

Figure 1. Growth curves and medium pH changes in each strain when konjac glucomannan (KGM) or glucose was the sole carbon source. (**A**) *Lactiplantibacillus plantarum* 17001, (**B**) *Lactiplantibacillus plantarum* SHY130, (**C**) *Limosilactobacillus fermentum* H131, (**D**) *Lacticaseibacillus rhamnosus* 127, (**E**) *Limosilactobacillus fermentum* 02001, (**F**) *Lacticaseibacillus casei* 01001, (**G**) *Lactiplantibacillus plantarum* S58, and (**H**) *Lacticaseibacillus rhamnosus* T53. ■, •, and ▲ indicate the change in $OD_{600 \text{ nm}}$; □, △, and ○ indicate the change in pH.

The strains that exhibited the ability to utilize KGM entered the stationary phase at a delayed stage when cultured in a KGM medium, which may be attributed to the intrinsic complexity of KGM metabolism and its relatively inefficient decomposition compared to glucose [17]. In addition, the changes in pH are consistent with the changes in $OD_{600\ nm}$, as pH changes are closely related to the growth of strains. The two strains with higher KGM utilization activity in our study are both *Lactiplantibacillus plantarum*, which is similar to the reported strong carbohydrate utilization ability of *Lactiplantibacillus plantarum* LP-F1 [18]. *Lactiplantibacillus plantarum* SHY130 is a probiotic isolated from traditional Chinese yak fermented yogurt. Our previous studies have reported that it can significantly improve body weight, hyperglycemia, glucose tolerance, and insulin resistance by regulating the enteroinsular axis [11]. Additionally, it can regulate liver metabolism and alleviate liver damage [19]. In this study, *Lactiplantibacillus plantarum* SHY130 showed stronger KGM utilization activity than the other strains and was therefore selected to be combined with KGM and added to the yogurt as a synbiotic.

3.2. Viable Bacterial Count, pH, and Titration Acidity of Yogurt during Storage

During the 14-day storage period, the dynamic changes in the number of LAB, pH value, and titration acidity in the yogurt were monitored. The results are shown in Figure 2. Throughout storage, the number of LAB in the yogurt of the control group and the SHY group gradually decreased. In contrast, the number of LAB in the yogurt with the synbiotics continued to increase over the 14-day storage period. On the 14th day, the number of viable bacteria in the synbiotic yogurt was approximately 10 log CFU/mL, while the amount of viable bacteria in the SHY group was only approximately 5 log CFU/mL. The change in the LAB numbers directly affects the pH value of the yogurt. As lactic acid continues to be generated, the pH value of the yogurt gradually decreases. Notably, the pH value of the control group decreased slowly, and on the 14th day, its pH value was significantly different from the other groups ($p < 0.05$). The change in the titration acidity of the yogurt is inversely proportional to the pH value.

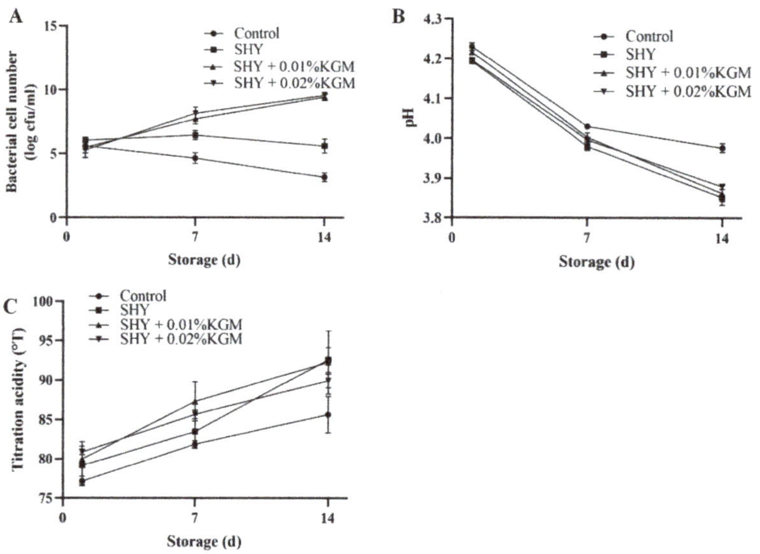

Figure 2. Bacterial cell numbers (**A**), pH (**B**), and titration acidity of yogurt (**C**). Control: the yogurt with only the addition of the starter culture; SHY: the yogurt with the addition of the starter culture and SHY130; SHY + 0.01% KGM: the yogurt with the addition of the starter culture, SHY130, and 0.01% KGM; SHY + 0.02% KGM: the yogurt with the addition of the starter culture, SHY130, and 0.02% KGM.

The growth and reproduction of lactobacilli depend on specific nutrients, such as lactose. Therefore, the decrease in the number of LAB in the control yogurt may be due to a reduction in the supply of the nutrients necessary for their proliferation and survival. In the yogurt with synbiotics, LAB can continue to grow, which may be due to the KGM in the synbiotics serving as an energy source, thus maintaining a high level of probiotic activity [20]. Similarly, Falah et al. found that when the prebiotic inulin was added to yogurt, the viability of the probiotic *Lactobacillus brevis* PML1 increased during storage [21]. Additionally, the production of lactic acid and the primary metabolite acetic acid may cause acid damage to organisms, and the addition of prebiotics can effectively resist this acid damage, thereby ensuring the reproduction of probiotics [22]. The decrease in the pH value and increase in the titration acidity in yogurt are closely related to the growth and metabolic activities of LAB. The fluctuations observed in the pH level and titration acidity in the control group are potentially associated with the initial concentration of the probiotic cultures introduced into the milk [23].

3.3. The Color of Yogurt

During storage, the color properties of yogurt undergo significant changes, as shown in Figure 3. For the lightness (L*) and red–greenness (a*) values of the yogurt in the four groups, no statistically significant changes were detected ($p < 0.05$) during the 14-day period. In contrast, the yellow–blue degree (b*) of yogurt in the control group decreased, while the b* value of the yogurt in the SHY130 + 0.01% KGM group increased. The b* values of the yogurt in the SHY130 group and the SHY130 + 0.02% KGM group remained unchanged during the 14 days of storage.

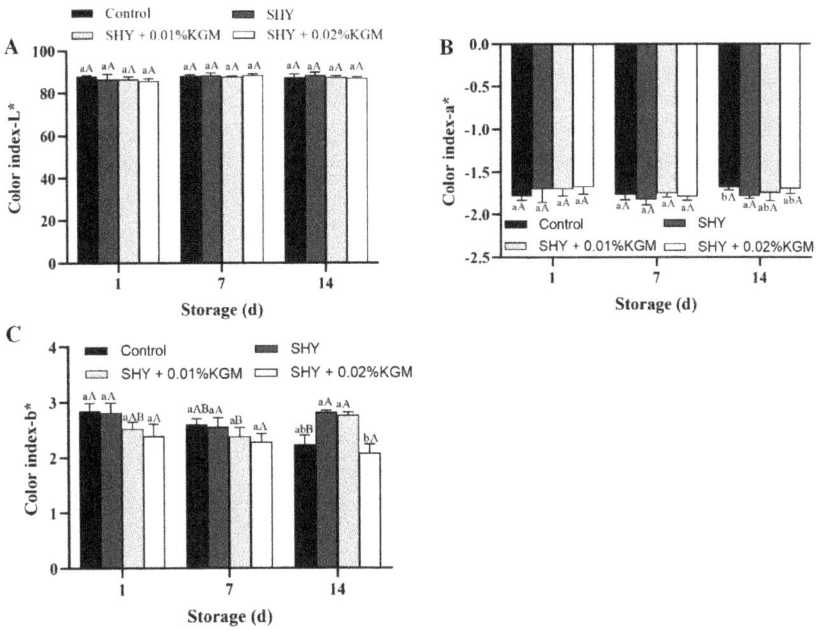

Figure 3. Changes in the color of yogurt during storage. (**A**) Color index L*, (**B**) color index a*, and (**C**) color index b*. Control: the yogurt with only the addition of the starter culture; SHY: the yogurt with the addition of the starter culture and SHY130; SHY + 0.01% KGM: the yogurt with the addition of the starter culture, SHY130, and 0.01% KGM; SHY + 0.02% KGM: the yogurt with the addition of the starter culture, SHY130, and 0.02% KGM. The means of varying storage times, indicated by varying uppercase letters (A,B), exhibit significant differences ($p < 0.05$); the means of varying yogurts, indicated by the lowercase letters (a,b), exhibit significant differences ($p < 0.05$).

The color of yogurt is an important indicator that influences its quality [24]. The alteration in the b* value during storage may be due to the utilization of riboflavin, β-carotene, and vitamin A present in milk by LAB. Changes in the number and species of LAB in different treatment groups lead to variations in the b* value of yogurt. Buchilina et al. reported that adding monk fruit sweetener to camel milk yogurt results in a decrease in b* value [25].

3.4. The Texture of Yogurt

The texture of the yogurt is shown in Table 1. During the initial 7 days of storage, the hardness, gumminess, and chewiness of the yogurt in all four groups increased. Subsequently, as the storage time was extended, the hardness, gumminess, and chewiness of the yogurt samples gradually decreased. Following the addition of *Lactiplantibacillus plantarum* SHY130 and KGM, the hardness, gumminess, and chewiness of the yogurt were markedly lower than those of the control group on the 1st and 14th day of storage ($p < 0.05$). However, the addition of KGM and probiotics did not result in significant alterations to the springiness and cohesiveness of the yogurt, and these textural parameters remained relatively stable throughout the storage duration.

Table 1. Changes in the texture parameters of the yogurt during storage.

Parameters	Treatment	Storage Period (Days)		
		1	7	14
Hardness (g)	Control	57.90 ± 2.55 ^{aB}	75.05 ± 2.01 ^{aA}	67.23 ± 1.02 ^{aAB}
	SHY	53.68 ± 1.62 ^{aB}	75.38 ± 0.92 ^{aA}	64.15 ± 4.14 ^{abB}
	SHY + 0.01%KGM	43.43 ± 2.42 ^{bB}	73.35 ± 2.08 ^{aA}	61.92 ± 3.69 ^{abB}
	SHY + 0.02%KGM	43.75 ± 0.25 ^{bC}	62.19 ± 0.46 ^{bA}	57.29 ± 1.16 ^{bB}
Springiness	Control	0.93 ± 0.00 ^{aA}	0.94 ± 0.00 ^{aA}	0.94 ± 0.00 ^{aA}
	SHY	0.93 ± 0.00 ^{aA}	0.93 ± 0.00 ^{aA}	0.92 ± 0.02 ^{aA}
	SHY + 0.01%KGM	0.93 ± 0.00 ^{aA}	0.93 ± 0.00 ^{aA}	0.93 ± 0.01 ^{aA}
	SHY + 0.02%KGM	0.93 ± 0.00 ^{aA}	0.93 ± 0.00 ^{aA}	0.95 ± 0.00 ^{aA}
Cohesiveness	Control	0.67 ± 0.01 ^{aA}	0.60 ± 0.00 ^{aA}	0.60 ± 0.01 ^{aA}
	SHY	0.67 ± 0.01 ^{aA}	0.61 ± 0.00 ^{aA}	0.60 ± 0.02 ^{aA}
	SHY + 0.01%KGM	0.68 ± 0.02 ^{aA}	0.63 ± 0.01 ^{aA}	0.60 ± 0.02 ^{aA}
	SHY + 0.02%KGM	0.69 ± 0.00 ^{aA}	0.63 ± 0.01 ^{aA}	0.60 ± 0.00 ^{aA}
Gumminess	Control	39.60 ± 1.28 ^{aB}	45.18 ± 1.05 ^{aA}	40.95 ± 0.35 ^{aB}
	SHY	36.10 ± 1.37 ^{aB}	46.08 ± 1.20 ^{aA}	37.62 ± 1.03 ^{abB}
	SHY + 0.02%KGM	29.70 ± 1.07 ^{bC}	45.59 ± 0.91 ^{aA}	35.43 ± 0.76 ^{bB}
	SHY + 0.02%KGM	30.23 ± 0.17 ^{bC}	40.50 ± 0.40 ^{bA}	35.63 ± 0.48 ^{bB}
Chewiness	Control	37.16 ± 1.19 ^{aB}	42.45 ± 0.98 ^{aA}	38.64 ± 0.56 ^{aB}
	SHY	33.81 ± 1.31 ^{aB}	43.13 ± 1.16 ^{aA}	34.80 ± 0.60 ^{abB}
	SHY + 0.01%KGM	27.82 ± 1.02 ^{bC}	42.57 ± 0.91 ^{aA}	33.01 ± 0.49 ^{bB}
	SHY + 0.02%KGM	28.24 ± 0.16 ^{bC}	38.15 ± 0.03 ^{bA}	33.67 ± 0.37 ^{bB}

Control: the yogurt with only the addition of the starter culture; SHY: the yogurt with the addition of the starter culture and SHY130; SHY + 0.01% KGM: the yogurt with the addition of the starter culture, SHY130, and 0.01% KGM; SHY + 0.02% KGM: the yogurt with the addition of the starter culture, SHY130, and 0.02% KGM. The means of varying storage times, indicated by different uppercase letters (A–C), exhibit significant differences ($p < 0.05$); the means of varying yogurts, indicated by different lowercase letters (a,b), exhibit significant differences ($p < 0.05$).

The texture of yogurt is characterized by several key indicators: hardness, cohesiveness, gumminess, springiness, and chewiness. Hardness directly correlates with the gel state of the yogurt; higher hardness indicates better coagulation. Cohesiveness reflects the yogurt's resistance to compression by a probe, with lower cohesiveness signifying smoother yogurt. Gumminess represents the energy needed to chew the yogurt until it is ready to be swallowed. Springiness measures how well the yogurt returns to its original height or volume after being deformed. Chewiness indicates the energy required to chew the yogurt. These parameters can be used to evaluate yogurt quality by simulating the jaw movement in the human oral cavity [26]. The early increase in hardness, gumminess, and chewiness in the synbiotic yogurt may be due to the formation of a three-dimensional network structure between KGM and protein. The latter decrease may be due to the hydrolysis and consumption of protein and KGM, making it difficult to form a larger three-dimensional network structure. This phenomenon is similar to findings from a study where increasing the supplementation of prickly pear decreased the hardness, adhesiveness, cohesiveness, and gumminess of probiotic yogurt [27]. Besides, the addition of synbiotics to yogurt appears to result in a reduction in hardness, gumminess, and chewiness, which can be attributed to KGM's water-binding capacity and its influence on the structural integrity and stability of the food matrix.

3.5. WHC, Syneresis, and Viscosity of Yogurt

The changes in the WHC, syneresis, and viscosity of each group of yogurt during storage are shown in Figure 4. The results indicate that the incorporation of probiotics and synbiotics into the yogurt did not have a significant impact on WHC and syneresis throughout the storage duration. WHC exhibited a decreasing trend as storage time progressed, whereas syneresis increased over the same period. Regarding viscosity, an initial increase was observed during the first 7 days of storage. Subsequent to this initial phase, the viscosity of yogurt in all the groups, except for the synbiotic group, began to decline.

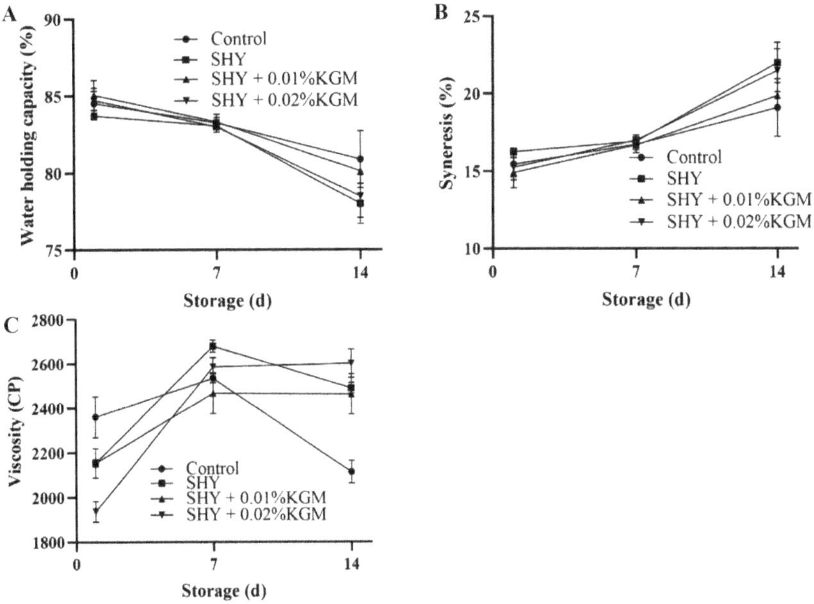

Figure 4. Changes in the water-holding capacity (**A**), syneresis (**B**), and viscosity (**C**) of yogurt during storage. Control: the yogurt with only the addition of the starter culture; SHY: the yogurt with the addition of the starter culture and SHY130; SHY + 0.01% KGM: the yogurt with the addition of the starter culture, SHY130, and 0.01% KGM; SHY + 0.02% KGM: the yogurt with the addition of the starter culture, SHY130, and 0.02% KGM.

The WHC of yogurt is an important measure of the denseness of the protein gel network and the texture of yogurt. The decrease in WHC during storage may be due to the rearrangement and shrinkage of the protein networks [28]. The increase in syneresis is related to the rearrangement of the casein network, which leads to increased interparticle binding, thus making the network prone to compression and shrinkage [29]. Furthermore, a sustained decrease in yogurt pH may also lead to the shrinkage of the casein network and higher levels of syneresis [30]. The increase in yogurt viscosity in the early stages of storage may be due to the continued formation of the protein network, while the decrease in viscosity in the later stages may be due to the destruction of the protein network by the pH value. This result is consistent with the findings of Gomes et al., who observed similar effects when adding different types of soluble fiber to high-protein yogurts [31]. The decrease in the viscosity of the synbiotic yogurts during the early stages of storage may be due to a bridging effect caused by the high concentration of KGM, which inhibits the formation of a continuous protein network, resulting in the highest viscosity in the control yogurt. Tseng et al. also found similar results, demonstrating that yogurt containing 3% fermented wine grape pomace had a lower viscosity [32]. In the middle stage, changes in viscosity may be due to the rearrangement of the protein network during storage; the effect of protein–protein interactions on viscosity exceeds that of synbiotics. As storage progresses, the consumption of KGM by microbial activity in the later stages of storage makes its contribution to the protein–polysaccharide network increasingly critical, maintaining the stability of yogurt viscosity [33].

3.6. Yogurt's Antioxidant Activity

During the storage period, the changes in the DPPH free radical scavenging ability and ABTS free radical scavenging ability of each group of yogurt are shown in Figure 5. The results indicate that the DPPH free radical scavenging activity in the yogurt samples progressively increased with the duration of storage. Specifically, the yogurt samples supplemented with synbiotics exhibited higher scavenging activity compared to the control group. The DPPH scavenging activity in the SHY130 + 0.02% KGM group reached the highest level on the 14th day (39.08 ± 1.17%). In contrast, the ABTS scavenging activity in the SHY130 + 0.02% KGM group peaked on the 7th day (25.21 ± 0.24%).

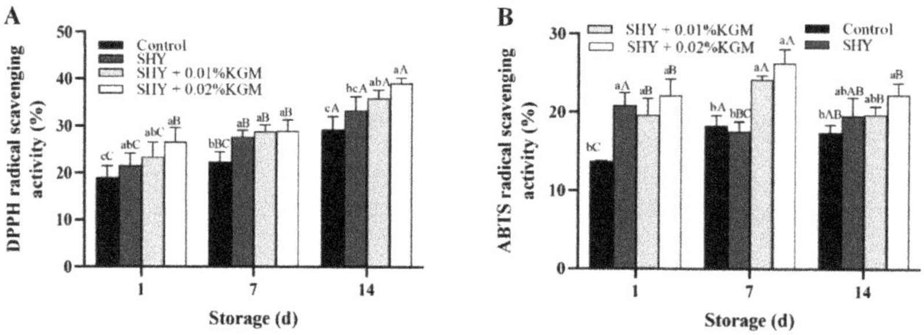

Figure 5. DPPH (**A**) and ABTS (**B**) free radical scavenging ability of yogurt. Control: the yogurt with only the addition of the starter culture; SHY: the yogurt with the addition of the starter culture and SHY130; SHY + 0.01% KGM: the yogurt with the addition of the starter culture, SHY130, and 0.01% KGM; SHY + 0.02% KGM: the yogurt with the addition of the starter culture, SHY130, and 0.02% KGM. The means of varying storage times, indicated by different uppercase letters (A–C), exhibit significant differences ($p < 0.05$); the means of varying yogurts, indicated by different lowercase letters (a–c), exhibit significant differences ($p < 0.05$).

Compared with the yogurt without *Lactiplantibacillus plantarum* SHY130, the yogurt containing *Lactiplantibacillus plantarum* SHY130 demonstrated more efficient scavenging activity in both the DPPH and ABTS free radical scavenging experiments. This phenomenon may be attributed to the ability of *Lactiplantibacillus plantarum* SHY130 to produce a variety of antioxidant enzymes (such as superoxide dismutase, catalase, glutathione reductase, and thioredoxin) that scavenge free radicals, thereby improving the overall antioxidant capacity of yogurt [34]. Additionally, the antioxidant activity of yogurt may originate from bioactive peptides produced during the fermentation process, with milk protein being an important source of these bioactive peptides. Sah et al. purified antioxidant peptides from probiotic yogurt supplemented with pineapple peel powder and identified two peptides derived from β-casein, namely 193YQEPVLGPVRGPFPPIV209 and 69SLPQNIPPLTQTPVVVPPF87 [35]. The addition of different concentrations of KGM enhanced the antioxidant capacity of yogurt, possibly because KGM can increase the activity and quantity of *Lactiplantibacillus plantarum* SHY130 in yogurt. Our results are consistent with the research results of Sha et al., who found that supplementing yogurt with pineapple peel or inulin increased its antioxidant activity [36]. The different antioxidant activities of synbiotic yogurts suggest that free radical scavenging activity may depend on strain characteristics and that there may be differences in reaction kinetics between different free radicals and antioxidants [37].

3.7. Sensory Properties of Yogurt

The sensory evaluation outcomes for the yogurt samples are presented in Table 2. The statistical tests revealed that the incorporation of the synbiotics did not result in statistically significant alterations in the sensory attributes of the yogurt, including the color, flavor, texture, and odor ($p > 0.05$). Additionally, the sensory evaluation panel identified a significant decline in the overall acceptance of the yogurt samples enriched with 0.02% KGM ($p < 0.05$). These data suggest that the synbiotic yogurts with 0.01% KGM may represent an optimal compromise, preserving the sensory appeal and quality attributes of the product.

Table 2. The sensory scores of yogurt after 7 days.

Sample	Color	Flavor	Texture	Odor	Overall Acceptability
Control	7.57 ± 1.08^{a}	7.15 ± 1.47^{a}	7.21 ± 1.04^{a}	7.15 ± 1.29^{a}	7.46 ± 0.81^{a}
SHY	7.54 ± 1.05^{a}	6.85 ± 1.61^{a}	7.33 ± 0.93^{a}	7.33 ± 1.44^{a}	7.17 ± 1.30^{ab}
SHY + 0.01%KGM	7.17 ± 1.30^{a}	6.78 ± 1.48^{a}	7.41 ± 1.21^{a}	6.85 ± 1.29^{a}	7.20 ± 1.21^{ab}
SHY + 0.02%KGM	7.20 ± 1.44^{a}	6.35 ± 1.78^{a}	7.13 ± 1.10^{a}	6.74 ± 1.71^{a}	6.67 ± 1.38^{b}

Note, control: the yogurt with only the addition of the starter culture; SHY: the yogurt with the addition of the starter culture and SHY130; SHY + 0.01% KGM: the yogurt with the addition of the starter culture, SHY130, and 0.01% KGM; SHY + 0.02% KGM: the yogurt with the addition of the starter culture, SHY130, and 0.02% KGM. Means within a column with different superscripts (a,b) differ significantly ($p < 0.05$).

4. Conclusions

In this study, we screened *Lactiplantibacillus plantarum* SHY130 for its high bioavailability activity towards KGM. This strain was combined with KGM and added to yogurt, and their effects on the quality and sensory properties of yogurt were evaluated. The addition of KGM increased the activity of probiotic bacteria, as evidenced by the viable bacterial count and pH of the yogurt. The synbiotics reduced the hardness, gumminess, and chewiness of the yogurt texture. Additionally, the synbiotic supplementation did not significantly impact the water-holding capacity and syneresis of the yogurt. However, it may have an adverse effect on the viscosity of yogurt in the early stages of storage, but it increases the viscosity in the later stages. Furthermore, the synbiotics significantly improved the antioxidant properties of the yogurt. Although the addition of the synbiotics did not affect the sensory properties, the overall acceptability of the synbiotic yogurt containing 0.02% KGM was low. Taken together, these results provide a new reference and approach for the development of new synbiotic yogurt.

Author Contributions: Conceptualization, H.S. and J.S.; Methodology, P.H. and Y.W.; Formal analysis, J.Y.; Investigation, Y.W.; Resources, H.S.; Data curation, Y.W.; Writing—original draft, P.H. and J.S.; Writing—review & editing, J.Y. and J.S.; Visualization, P.H.; Supervision, J.S.; Project administration, J.S. All authors have read and agreed to the published version of the manuscript.

Funding: This work was supported by the Natural Science Foundation of Chongqing (CSTB2024NSCQ-MSX0192), the Sichuan Province Key Research and Development Program (2024YFHZ0077), the Science and Technology Research Program of Chongqing Municipal Education Commission (KJQN202300204), and the Innovation and Entrepreneurship Training Program for College Students at Southwest University (X202310635033).

Institutional Review Board Statement: This study involves only a one-time, short-term evaluation of the sensory attributes of yogurt and does not constitute a long-term functional product test. The study has not involved ethical reviews for human trials.

Informed Consent Statement: Informed consent was obtained from all subjects involved in the study.

Data Availability Statement: The original contributions presented in the study are included in the article, further inquiries can be directed to the corresponding author.

Conflicts of Interest: The authors declare no conflicts of interest.

References

1. Baltova, K.; Dimitrov, Z. Probiotic and cultural characteristic of strain Lactobacillus gasseri 4/13 of human origin. *Biotechnol. Biotechnol. Equip.* **2014**, *28*, 1084–1088. [CrossRef] [PubMed]
2. Han, X.; Yang, Z.; Jing, X.; Yu, P.; Zhang, Y.; Yi, H.; Zhang, L. Improvement of the Texture of Yogurt by Use of Exopolysaccharide Producing Lactic Acid Bacteria. *BioMed Res. Int.* **2016**, *2016*, 7945675. [CrossRef] [PubMed]
3. Meybodi, N.M.; Mortazavian, A.M.; Arab, M.; Nematollahi, A. Probiotic viability in yoghurt: A review of influential factors. *Int. Dairy J.* **2020**, *109*, 104793. [CrossRef]
4. Jooyandeh, H.; Momenzadeh, S.; Alizadeh Behbahani, B.; Barzegar, H. Effect of Malva neglecta and lactulose on survival of Lactobacillus fermentum and textural properties of synbiotic stirred yogurt. *J. Food Sci. Technol.* **2023**, *60*, 1136–1143. [CrossRef] [PubMed]
5. Mudgil, D.; Barak, S.; Patel, A.; Shah, N. Partially hydrolyzed guar gum as a potential prebiotic source. *Int. J. Biol. Macromol.* **2018**, *112*, 207–210. [CrossRef]
6. Gao, F.; Zhang, W.; Cao, M.; Liu, X.; Han, T.; He, W.; Shi, B.; Gu, Z. Maternal supplementation with konjac glucomannan improves maternal microbiota for healthier offspring during lactation. *J. Sci. Food Agric.* **2024**, *104*, 3736–3748. [CrossRef]
7. Maeda, M.; Shimahara, H.; Sugiyama, N. Detailed examination of the branched structure of konjac glucomannan. *Agric. Biol. Chem.* **1980**, *44*, 245–252. [CrossRef]
8. Zhang, Y.; Zhao, Y.; Yang, W.; Song, G.; Zhong, P.; Ren, Y.; Zhong, G. Structural complexity of Konjac glucomannan and its derivatives governs the diversity and outputs of gut microbiota. *Carbohydr. Polym.* **2022**, *292*, 119639. [CrossRef]
9. Mao, Y.-H.; Xu, Y.; Song, F.; Wang, Z.-M.; Li, Y.-H.; Zhao, M.; He, F.; Tian, Z.; Yang, Y. Protective effects of konjac glucomannan on gut microbiome with antibiotic perturbation in mice. *Carbohydr. Polym.* **2022**, *290*, 119476. [CrossRef]
10. Zhang, Y.; Tong, C.; Chen, Y.; Xia, X.; Jiang, S.; Qiu, C.; Pang, J. Advances in the construction and application of konjac glucomannan-based delivery systems. *Int. J. Biol. Macromol.* **2024**, *262*, 129940. [CrossRef]
11. Wang, G.; Song, J.; Huang, Y.; Li, X.; Wang, H.; Zhang, Y.; Suo, H. Lactobacillus plantarum SHY130 isolated from yak yogurt attenuates hyperglycemia in C57BL/6J mice by regulating the enteroinsular axis. *Food Funct.* **2022**, *13*, 675–687. [CrossRef]
12. Fan, X.; Du, L.; Xu, J.; Shi, Z.; Zhang, T.; Jiang, X.; Zeng, X.; Wu, Z.; Pan, D. Effect of single probiotics Lacticaseibacillus casei CGMCC1.5956 and *Levilactobacillus brevis* CGMCC1.5954 and their combination on the quality of yogurt as fermented milk. *LWT* **2022**, *163*, 113530. [CrossRef]
13. Naibaho, J.; Butula, N.; Jonuzi, E.; Korzeniowska, M.; Laaksonen, O.; Föste, M.; Kütt, M.-L.; Yang, B. Potential of brewers' spent grain in yogurt fermentation and evaluation of its impact in rheological behaviour, consistency, microstructural properties and acidity profile during the refrigerated storage. *Food Hydrocoll.* **2022**, *125*, 107412. [CrossRef]
14. Kariyawasam, K.M.G.M.M.; Lee, N.-K.; Paik, H.-D. Synbiotic yoghurt supplemented with novel probiotic *Lactobacillus brevis* KU200019 and fructooligosaccharides. *Food Biosci.* **2021**, *39*, 100835. [CrossRef]
15. Du, H.; Yang, H.; Wang, X.; Zhu, F.; Tang, D.; Cheng, J.; Liu, X. Effects of mulberry pomace on physicochemical and textural properties of stirred-type flavored yogurt. *J. Dairy Sci.* **2021**, *104*, 12403–12414. [CrossRef]
16. Zhang, T.; Jeong, C.H.; Cheng, W.N.; Bae, H.; Seo, H.G.; Petriello, M.C.; Han, S.G. Moringa extract enhances the fermentative, textural, and bioactive properties of yogurt. *LWT* **2019**, *101*, 276–284. [CrossRef]
17. Wan, X.; Wei, L.; Zhang, W.; Lei, Y.; Kong, Q.; Zhang, B. Production, characterization, and prebiotic activity of oligosaccharides from konjac glucomannan by Bacillus amyloliquefaciens WX-1. *J. Funct. Foods* **2022**, *88*, 104872. [CrossRef]
18. Cui, Y.; Wang, M.; Zheng, Y.; Miao, K.; Qu, X. The Carbohydrate Metabolism of Lactiplantibacillus plantarum. *Int. J. Mol. Sci.* **2021**, *22*, 13452. [CrossRef]
19. Wang, G.; Song, J.; Wang, C.; Chen, X.; Suo, H. Metabolomics reveals the role of Lactobacillus plantarum SHY130 in hepatic metabolic regulation in a mouse model of type 2 diabetes. *J. Sci. Food Agric.* **2023**, *103*, 6406–6415. [CrossRef]
20. Li, H.; Zhang, T.; Li, C.; Zheng, S.; Li, H.; Yu, J. Development of a microencapsulated synbiotic product and its application in yoghurt. *LWT* **2020**, *122*, 109033. [CrossRef]

21. Falah, F.; Vasiee, A.; Yazdi, F.T.; Behbahani, B.A. Preparation and Functional Properties of Synbiotic Yogurt Fermented with Lactobacillus brevis PML1 Derived from a Fermented Cereal-Dairy Product. *BioMed Res. Int.* **2021**, *2021*, 1057531. [CrossRef] [PubMed]
22. Donkor, O.N.; Nilmini, S.L.I.; Stolic, P.; Vasiljevic, T.; Shah, N.P. Survival and activity of selected probiotic organisms in set-type yoghurt during cold storage. *Int. Dairy J.* **2007**, *17*, 657–665. [CrossRef]
23. Li, H.; Song, W.; Liu, T.; Xu, S.; Zhang, S.; Zhang, Y.; Liu, D.; Li, H.; Yu, J. Developing novel synbiotic yoghurt with Lacticaseibacillus paracasei and lactitol: Investigation of the microbiology, textural and rheological properties. *Int. Dairy J.* **2022**, *135*, 105475. [CrossRef]
24. Popescu, L.; Cesco, T.; Gurev, A.; Ghendov-Mosanu, A.; Sturza, R.; Tarna, R. Impact of Apple Pomace Powder on the Bioactivity, and the Sensory and Textural Characteristics of Yogurt. *Foods* **2022**, *11*, 3565. [CrossRef] [PubMed]
25. Buchilina, A.; Aryana, K. Physicochemical and microbiological characteristics of camel milk yogurt as influenced by monk fruit sweetener. *J. Dairy Sci.* **2021**, *104*, 1484–1493. [CrossRef] [PubMed]
26. Kim, S.Y.; Hyeonbin, O.; Lee, P.; Kim, Y.-S. The quality characteristics, antioxidant activity, and sensory evaluation of reduced-fat yogurt and nonfat yogurt supplemented with basil seed gum as a fat substitute. *J. Dairy Sci.* **2020**, *103*, 1324–1336. [CrossRef] [PubMed]
27. Azeez Khalid Albayati, A.; Ağçam, E.; Karaca, O.B.; Ozogul, F. Effects of prickly pear supplementation on physico-chemical, textural, microbiological and sensory characteristics of probiotic set yoghurts. *Food Biosci.* **2024**, *60*, 104513. [CrossRef]
28. Ghaderi-Ghahfarokhi, M.; Yousefvand, A.; Ahmadi Gavlighi, H.; Zarei, M.; Farhangnia, P. Developing novel synbiotic low-fat yogurt with fucoxylogalacturonan from tragacanth gum: Investigation of quality parameters and Lactobacillus casei survival. *Food Sci. Nutr.* **2020**, *8*, 4491–4504. [CrossRef]
29. Ramirez-Santiago, C.; Ramos-Solis, L.; Lobato-Calleros, C.; Peña-Valdivia, C.; Vernon-Carter, E.J.; Alvarez-Ramírez, J. Enrichment of stirred yogurt with soluble dietary fiber from Pachyrhizus erosus L. Urban: Effect on syneresis, microstructure and rheological properties. *J. Food Eng.* **2010**, *101*, 229–235. [CrossRef]
30. Senadeera, S.S.; Prasanna, P.H.P.; Jayawardana, N.W.I.A.; Gunasekara, D.C.S.; Senadeera, P.; Chandrasekara, A. Antioxidant, physicochemical, microbiological, and sensory properties of probiotic yoghurt incorporated with various Annona species pulp. *Heliyon* **2018**, *4*, e00955. [CrossRef]
31. Gomes, E.R.; Barroso dos Anjos Pinto, C.; Stephani, R.; Fernandes de Carvalho, A.; Perrone, Í.T. Effect of adding different types of soluble fibre to high-protein yoghurts on water holding capacity, particle size distribution, apparent viscosity, and microstructure. *Int. Dairy J.* **2023**, *141*, 105609. [CrossRef]
32. Tseng, A.; Zhao, Y. Wine grape pomace as antioxidant dietary fibre for enhancing nutritional value and improving storability of yogurt and salad dressing. *Food Chem.* **2013**, *138*, 356–365. [CrossRef]
33. Salgado, M.J.G.; Ramos, M.S.; Assis, D.J.; Otero, D.M.; Oliveira, R.L.; Ribeiro, C.V.D.M.; Costa, M.P.; Oliveira, C.A.A. Impact of fiber-rich donkey milk yogurt on apparent viscosity and sensory acceptance. *LWT* **2021**, *145*, 111494. [CrossRef]
34. Ianniello, R.G.; Ricciardi, A.; Parente, E.; Tramutola, A.; Reale, A.; Zotta, T. Aeration and supplementation with heme and menaquinone affect survival to stresses and antioxidant capability of Lactobacillus casei strains. *LWT-Food Sci. Technol.* **2015**, *60*, 817–824. [CrossRef]
35. Sah, B.N.P.; Vasiljevic, T.; McKechnie, S.; Donkor, O.N. Antioxidant peptides isolated from synbiotic yoghurt exhibit antiproliferative activities against HT-29 colon cancer cells. *Int. Dairy J.* **2016**, *63*, 99–106. [CrossRef]
36. Sah, B.N.P.; Vasiljevic, T.; McKechnie, S.; Donkor, O.N. Effect of refrigerated storage on probiotic viability and the production and stability of antimutagenic and antioxidant peptides in yogurt supplemented with pineapple peel. *J. Dairy Sci.* **2015**, *98*, 5905–5916. [CrossRef]
37. Sah, B.N.P.; Vasiljevic, T.; McKechnie, S.; Donkor, O.N. Effect of probiotics on antioxidant and antimutagenic activities of crude peptide extract from yogurt. *Food Chem.* **2014**, *156*, 264–270. [CrossRef]

MDPI AG
Grosspeteranlage 5
4052 Basel
Switzerland
Tel.: +41 61 683 77 34

Foods Editorial Office
E-mail: foods@mdpi.com
www.mdpi.com/journal/foods

www.ingramcontent.com/pod-product-compliance
Lightning Source LLC
Chambersburg PA
CBHW061522210926
51850CB00008B/2000
9783725869527